PALGRAVE STUDIES IN CULTURAL AND INTELLECTUAL HISTORY

Series Editors

Anthony J. La Vopa, North Carolina State University.

Suzanne Marchand, Louisiana State University.

Javed Majeed, Queen Mary, University of London.

The Palgrave Studies in Cultural and Intellectual History series has three primary aims: to close divides between intellectual and cultural approaches, thus bringing them into mutually enriching interactions; to encourage interdisciplinarity in intellectual and cultural history; and to globalize the field, both in geographical scope and in subjects and methods. This series is open to work on a range of modes of intellectual inquiry, including social theory and the social sciences; the natural sciences; economic thought; literature; religion; gender and sexuality; philosophy; political and legal thought; psychology; and music and the arts. It encompasses not just North America but Africa, Asia, Eurasia, Europe, Latin America, and the Middle East. It includes both nationally focused studies and studies of intellectual and cultural exchanges between different nations and regions of the world, and encompasses research monographs, synthetic studies, edited collections, and broad works of reinterpretation. Regardless of methodology or geography, all books in the series are historical in the fundamental sense of undertaking rigorous contextual analysis.

Published by Palgrave Macmillan

Indian Mobilities in the West, 1900–1947: Gender, Performance, Embodiment
By Shompa Lahiri

The Shelley-Byron Circle and the Idea of Europe
By Paul Stock

Culture and Hegemony in the Colonial Middle East
By Yaseen Noorani

Recovering Bishop Berkeley: Virtue and Society in the Anglo-Irish Context
By Scott Breuninger

The Reading of Russian Literature in China: A Moral Example and Manual of Practice
By Mark Gamsa

Rammohun Roy and the Making of Victorian Britain
By Lynn Zastoupil

Carl Gustav Jung: Avant-Garde Conservative
By Jay Sherry

Law and Politics in British Colonial Thought: Transpositions of Empire
Edited by Shaunnagh Dorsett and Ian Hunter

Sir John Malcolm and the Creation of British India
By Jack Harrington

Nature Engaged

Science in Practice from the Renaissance to the Present

Edited by

Mario Biagioli and Jessica Riskin

NATURE ENGAGED

Copyright © Mario Biagioli and Jessica Riskin, 2012.

All rights reserved.

First published in 2012 by
PALGRAVE MACMILLAN®
in the United States—a division of St. Martin's Press LLC,
175 Fifth Avenue, New York, NY 10010.

Where this book is distributed in the UK, Europe and the rest of the world,
this is by Palgrave Macmillan, a division of Macmillan Publishers Limited,
registered in England, company number 785998, of Houndmills,
Basingstoke, Hampshire RG21 6XS.

Palgrave Macmillan is the global academic imprint of the above companies
and has companies and representatives throughout the world.

Palgrave® and Macmillan® are registered trademarks in the United States,
the United Kingdom, Europe and other countries.

ISBN: 978–0–230–10276–7

Library of Congress Cataloging-in-Publication Data is available from the
Library of Congress.

A catalogue record of the book is available from the British Library.

Design by Newgen Imaging Systems (P) Ltd., Chennai, India.

First edition: December 2012

10 9 8 7 6 5 4 3 2 1

This book is dedicated
by all the authors
to J. L. Heilbron,
scholar, mentor, colleague, friend.

Contents

Part III   Histories

Part IV   Things

Figures

Acknowledgments

We offer heartfelt thanks to Suzanne Marchand for her careful readings of the manuscript and incisive editorial suggestions. We also thank her and her two coeditors, Anthony La Vopa and Javed Majeed, for finding a place for our volume in such an excellent company in the *Palgrave Studies in Cultural and Intellectual History*.

We are grateful to Nina Bell, Lori Kelley, and Linda Schneider for their expert assistance in putting together the manuscript; to Chris Chappell and Sarah Whalen for their stewardship of the manuscript through the editorial process; to the anonymous reviewers for their comments; to Rich Bellis and Deepa John for turning the edited manuscript into a book; to Sukanya Sai Satheesh for her skillful and thorough copyediting; and to Laura Bevir for the excellent index.

Introduction: A Mingled Yarn

Jessica Riskin

The web of our life is a mingled yarn.

—William Shakespeare, *All's Well that Ends Well,* c. 1602

The history of science has become crucial to the larger discipline of history, and the essays gathered in this volume demonstrate why. They reveal the entanglements of the history of science with all histories: cultural, intellectual, legal, religious, military, institutional, architectural, social, quantitative, colonial, and environmental. Moreover, these essays instantiate the crucial methodological function that the history of science has served, over the past two decades, for the larger discipline of history: the history of science has been the key site for discussions of major epistemological questions, and more broadly, questions about knowledge-making. How is knowledge dependent on its context? Are all truths local and contextual? Are some truths more transcendently true than others? How much of the world can we know? The history of science has been the key site, too, for posing and trying to answer moral questions about knowledge-making, and for understanding the interfaces among institutions, individuals, and the marketplace.[1] You will witness all of this in the following pages.

Historians of science, in our turn, need and use all the other branches of history. Ours is perhaps the most intradisciplinary field within the discipline, and our borrowings and excursions are necessary, not elective. As historians of science started tracing the web of scientific practices beyond the evidence provided by scientific publications and the manuscript drafts leading to them, it became obvious that practitioners of the sciences have also been instrument-makers, astrologers, inventors, artists, doctors, travelers, colonialists, lawyers, institution-builders, alchemists, political actors, and that all these roles (and many others) have been integral to their making of science. The image of science became quickly more complicated, and to understand the wider range of evidence that historians of science were finding, they had to become real historians.

1

Although this more expansive focus in the history of science has often carried announcements of changing methodologies and political outlooks—from internalism to externalism, from science to its cultural context, from science celebrators to science critics—we think that the broadened focus was in fact primarily the effect of something more mundane: research. As the historian John Heilbron teaches his graduate students (who have included this volume's editors and several of its authors), you must follow where your subject leads you, into any area of history—institutional as well as intellectual, practical as well as philosophical, technical as well as theoretical. We believe he is right. The demise of the so-called internalist view of science may have resulted less from methodological revolutionizing than from empirical overwhelming: it burst at the seams due to vast and mounting evidence that science is inseparably woven into the rest of life.

By following its protagonists and tracing their practices, the history of science has eventually penetrated anywhere knowledge is made, used, arranged, taught, displayed, debated, censored, communicated, made secret, exposed, or simply claimed, thus providing windows into virtually every social and cultural phenomenon. The narrow focus on formalized knowledge of earlier history of science has given way to large vistas.[2] As part of this work of complication and contextualization, historians of science have tended to leave behind the notion of a monolithic science with a single method, finding instead many kinds of scientific activity. The sciences have come to figure in the history of science as varieties of a distinctive mode of human engagement with natural phenomena—a combination of theories and practices that emerged from a host of older modes of engagement (philosophical, artisanal, magical) during the early modern period and has been a central force in organizing human society and understanding ever since.

The earlier notion of a single science helped to fuel arguments that the epistemic status of science must necessarily be undermined by historicization, contextualization, and localization, in fact, the very kinds of analyses in which historians engage. Indeed, several generations of historians of science during the 1960s through the 1990s labored to analyze and qualify the epistemic power of the sciences. Thomas Kuhn effectively launched the revolution with his 1962 manifesto, *The Structure of Scientific Revolutions*. Here he demonstrated by means of historical examples the crucial importance of what he called "paradigms," conceptual frameworks, in science. Paradigms were permeable to every aspect of human intellectual, cultural, and social life.[3] Kuhn was both a philosopher and a historian of science; he and others of his generation built close relations between the two disciplines. The history of science adopted the linguistic and theoretical focus of philosophy during the 1960s and 1970s, while historians and philosophers alike used these philosophical tools to analyze the role of historical context in natural science.[4]

There followed a series of challenges from the history of science and science studies leveled at the sciences, as historians turned to cultural history and the sociology and ethnography of scientific practices during the 1980s and 1990s.

Sociologists of science connected scientific ideas and theories to social interests.[5] Feminist scholars examined the role of assumptions about gender in shaping practice and theory.[6] Historians of economics analyzed economists' claims to produce scientific representations of economic processes.[7] Science studies, an interdisciplinary field comprised primarily of sociology, anthropology, history, and philosophy, saw a proliferation of theories describing themselves as "constructivist" (i.e., devoted to revealing the many forces—social, cultural, economic, and institutional—at work in constructing scientific knowledge and practice).[8]

The best thing to emerge from these writers' collective work was a sense that scientific knowledge, like any human product, has human contours: linguistic, theoretical, cultural, social, political, and institutional. As with all intellectual movements, this one also had its excesses, which provoked correspondingly extreme reactions. There was rancor. Alan Sokal, a physicist at New York University, was angered by the relativist turn of the humanities, and particularly of science studies, which he saw as an abdication of humanists' proper role. Humanists, Sokal thought, should be critics who question established beliefs to correct them; relativists criticized, but not to correct. In the spring of 1996, Sokal published a satirical science studies article in the journal *Social Text,*[9] and simultaneously published an acknowledgement of the spoof, laying out his motivations, in *Lingua Franca.*[10]

During the ensuing "Science Wars" many combatants, including many scientists, angrily rejected certain of these contemporary trends in the history of science and science studies.[11] The Science Wars pitted the "rationalists" or "realists" against the "relativists" or "constructivists." Ironically, the deadlock resulted from a shared assumption: radicals on both sides imagined that truth, whether historical or scientific, must be all or nothing, transcendent or else fraudulent.

By seeking a third possibility, we have arrived at a kind of peace accord. We agree that there is no view from nowhere, and also that there are better and worse vantage points, more and less useful places to stand while looking. Science describes and explains the world, not from outside, but from privileged as well as compromised places within it. Science is neither natural fact nor social process, neither discovery nor invention. It is neither reflection nor application, neither intellectual nor embodied, but all of these. It is not morally neutral; nor is it a pure exercise of power. Science is a mingled yarn, no more separable into discrete parts (natural vs. social, objective vs. subjective) than the thread of life. These days among historians, philosophers, scientists, and cultural critics, you would be hard put to find a radical relativist, or a naïve realist.

If this were a fairy tale, it might go like this. Once upon a time, in the days when internalists and externalists roamed the earth, when objectivists and constructivists ate one another for lunch, a generation of scholars struggled to find a third way through the scorching deserts, the tangled jungles, the foggy marshes, and the grassy plains of history. Kuhn, for example, spent a tormented career laboring to rewrite his accidental masterpiece, persuaded that no one, finally, had understood that his answer to realism was never meant to be relativism. But what was it meant to be? We think we have arrived at an answer.

The struggles of those decades did reveal a third way. Now that the Science Wars of the late twentieth century are ended and we have learned both the lessons and the limits of social construction, we are ready to move on. From the wars, we have brought back one major lesson: that science is in and of history, that nature and culture are different words for the same world. We might call the resulting approach a kind of historical pragmatism, for it is one deeply rooted in the practice of history. When pursuing specific and complex historical questions, there is little explanatory power to be gained by invoking "society" or "nature," or arguing about which one is the *explanandum* and which the *explanans*. Specific answers to specific questions can never take so general a form: the God of history is truly in the details.

Manifestations of this methodological transition include an increased attention to instruments, apparatus, and scientific practices;[12] a concern with institutions, communities, and infrastructure;[13] a heightened interest in the communication of knowledge and the history of the book;[14] and a currently thriving history of "things."[15] In retrospect, though, historians of science, through the practice of their craft and the suspension of theoretical preconceptions about the boundaries of their field, had arrived at these pragmatically motivated projects well before the Science Wars. For example, Heilbron's monumental study of the history of electricity, published in 1979, pioneered the close consideration of instruments, apparatus, and practices; while his studies of the Royal Society of London under Newton's presidency and of the Lawrence Berkeley Laboratory, published during the 1980's, were pivotal contributions to the emerging history of scientific institutions.[16] Recent and current work in these burgeoning areas thus reaches back to a longer-standing tradition.

The following chapters exemplify and embody this current pragmatism in the history of science. They recover the historical conjunctions in modern scientific ideas and practices, revealing the deep involvement of science in the major institutional bases of modern social life, and their reciprocal involvement in the theory and practice of science: law, market, church, school, and nation. The work of three generations of historians gathered here demonstrates the everyday mingling of theories, practices, instruments, institutions, nature, and society. With a chronological span reaching from the Renaissance to the present, our topics range from sundials to genetic sequences, from calculating instruments to devices that simulate human behavior, from early cartography to the origins of operations research in World War II Britain.

Nature Engaged is not aimed at calling the attention of other history of science specialists to a new topic or research problem within the discipline. Rather, we invite the broader readership of the Palgrave Studies in Cultural and Intellectual History series into the history of science by showing the many perspectives and engagements it has to offer. We have therefore chosen variety over focus, gathering work from prominent scholars who have been instrumental in challenging the boundaries of the discipline. These chapters reveal disciplinary engagements with both theoretical and practical fields: the sciences, intellectual property, political theory, religion, museum studies, legal and forensic evidence, philosophy, warfare, technology, and business. Each of the essays relates a story of mutual permeation: facts and values, nature and culture, science and society, physics (in Aristotle's encompassing sense) and history. The book is divided into four parts, corresponding with four areas of mutual engagement and permeation: "Conventions," "Laws," "Histories," and "Things."

The essays in "Part I: Conventions" treat the engagement of science in the formation and promulgation of conventions of all kinds—instrumental, social, pedagogical, military—and the reciprocal engagement of these conventions in the formation and promulgation of scientific ideas and theories. Ken Alder opens our book with a pun and its consequential implications: "Convention," of course, means a kind of gathering and also the system of agreements that such a gathering both requires and generates. Alder presents the nineteenth-century emergence of scientific internationalism through the now-familiar ritual of the international scientific convention and associated conventions. This apparently tame topic, Alder shows, hides a radical past—the very idea of international scientific meetings once carried more than a whiff of danger. Scientific internationalism as a mode of interaction replaced an older form that Alder calls cosmopolitanism, in which eighteenth-century savants corresponded with their counterparts in other countries, visited one another, and joined one another's national philosophical academies as foreign members.

By traversing state and religious boundaries, scientific cosmopolitanism already posed a threat to the monarchical and church powers that fostered it in the paradoxical hope of domesticating it. But cosmopolitanism also relied on the social systems of those monarchies: systems governed by personal interactions and agreements on standards and terminologies. This older mode of cosmopolitan engagement could not outlive the world in which it operated: the sociable and intimate world of the Republic of Letters. It gave way, by the mid-nineteenth century, to a formal system in which nations jointly determined, by means of international treaties, such matters as the length of the meter.

This giving way, Alder shows, was necessarily violent. Conventions in both of their conjoined senses, technical and social, do not succumb easily. Their remaking was required for and also helped to effect the breaking down of local markets and the ruptures in sovereignty of the mid- to late nineteenth century: revolutions, conquests, unifications, and decolonizations. The conventions of scientific internationalism that arose from these struggles both supported and limited the new sovereign powers, as they enabled and delimited the evolving sciences. The generation of scientific positivists who lived through these developments, indeed, arrived at the strong view that scientific laws themselves were essentially conventional.

Take, for example, the law that water boils at 212 degrees Fahrenheit. This law is surprisingly conventional, Hasok Chang shows in chapter 2. Not only is it not (necessarily) the case but the discoverers (or inventors?) of the boiling point of water also knew it very well. Indeed, eighteenth-century experimenters in physics and chemistry such as the Genevan naturalist Jean-André De Luc carefully recorded the variability of the boiling point of distilled water under a given pressure depending upon the sort of vessel containing it and the source of heat one employed. Others, including the Paris chemistry and physics professors Joseph-Louis Gay-Lussac and Jean-Baptiste Biot, affirmed that water boiled at a lower temperature in metal than in glass.

Chang's research, an amalgam of historically informed experimental science and experimentally informed historical investigation, further revealed that water boils at a lower temperature in a scratched beaker than in a pristine, smooth one. Boiling behavior also occurs differently in different-shaped vessels and changes over time. By exploiting such variabilities, De Luc reported having been able to bring water to 112°C without boiling, and two centuries on, Chang has confirmed the finding.

To say that distilled water under standard pressure boils at 100°C is therefore not precisely true. Rather, under a very circumscribed set of conditions, one can get this to happen. One would think this variability in the boiling point of water would be utterly common knowledge: after all, who has never boiled water? Who, indeed, has not done so in variously shaped vessels, some metal, some glass, some more scratched, some smoother? We are not talking about specialized procedures. But the conventional dimension of even so elementary a scientific fact disappears in both historical and scientific retellings. Restoring it, Chang argues, makes for truer history and truer science by the same token.

The length of the meter and the boiling point of water, two basic elements not only of science but also of daily life, both apparently supremely simple, the stuff of elementary-school science classes—each hides tangles of technical, social, and physical complexity in its history. The same is true of another part of the basic science curriculum, the periodic table. In his chapter Michael Gordin examines the complicated origins of this convention: initially, there were two competing periodic systems of elements—one by the German chemist Julius Lothar Meyer and the other by the Russian Dmitri Mendeleev. Although Mendeleev's name is now the one primarily associated with the invention, in 1882 the Royal Society of London saw the two as coinventors, awarding them both the Davy Medal.

Mendeleev and Meyer appear the most unlikely of coinventors, having converged on the periodic table of the elements from completely different theoretical commitments (or lack thereof). Mendeleev was a theoretical conservative who had opposed the latest chemical theories of atomic weight and valency, whereas Meyer had been one of their supporters. However, Mendeleev did use the structural patterns emerging from his table to predict the existence of unknown elements—predictions that proved correct, earning him much recognition, and going a long way toward sealing his status as the inventor of the periodic system. Meyer, in contrast, did not venture predictions (despite his more open stance toward theory), a caution that may have cost him a greater share of posterity. Gordin shows that these two radically different chemists shared a deep interest in and commitment to pedagogy, which can help to explain both their convergences and their divergences.

Each of them came to think of a periodic arrangement of the elements as a good way to convey their features to students in a textbook. At the same time, the differences in their approaches reflected their distinct pedagogical traditions. The long tradition of historiography of the periodic table, by celebrating a single discoverer, has obscured the many, important contingencies in its invention, which Gordin restores in his retelling.

The divisive power of scientific conventions is the theme of Dominique Pestre's essay on early information-processing practices developed at the interface between British science and the military during World War II, often with Churchill's direct prodding and supervision. The war, Pestre argues, was won not only with soldiers and weapons, but also with paper, statistics, diagrams, efficient protocols and administrative rules. Modern war, as he shows it, was a paper war—a war of offices and reports. He sets out a particularly bureaucratic history of scientists' involvement in World War II, intended as a corrective to popular narratives of scientific geniuses developing the ultimate weapon. The engagement of scientists in organizing the war yielded management and data-processing techniques that became part of postwar everyday governmental and corporate administration.

Based on extensive research in British archives, Pestre shows that the British (not the Americans) developed these techniques, and they did so because they could do little else. In the UK, German attacks created an ongoing emergency situation that precluded the more long-term weapons development programs one saw in the United States, such as the Manhattan project and cybernetics-based automatic firing controls. In the UK, time was measured in weeks, not months, let alone years. The British had to learn how to do with what they had. This meant an accelerated development of management and decision-making techniques based on new data-collection and processing; it also meant a local and problem-specific, rather than a global and systematic, approach. The results included systems for radar-based antiaircraft batteries and air response to German air attacks; for detecting enemy submarines; for designing naval convoys to minimize losses; for statistically monitoring resources and losses; and for using diagrams, charts, and maps of the information to enable prompt strategic and political decisions. Pestre shows how the British military and collaborating academic scientists not only won the war but also actually created the science of management.

If the chapters in Part I canvass various conventional dimensions of scientific practice—metric, instrumental, taxonomic, pedagogical, systematic, military—those in "Part II: Laws" focus on a particular category of conventions, legal ones, in their engagements with science. These chapters examine how legal and scientific concepts and practices have shaped one another.

Part II opens with Matthew Jones's analysis of the origins of modern patent law through the lens of a landmark example of technological innovation. Examining G. W. Leibniz's negotiations with the Académie royale des sciences concerning his calculating machine during the 1670s, Jones shows how these represented a turning point in early modern practices concerning intellectual property. Unlike modern patents, Jones argues, early modern privileges for inventions (which are commonly but anachronistically also referred to as "patents") did not provide property rights in inventions construed as ideas embodied in machines. Rather, the privileges were more like gifts that the sovereign offered a subject as a reward for the production of a useful (and typically working) machine.

Early modern calculating machines challenged the legal status quo. Their inventors, first Pascal but more so Leibniz, cast their inventions as essentially ideas, of which the material machine, the traditional object of the privilege, was merely the execution. This execution, however, required skills that the inventors recognized they did not have, leaving them tethered to skilled artisans who could represent themselves as coinventors or at least necessary links in the inventive chain. Therefore, the transition from early modern privileges that rewarded machines to modern patents creating rights in the inventors' ideas, Jones concludes, hinged on more than a transition from political absolutism to representative democracies. It also required a reconfiguration of inventive skills and relations between the philosopher and the artisan.

Next, Mario Biagioli revisits the long-standing debate in the history and sociology of science about the epistemic role of eye-witnessing by reconstructing Kepler's sophisticated and innovative engagement with legal practices of testimony. There have always been exchanges between law and science around evidential practices—the law introducing scientific evidence and scientists as expert witnesses in court, and science invoking legal standards of testimony to back up the epistemic value of collective witnessing practices in the experimental sciences. Kepler's engagement with the law—specifically inquisitorial law—is striking in that it frames the role of eyewitnessing in ways that are radically different from those developed by English experimental communities who relied instead on common law tradition.

What transpires from these comparisons is an intriguing analogy between scientific styles of argumentation and evidentiary practices and the legal traditions the scientists are borrowing from. Furthermore, Kepler's use of testimony during the telescopic observations he conducted in response to Galileo's discoveries exemplifies the use of eyewitnessing for rhetorical rather than epistemic purposes. Biagioli argues that, surprisingly, Kepler did not take witnessing to produce the kind of evidence natural philosophers should care about but evidence that was nevertheless useful to convince nonprofessional audiences. It was, so to speak, cheap evidence that, thanks to its aura of legal credibility, could silence uninformed skeptics. It could not, however, be binding to true philosophers.

An ongoing episode in the history of intellectual property is the focus of Daniel J. Kevles's chapter; he analyzes the regulation of property rights to genes. Kevles draws upon a surprising historical analogy: property rights to the railroads in nineteenth-century America. Though railroads are huge and genes tiny, railroads built and genes discovered, railroads an emblem of the industrial revolution and genes of the information revolution, Kevles argues that the struggles over the ownership of each have been importantly similar and that there are therefore historical lessons to be drawn from the railroads and applied to genes.

In particular, conflicts between the private interests of the railroads and the public interests of those whose lives they increasingly shaped—shippers, suppliers, consumers, farmers—led during the 1870s to the establishment of a regulatory regime. This regime consisted of railway commissions with the authority, for example, to set maximum rates and prohibit price discrimination. Lawmakers and, ultimately, the Supreme Court justified these laws on the ground that private property, when important to public interests, must submit to governmental control for the common good.

Kevles argues that the science of human genes is as essential to health and medicine as were the nineteenth-century railroads to the economy they propelled. Moreover, human genes are like railroads in being "natural monopolies": "monopolies created by nature or circumstance." Railroads tended to be natural monopolies because of the limited number of geographically and economically viable routes through a given region. Genes are natural monopolies because of their specialized functions: a gene that disposes a person to a disease is unique in that regard.

The question what a gene is figures importantly in arguments about whether and how they should be patented. Is a gene a particular substance or a sequence of information or both? Does it include any or all of its possible mutations? Is a gene also the sum of all its functions, including those that are currently unknown? Those who oppose gene patenting have argued that a gene is a product of nature and therefore not patentable. Patent owners have responded that genes are chemical structures discovered in laboratories just like pharmaceuticals or paints or dyes.

The scientific, legal, and practical answers to these questions are, Kevles shows, conjoined. Rather than trying to distinguish them, he draws upon historical precedents to propose an answer that instead acknowledges their entanglement, an answer whose logic is fundamentally not scientific but moral. Precisely because genes constitute a convergence of nature and artifice, public and private interests, science and commerce, there is a clear moral and social imperative for the regulation of gene patenting.

Next, taking up the engagements of science and law from the law side, by examining the penetration of scientific ideas and practices into legal settings, Tal Golan recounts the early career of epidemiology in American law. Epidemiological evidence entered American courtrooms in the 1970s, where it proliferated wildly during the 1980s, provoking a series of legal innovations to screen and control it by the early 1990s. Golan argues that the fortunes of epidemiological evidence accompanied the incidence of mass tort litigation and, more generally, that the collective plaintiffs and statistical arguments of mass tort litigation remade the relations of science to law in late twentieth-century America.

In the convergence of epidemiology and mass tort claims, causal arguments gave way to correlative ones, risk factors took the place of biological mechanisms and statistical studies replaced experimental reports. These developments were controversial among medical scientists, many of whom argued that epidemiology was not a true science in itself, but rather a technique for generating causal hypotheses. Golan shows that it was really in legal and regulatory settings that epidemiological methods, for a time, most easily flourished. Here emerged a kind of "black-box" epidemiology, focusing on evaluations of risk and putting aside questions of causal mechanism.

In the early 1990s, responding to the escalating frequency of black-box epidemiological arguments, and also to growing worries about junk science in the courtroom, epidemiologists, judges, and legal scholars introduced various techniques for classifying and evaluating the causal implications of statistical correlations. These constituted the methods for a new mode of scientific analysis and argumentation, Golan's story reveals, one that was at once and in equal measure scientific and legal.

Thus the chapters of parts I and II examine how scientific and social conventions have continually shaped and reshaped one another. Those of "Part III: Histories" explore how modern scientific and modern historical ideas and practices have helped to constitute and reconstitute one another.

An episode in the history of the science of history is Anthony Grafton's topic. He traces sixteenth-century transformations in the technical field of chronology, in particular at the hands of a man better known for his mappings of space rather than time, the cartographer Gerardus Mercator. Chronology involved the use of various methods to identify and order dates in ancient history. These traditionally centered upon literal readings of the biblical text. To these, Mercator added other methods. For example, from Jewish exegetical tradition he borrowed the kabbalistic manipulation of Hebrew words and letters. To grapple with the problem posed by the Egyptian dynasties, namely that they indicated a history extending back long before the Flood, Mercator consulted the neo-Platonic commentaries used by his predecessor, the Italian humanist Marsilio Ficino, and hesitantly admitted the possibility of a long, antediluvian Egyptian history.

Mercator's map of time included both the sacred time of Jesus's mission on earth, as narrated by the four Evangelists, and the more general time of world history, drawn from Egyptian and other sources. Mercator arranged these on a single, uniform scale, which meant leaving large, empty spaces. It also meant devising an intricate tabular scheme for correlating different narratives within a given tradition (the Gospels), different chronological traditions (Egyptian, Babylonian, Greek, Jewish, Christian) and their corresponding eras (the era of Nabonassar, the era of the Olympic Games, the era of Solomon's Temple).

This work of correlation involved moving among solar and lunar calendars. It relied heavily, in other words, upon astronomical knowledge and calculation. Moreover, to have fixed, universal points of reference, Mercator followed the example of the German astronomer Petrus Apianus, a generation senior to him, and turned to a readily available source of dateable celestial events, namely eclipses. Mercator drew his astronomical data especially from Ptolemy.

Something surprising took place as Mercator carried out his work of applying astronomical data to the determination of historical dates; he discovered what he believed to be a hitherto unremarked anomaly in the motion of the moon. In other words, having started out to use astronomy to inform history, he found himself doing the reverse as well: using history to inform astronomy. He thereby left himself at sea in chronological space, without the temporal bedrock he had sought in astronomy, but having decisively demonstrated the profound interdependence of historical and astronomical science, narrative and natural time.

Modern scientific and modern historical understanding have informed one another from their (conjoined) inception; here is a central theme, too, of Paula Findlen's chapter. She analyzes the rivalry to tell the definitive story of Galileo's trial, arguably the leading exemplar of a historical watershed around a scientific development. In her investigation, Findlen goes beyond the best-known and most authoritative early account of Galileo's life, *Racconto istorico della vita del Signor Galileo Galilei*, written by his last living disciple, Vincenzo Viviani. Findlen instead focuses on the competition to produce an authorized Jesuit biography of Galileo during the 1670s. Viviani's account and the previous and subsequent ones appear in her reading as instances in a general examination of how Galileo's first biographers understood the significance of their project.

The first generation of biographers for the most part cautiously omitted any mention of the trial, either stopping before or leaping over it. Then, during the late 1650s and 1660s, the project to produce a definitive account of Galileo's life became the shared preoccupation of the community people connected with the Galilean experimental science society, the Accademia del Cimento. Authors including Viviani began to write more extensive accounts of Galileo's life and to include tentative and apologetic mentions of the trial.

In 1678, Viviani received a letter soliciting his help from a young Jesuit mathematics professor, Antonio Baldigiani, who was editing Athanasius Kircher's *Etruria Illustrated*. Kircher, illustrious German Jesuit polymath and great nexus in the world of letters, was now old and frail, leaving Baldigiani some freedom as editor to shape the work, which was to include a biography of Galileo—here would be the first authoritative Jesuit account of Galileo's life. Viviani agreed to help, and the subsequent correspondence between the two men displays the extreme sensitivity of their project. Their exchanges are especially telling, since they disagreed about what would be the best strategy. Findlen's analysis of their extraordinary correspondence shows that the trial, which has long masqueraded as a watershed dividing science from culture and authority, in fact had irreducibly mixed meanings from its very first retellings. How one understood the historical import of this landmark event was inseparable from how one understood its scientific meaning, and vice versa.

Parts I, II, and III then, are about the entanglements of scientific ideas and practices with social, legal, and historical conventions. The final section,

"Part IV: Things," examines the entanglements of science with the stuff of the world. This section tours the material culture of science since the Renaissance and shows the inseparability of ideas and objects, theories and devices. In Jessica Riskin's piece, the devices in question were the moving, mechanical figures ubiquitous first in early modern churches and cathedrals, and then on the grounds of wealthy estates. These automata closely informed Descartes's philosophical notion of living bodies as machines (as well as this notion's momentous counterpart, the disembodied human mind).

To approach Descartes's revolutionary philosophy in terms of the devices that informed it is to see it differently in a number of ways. First, the devices indicate what was *not* very new or radical in Descartes's revolution: neither the philosophical idea of animal-machinery, in itself, nor the actual images of life in mechanism ran counter to established Christian practice or doctrine. Quite the contrary, automata appeared first and most commonly in churches and cathedrals; moreover, the idea as well as the technology of human-machinery was also indigenously Catholic. The Church was a primary sponsor of the literature that accompanied the technology of lifelike machines, and the body-machine was also a recurrent motif in Scholastic writing.

The devices show the roots of Descartes's idea in the world he inhabited, and in so doing, they also help to indicate a kind of instability, a fault line running through the very core of his program. His idea of the mechanical body took on an array of meanings, such as passivity, unresponsiveness, even lifelessness, that it did not initially hold. Indeed, in the first instance, his animal-machine model meant something like the opposite: responsiveness, feeling, vitality. Looking at the actual life-like machines to which Descartes referred reveals this fundamental instability in his idea of living machinery, and so reopens an older, perhaps only temporarily eclipsed, set of possibilities for what it can mean to be both mechanical and alive.

Jim Bennett's objects are Renaissance sundials: instruments we have come to misunderstand as mere time-telling devices that decorate old public buildings and monuments. Bennett reconstructs the early modern discipline of dialing, or sundial-design, which enjoyed an enormous following across all levels of society. He examines the intellectual concerns of their designers, which included some of the leading astronomers of the time, and the technical backgrounds of their users, including the books they read and the other instruments they knew. Bennett shows that telling time was only the tip of the iceberg of Renaissance and early modern dialing.

Sundials expressed the concerns and ingenuity of practitioners of cosmography, "the whole and perfect description of the heavenly, and also elementall parte of the world," as the Elizabethan mathematician John Dee expressed it in 1570. Cosmography was indeed concerned with time, but not as the answer to the question: "What time is it?" To a cosmographer, time was an index of geographical location, the determination of which was the discipline's raison d'être. Complementary to astronomy, cosmography traced the motions of the sun and the celestial sphere, dividing and organizing the earth's surfaces with geometrical lines—equator, tropics, and lines of latitude —in relation to those motions. In that framework, time was the most immediate link between the heavens and the earth, and the dials captured it.

Finally, Giuliano Pancaldi looks at a roomful of things: William Thomson's apparatus room at the University of Glasgow, where he taught natural philosophy from 1846 until his retirement more than a half-century later. At Glasgow, Thompson conducted research in electricity and magnetism that not only made him famous among fellow scientists, but also gained him key telegraph patents and a seat on the board of directors of the Atlantic Telegraph Company. Around 1880, electricity and magnetism developed from subjects of primarily pure science to key areas of engineering. Pancaldi chooses Thomson as a pivotal figure during the period in which the study of electromagnetism began to travel outside the academy, but had not yet fully arrived in the industrial laboratory. And he chooses one of Thomson's devices, the mirror galvanometer, as the emblem of that hybrid stage.

Invented in the apparatus room at Glasgow, the mirror galvanometer was a laboratory instrument that quickly became an extraordinarily useful detector of telegraphic messages, allowing the telegraph industry to proceed with the development of transoceanic networks despite a patchy scientific understanding of the problems of signal transmission over long and submerged cables. The mirror galvanometer straddled pure science and industrial application, as did the room in which it originated: a university setting permeable to industrial concerns, structured as a cross between a laboratory and an apparatus-building shop. Thomson's approach to his career also moved between both worlds. Pancaldi shows that Thomson initially sought a scientific entry into high-stakes telegraphic problems, applying for a patent for a theoretically informed design for transatlantic cables in 1854. But when that strategy failed, he achieved great success with the humble galvanometer, and by 1858 he had built a central position for himself in both the science and the industry of the telegraph.

Pancaldi's analysis of Thomson's hybrid strategies, located between academia and industry, and hybrid tools, simultaneously theoretical and practical, closes our book by epitomizing the shared outlook of the chapters we have assembled here.

Thus, the stories told in these chapters, from Renaissance to contemporary science, are about the worldly engagements of the sciences in legal, economic, pedagogical, religious, military, and political institutions, and in the evolving landscape of daily objects and devices. Through these connections, the chapters come together as instances of what we have called a third way, a new pragmatic consensus in action. We assume the connection of part to whole, event to institution, meaning to context, and so sketch the outlines of a pragmatic genealogy of science in the mingled yarn of modern history.

Notes

1. See the recent discussion of the ubiquitous place of the history of science in early modern history in Pamela Smith, "Science on the Move: Recent Trends in the History of Early Modern Science," *Renaissance Quarterly* 62 (2009): 345–375.

2. For a discussion of the field's cross-disciplinary engagements, see Mario Biagioli, "Post-Disciplinary Liaisons," *Critical Inquiry* 35 (2009): 816–833, which discusses how, by participating in the broader field of science studies, the history of science has found it natural to develop important intellectual linkages also with sociology, philosophy, and anthropology as well as with history, visual studies, museum studies, philosophy, religion, political theory, literary studies, communication and media studies, science policy, and of course the sciences themselves.

3. T. S. Kuhn, *The Structure of Scientific Revolutions* (Chicago: Chicago University Press, 1962).

4. For landmark texts, see Georges Canguilhem, *Le normal et le pathologique* (Paris: Presses universitaires de France, 1972); Michel Foucault, *Les mots et les choses* (Paris: Gallimard, 1966); Ludwik Fleck, *The Genesis and Development of a Scientific Fact* (Chicago: University of Chicago Press, 1979 [first published in German in 1935]); and Paul Feyerabend, *Against Method: Outline of an Anarchistic Theory of Knowledge* (London: Verso, 1975).

5. S. Barry Barnes, *Scientific Knowledge and Sociological Theory* (London: Routledge and Kegan Paul, 1974); Barnes and David Edge, eds., *Science in Context: Readings in the Sociology of Science* (Cambridge: MIT Press, 1982); Barnes and David Bloor, "Relativism, Rationalism and the Sociology of Knowledge," in Martin Hollis and Steven Lukes, eds., *Rationality and Relativism* (Cambridge: MIT Press, 1982).

6. For classic examples see Carolyn Merchant, *The Death of Nature: Women, Ecology and the Scientific Revolution* (New York: HarperCollins, 1980); Margaret Rossiter, *Women Scientists in America* (Baltimore: Johns Hopkins University Press, 1982); Evelyn Fox Keller, *A Feeling for the Organism: The Life and Work of Barbara McClintock* (New York: Freeman, 1983); *Reflections on Gender and Science* (New Haven: Yale University Press, 1985); Sandra Harding, *The Science Question in Feminism* (Ithaca: Cornell University Press, 1986); Donna Haraway, *Primate Visions: Gender, Race and Nature in the World of Modern Science* (New York: Routledge, 1989).

7. Philip Mirowski, *Against Mechanism: Protecting Economics from Science* (Lanham, MD: Rowman and Littlefield, 1988); *More Heat than Light: Economics as Social Physics, Physics as Nature's Economics* (Cambridge: Cambridge University Press, 1991); Donald MacKenzie, *An Engine, Not a Camera: How Financial Models Shape Markets* (Cambridge: MIT Press, 2006); *Material Markets: How Economic Agents are Constructed* (Oxford: Oxford University Press, 2009).

8. On constructivism, see Jan Golinksi, *Making Natural Knowledge: Constructivism and the History of Science* (New York: Cambridge University Press, 1998). For key texts in this tradition, see Donna Haraway, "A Manifesto for Cyborgs: Science, Technology, and Socialist Feminism in the 1980s," *Socialist Review* 15 (March–April 1985): 65–107; Michel Callon, "Some Elements of a Sociology of Translation: Domestication of the Scallops and the Fishermen of St Brieuc Bay," in John Law, ed., *Power, Action and Belief: A New Sociology of Knowledge,* (London: Routledge & Kegan Paul, 1986), 196–233; Bruno Latour, *Science in Action: How to Follow Scientists Through Society* (Cambridge: Harvard University Press, 1988); Hans-Joerg Rheinberger, *On Historicizing Epistemology: An Essay,* tr. David Fernbach (Stanford: Stanford University Press, 2010); and Andrew Pickering, *The Mangle of Practice: Time, Agency and Science* (Chicago: University of Chicago Press, 1995).

9. Alan Sokal, "Transgressing the Boundaries: Towards a Transformative Hermeneutics of Quantum Gravity," *Social Text* 46, no. 37 spring/summer 1996), 217–252.

10. Alan Sokal, "A Physicist Experiments With Cultural Studies," *Lingua Franca* (May/June 1996), 62–64.

11. Key texts of the Sciences Wars include: Andrew Ross, ed., *Science Wars* (Durham: Duke University Press, 1996); Paul Ross and Norman Levitt, eds., *Higher Superstition: The Academic Left and Its Quarrels with Science* (Baltimore: Johns Hopkins University

Press, 1997); Mario Biagioli, "Introduction," in Biagioli, ed., *The Science Studies Reader* (New York: Routledge, 1998), xi–xvi; Noretta Koertge, ed., *A House Built on Sand: Exposing Postmodernist Myths About Science* (Oxford: Oxford University Press, 1998); Alan Sokal and Jean Bricmont, *Fashionable Nonsense: Postmodern Intellectuals Abuse of Science* (New York: Picador, 1999); The Editors of Lingua Franca, eds., *The Sokal Hoax: The Sham that Shook the Academy* (New York: Bison Books, 2000); and Bruno Latour, "The Invention of the Science Wars: The Settlement of Socrates and Callicles," in Latour, *Pandora's Hope: Essays on the Reality of Science Studies* (Cambridge: Harvard University Press, 1999), Ch. 7.

12. Ken Alder, *Engineering the Revolution: Arms and Enlightenment in France, 1763–1815* (Chicago: University of Chicago Press, 1997); Hasok Chang, *Inventing Temperature: Measurement and Scientific Progress* (Oxford: Oxford University Press, 2004); Harry Collins, *Tacit and Explicit Knowledge* (Chicago: University of Chicago Press, 2010); J. L. Heilbron, *The Sun in the Church: Cathedrals as Solar Observatories* (Harvard: Harvard University Press, 1999); and Simon Schaffer, "Experimenters' Techniques, Dyers' Hands, and the Electric Planetarium" in *Isis* 88 (1997): 456–483.

13. Paula Findlen, *Possessing Nature: Museums, Collecting, and Scientific Culture in Early Modern Italy* (Berkeley: UC Press, 1996); Heilbron, *The Sun in the Church*; Lewis Pyenson and Susan Sheets-Pyenson, *Servants of Nature: A History of Scientific Institutions, Enterprises, and Sensibilities* (New York: Norton, 1999).

14. Timothy Lenoir, ed., *Inscribing Science: Scientific Texts and the Materiality of Communication* (Stanford: Stanford University Press, 1998); Adrian Johns, *The Nature of the Book: Print and Knowledge in the Making* (Chicago: University of Chicago Press, 1998); and Simon Schaffer, Lissa Roberts, Kapil Raj and James Delbourgo, eds., *The Brokered World: Go-Betweens and Global Intelligence, 1770–1820* (Sagamore Beach, MA: Science History Publications/USA, 2009).

15. Steven Lubar and W. David Kingery, eds., *History From Things: Essays on Material Culture* (Washington, DC: Smithsonian Books, 1995); Heilbron, *The Sun in the Church*; Lorraine J. Daston, ed., *Biographies of Scientific Objects* (Chicago: University of Chicago Press, 2000); Annemarie Mol, *The Body Multiple: Ontology in Medical Practice* (Durham, NC: Duke University Press, 2002); Bill Brown, ed., *Things* (Chicago: University of Chicago Press, 2004); Lorraine J. Daston, ed., *Things that Talk: Object Lessons from Art and Science* (Cambridge: Zone Books, 2004); Ian Hacking, *Historical Ontology* (Cambridge: Harvard University Press, 2004); and Ursula Klein and Wolfgang Lefevre, *Materials in Eighteenth-Century Science: A Historical Ontology* (Cambridge: MIT Press, 2007).

16. J. L. Heilbron, *Electricity in the 17th and 18th Centuries: A Study of Early Modern Physics* (Berkeley: UC Press, 1979); Heilbron, *Physics at the Royal Society during Newton's Presidency* (Los Angeles: Clark Library, 1983); Heilbron and Robert Seidel, *Lawrence and His Laboratory: A History of the Lawrence Berkeley Laboratory* (Berkeley: UC Press, 1989). For other landmark texts in these traditions, see Roger Hahn, *The Anatomy of a Scientific Institution: The Paris Academy of Sciences, 1666–1803* (Berkeley: UC Press, 1971); Daniel J. Kevles, *The Physicists: The History of a Scientific Community in Modern America* (New York: Knopf, 1977); Ian Hacking, *Representing and Intervening: Introductory Topics in the Philosophy of Natural Science* (Chicago: University of Chicago Press, 1983); Steven Shapin and Simon Schaffer, *Leviathan and the Air-pump: Hobbes, Boyle and the Experimental Life* (Princeton: Princeton University Press, 1985); Peter Galison, *How Experiments End* (Chicago: University of Chicago Press, 1987); David Gooding, Trevor Pinch, and Simon Schaffer, eds., *The Uses of Experiment: Studies in the Natural Sciences* (Cambridge: Cambridge University Press, 1989); and Jed Z. Buchwald, *Scientific Practice: Stories and Theories of Doing Physics* (Chicago: University of Chicago Press, 1995).

Part I
Conventions

1

Scientific Conventions: International Assemblies and Technical Standards from the Republic of Letters to Global Science

Ken Alder

The Republic of Letters never assembled. The cosmopolitan savants of early modern Europe corresponded with fellow natural philosophers across the Continent and around the globe, and many traveled great distances to study alongside colleagues in foreign lands. But the Republic of Letters itself was never more than a virtual community.[1] More than that: no one even *proposed* that savants from different nations collectively assemble to discuss matters of common concern. Or rather, no such meeting was proposed or assembled until the Republic of Letters was in its death throes at the very end of the eighteenth century, when the rise of a very different kind of republic provoked a devastating series of nationalist wars across Europe. And even then, transnational science had to wait out a sixty-year gestation—until the second half of the nineteenth century—before conferences attended by scientists from different nations became an acceptable feature of scientific life and a new form of transnational science—call it international science—was fitfully born.

In the 150 years since that birth, such international conferences have become a banal—yet much appreciated—lubricant of scientific life, thanks to the sponsorship of transnational professional societies, supranational agencies, and now global NGOs. Indeed, such junkets—as they are sometimes called—have become so common that attendees may be forgiven for assuming that such meetings have been convened for as long as science itself has been transnational. And given the cosmopolitan roster of early modern scientific academies—and the transnational character of their endeavors—such an assumption might seem warranted. But as figure 1.1 demonstrates, the "tradition" of international scientific gatherings only took off in the later half of the nineteenth century, after science had already been a cosmopolitan

19

endeavor for nearly three centuries and the scientific academies had been meeting for two.[2] Why did it take so long?

One short, easy answer is "trains," and like many short, easy answers, it is not wrong. But it is not the whole story either, and in any case, not for the simple reason supposed. Savants had been visiting foreign colleagues for centuries before railways eased travel, and on two exceptional occasions in the 1790s a number of savants from different nations did gather for coordinated conventions—the first one in Germany, the second in France. This chapter will discuss these early, precocious efforts of cosmopolitan savants to assemble amid the wars of the French Revolution and compare them with the approach taken some 60 years later during the protracted birth of international scientific conferences amid the renewed warfare of the late nineteenth century. In the process we will discover that, like many other things that today seem banal or even frivolous, the idea of an international scientific convention once carried more radical overtones. Doing so will also help us understand how social and scientific norms are cocreated in distinct ways in eras governed by distinct systems of national/international law.

For the early period, I will consider three different episodes, each with its own vision of what might constitute scientific cooperation across national boundaries—(1) the first transnational scientific conference, which took place in August 1798 in Gotha with the ambivalent sanction of its duke, and convened by the astronomer Franz Xaver von Zach to discuss astronomical and metrological standards; (2) the second transnational scientific conference, which met in Paris later in the fall of 1798 under the auspices of the French Revolutionary state, with the goal of calculating the length of the meter; and (3) the never-implemented proposal made in 1816, after the collapse of

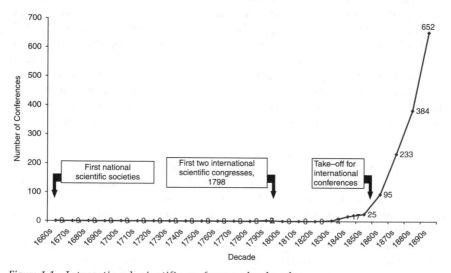

Figure 1.1   International scientific conferences by decade.

Source: Data adapted from Union des Associations Internationales, *Les Congres Internationaux de 1681 a 1899* (Brussels, 1960) 3.

Napoleon's empire, by Abbé Henri Grégoire, then in exile, for what he called an International Congress of Savants to be held (prospectively) in Frankfurt-am-Main, to which all the savants of the universe would be invited. For the later period, I consider the series of international assemblies held in Paris beginning in 1869, meetings that were meant to resolve the discrepancies in the metric system, but were interrupted by the Franco-Prussian War, only to (apparently) culminate in the Metrical Convention of 1875—although final resolution was delayed while a series of further international scientific meetings persisted through the late 1880s.

In both these periods, it was the perceived need for transnational metrical standards that was used to justify the assemblies. As the organizers of these meetings all noted, the exchange of scientific ideas and practices transcended local and national boundaries and only common standards of measurement would aid communication and the comparison of results. More to the point, the consolidating nation-states of Europe realized that the expansion of both domestic and long-distance commerce would be eased by the use of common technical and measurement standards. Hence, all these conventions were devoted to conventions. This is more than a pun. Assembling people proved essential to assembling norms. It may well be that many of these conferences were junkets, more beneficial to hotel-keepers than knowledge—as was often noted at the time. It may also be that the assemblies only provided a diplomatic cover for the hard work done by private industry, academic centers, and national laboratories that defined workable standards. Yet coordinating such standards often seems to have required the kind of bargaining facilitated by face-to-face meetings. Moreover, getting nation-states to adopt such standards required even more intense bargaining, because surrendering parochial measures meant giving up a degree of sovereign control in return for the promise of more efficient administration and long-distance trade. And this required the bodily presence and assent, not just of scientists, but of ministers plenipotentiary.

In short, these foreign "junkets" enabled participants with divergent interests to mix science with sociability, private wrangling with public ceremony, and collaborative experimentation with committee draft-writing in such a way as to produce consensus. These scientific conventions were modeled on assemblies for international diplomacy, and like them, involved chandelier festivities, back-room back-stabbing, and reams of bureaucratic prep-work. And like them, their outcome depended on prevailing understandings of sovereignty. No wonder they were so hard to pull off.

Cosmopolitan science...and its limits

For at least two centuries prior to 1800, a scattered band of European natural philosophers—along with their colleagues overseas—had imagined themselves to be simultaneously loyal subjects of their sovereign and citizens of something they called the Republic of Letters, a communications network that linked investigators of nature (and other kinds of scholars) to similarly inclined investigators around the globe. Even Isaac Newton, famous for working in splendid isolation, relied in his *Principia* on colleagues located at the remote edges of empire for reports on such matters as tides in Siam, pendulums in the Caribbean, and comets in South Asia.[3] Yet this global Republic of Letters was never more than a virtual republic; it had no territorial ambitions, no citizen army, no formal laws, no deliberative assemblies. And for just these reasons, the sovereign princes of Europe tolerated this divided loyalty. After all, in the natural law conception of monarchy—which dominated political thought in this period—all the prince's subjects were understood to owe multiple allegiances: to their prince, yes; but also to patrons, seigniorial authorities, and corporate bodies, such as guilds or universities. What *did* vex the nightmares of the consolidating powers of early modern Europe were *unregulated* assemblies of its subjects— particularly the urban poor, aggrieved peasants, and unlicensed theologians or natural philosophers. So beginning in the seventeenth century, European powers had begun to charter local scientific meeting clubs as formal academies, convening assemblies of these savants under the auspices of the nation-state. The most famous of these academies were the Royal Society in London and the Academy of Sciences in Paris.[4] To be sure, these national academies still honored the transnational scope of scientific exchange. They published the works of foreign savants, held prize competitions open to investigators from all lands, and elected "corresponding" members from abroad. In 1753 the Royal Society's president, George Parker—Earl of Macclesfield, astronomer, and MP—expressed this cosmopolitan ethos in grand terms:

> Learned men and Philosophers of all Nations... [s]hould consider themselves and each other as Constituent parts and Fellow Members of one and the same illustrious Republick; and look upon it to be beneath Persons of their character, to betray. . . a fond partiality for this or that particular district, where it happened to be their lot either to be born or reside.[5]

It is one of the tragedies of our modern world that this sort of cosmopolitanism commonplace from the age of the Republic of Letters increasingly came into conflict with public sentiment in the late eighteenth century, when citizenship increasingly became defined in terms of an indivisible allegiance to a unitary sovereign power lodged in the nation-state. And the history of the late eighteenth century documents how the new nationalistic sovereign powers increasingly viewed the foreign connections of their cosmopolitan citizens with a jaundiced eye.

Consider what happened in the decades prior to the American Revolution when the anti-parochialism of North American savants meant that they began to collaborate across colonial boundaries in a cause increasingly antithetical to the interests of their king. Throughout the course of the eighteenth century scientific societies had been founded from Boston to Georgia largely for the purpose of increasing "the common stock of knowledge" for the benefit of the commonweal. But increasingly, eminent members of these various societies—Benjamin Franklin, Thomas Paine, and Alexander Hamilton, among others—believed that the imperial government was incapable of fostering such public improvements, and the public interest could only be served by intercolonial collaboration *among* such scientific societies. As one booster put it: "The improvement of natural knowledge will be a means of uniting ingenious men of all societies who will begin to [wear] away by degrees any harsh opinions parties may have conceived of each other."[6] At a time when the residents of the various colonies were otherwise divided by religion, custom, and economic interests, this new continent-wide collaboration—as Michael Guenther has recently shown—taught colonial elites how they could unite around a common project.[7] Indeed, both the Revolutionary Continental Congress and even the later Constitutional Convention can in some sense be described as "international" assemblies in which gentlemen representing 13 sovereign states negotiated their way toward a common cause, in part by leaning on personal connections and collaborative skills many had honed through their membership in pre-revolutionary scientific societies. Of course, this cross-colonial collaboration ultimately forged a new sovereign nation.

Another threat that cosmopolitanism could pose to sovereign powers was that savants might refuse to go to war when their homelands did. The eighteenth-century catch-phrase "the sciences are never at war" was widely invoked by French and British savants to justify their collegiality through the bitter conflicts of the era. But did savants really forbear from war-work? For decades, commentators have pointed to the persistence of the war-time correspondence between London and Paris as proof that this cosmopolitan ideal survived the world's first total ideological war intact. This sanguine view of irenic science is mistaken, however. In fact, savants did go to war on behalf their homelands, as the Revolutionary Wars placed new demands on the patriotic contributions of all productive citizens, including savants, who, for the first time, labored in secret weapons labs, managed munitions factories, and organized the transfer of scientific booty. The research of Elise Lipkowitz shows how this new type of war also transformed the flow and content of scientific communication.[8] Whereas savants had once written directly to their colleagues via the networks typical of the Republic of Letters, war-time disruption meant that savants in conquered nations were increasingly obliged to communicate via London and Paris. In other words, the war-time persistence of the Paris-London axis does not prove that savants in the capitals transcended political divisions, but that the war actually strengthened imperial networks. So here too, cosmopolitanism proved unable to withstand the intensified demands of national allegiance.

It is in this context of consolidating nationalism that we must consider the two precocious international conferences that pre-date the "take off" of such meetings in the mid-nineteenth century. Both meetings took place in 1798, and both concerned the metric system then being designed in France. The context of these meetings—and their shortcomings—help explain both the demise of cosmopolitan science and the character of the new international science that would later emerge in its stead.

Gotha, August 1798

The world's first international scientific meeting—a "Congress," as it was initially called—took place in Gotha in August 1798. Its host would later insist that the assembly had been an impromptu affair. Franz Xaver von Zach, director of the Gotha Observatory, claimed that he had simply taken advantage of the impending visit of the eminent French astronomer, Jérôme Lalande, to invite a number of central European astronomers to meet the "dean" of their field. Yet in fact this gathering actually advanced the longstanding scientific and political aims of Zach, Lalande, and the Duke of Saxe-Gotha—at least at first.

Close ties bound Paris to tiny Gotha, an otherwise undistinguished duchy not far from Weimar. The duke's mother had been a pupil of the philosopher Christian Wolff and a correspondent of Voltaire, who had used her as his model for the character of Professor Pangloss.[9] Gotha's elite considered themselves to be living in a happy German outpost of the Republic of Letters. The duke himself was an amateur astronomer, who decreed that his principal legacy would be a world-class observatory. Zach, its Hungarian-born director, was a new breed of scientific entrepreneur, who put Gotha on the scientific map by corresponding with the world's leading astronomers and publishing their letters in his journals.[10] To further boost Gotha's visibility, Zach had long cultivated the friendship of Lalande, urging him to visit the town, where he promised that the duke would receive him "like a god"—that is to say, in a manner commensurate with his "immortal merits."[11]

For Lalande, the trip was both a victory lap and an opportunity to demonstrate that cosmopolitan fellowship had survived the Revolutionary Wars. Some 50 years earlier, at the age of 19, he had been sent to Germany by the Parisian Academy to conduct lunar observations in coordination with Lacaille's simultaneous observations at Cape Town. In Berlin, Lalande had dined at the high table of the Enlightenment, exchanging witticisms with Frederick the Great and Voltaire, and performing calculations alongside Pierre-Louis Maupertuis and Leonhard Euler. Then, in the 1760s, he had coordinated worldwide efforts to track the transit of Venus. By the 1790s, he had former students and correspondents around the globe. Initially, Lalande had welcomed the French Revolution for its libratory and universal principles—including the promise of universal weights and measures—but had become disenchanted by the country's violent and xenophobic turn as it went to war with the surrounding kingdoms. Worse, French conquests had engendered a new hyperpatriotic nationalism abroad. As Zach sadly informed Lalande: after five years of war, most ordinary Germans considered the French "drinkers of blood" and viewed all French ideas as "propaganda."[12] Among these reviled ideas was the revolutionary proposal for an international metric system. Zach noted that he had himself been branded a Jacobin and "democrat" simply for honoring French science.

Zach and Lalande decided that the best way to counteract such mutual mistrust was to personally demonstrate the pacific accomplishments of collaborative cosmopolitan science. Lalande agreed to visit Gotha as soon as a lull in the war made travel possible. So no sooner had Napoleon's Italian victories of 1797 imposed a new peace on Europe than Lalande set his plans in motion and Zach sent invitations to Copenhagen, Prague, Basel, and various German-speaking lands.[13] Lalande secured a passport from the French ministry, and in July 1798 he was on his way.

This début international conference combined science and sociability in ways that foreshadow such meetings today. The participants ranged from eminences to students, and the agenda focused on issues of coordination. Among the dozen-plus astronomers in attendance were Johann Bode, director of the Berlin observatory; an advanced student from Cambridge University; and Lalande's illegitimate daughter, Amélie, who performed all his astronomical calculations. Once assembled, the astronomers calibrated their state-of-the-art timepieces and agreed to henceforth report all observations in mean time. They even sought to negotiate a separate peace in the heavens; when Lalande proposed a new constellation named "Globus aerostaticus" in honor of the French Montgolfiers, Bode countered with "Officina typographica," in honor of the printing press's great German inventor, Guttenberg. On an excursion to Inselsberg to test a new sextant, they also enjoyed a festive dinner hosted by the duchess. The cannon salute shattered three windows and the astronomers drank champagne and danced until dawn. This and the other side junkets were organized by Zach's young Swiss adjutant, who complained privately about having to do all the scut work. In letters to his family, he called Lalande an "old vain fop" and his daughter, "wild, impertinent, and pretentious."[14] His family, however, was not to breathe a word about this; if anyone asked them, they were to say that several important astronomers had met in Gotha to discuss the important question of weights and measures.

In the end, however, the attendees could not reach a consensus on this key issue (which many suspected was the real reason that the Parisian authorities had allowed Lalande to travel to Germany in the first place). Despite Lalande's pleas—if ever any reform ought to command the consent of all rational people, this was it—his foreign colleagues said they were not licensed to urge new measures on their governments, lacking any authority to intervene on political questions.[15] The most they could promise was that wherever they had once used the French royal foot they would henceforth report their data in meters.

Still, Zach and Lalande initially judged the meeting a success and announced their intention to hold a second. But Zach took fright when the German press ascribed conspiratorial motives to the assembly.[16] Rumors circulated in Vienna and London that Lalande had come on a cloaked mission from the French government and that this scientific congress was the seed of a revolutionary plot to supplant the region's legitimate rulers.[17] The rumors insinuated that at the meeting a cabal had been hatched by Lalande, the atheist, and Zach, the puppet-master of the Francophile duchess, to join forces with the Illuminati, a secret society of "free-thinkers."[18] As it happened, Lalande and Zach *were* prominent Free Masons; the duke *was* a protector of the Bavarian Illuminati; and Zach *was* the duchess's lover. But in the new populist-nationalist dawn, these polite affiliations, typical of the Republic of Letters, had become suspect.

Lalande in Paris might laugh off these prattlers, but in Gotha the duke felt threatened by the rising tide of German patriotism. He retrospectively decreed that the meeting no longer be referred to as a "Congress"—apparently because the term had democratic (i.e., subversive) overtones.[19] And apparently other rulers shared his concerns. According to one eminent astronomer, the gathering had set the "little German princes...trembling and shaking."[20] The Austrians had forbid their savants to attend in the first place.[21] In the end, even the participants turned defensive. Zach suppressed all reports of the proceedings in his *own* journal.[22] And in his account of his visit, Bode dismissed his trip to Gotha as a "jaunt," even though it had been approved and paid for by the Prussian monarch.[23] The new constellations were never adopted. Zach and Lalande's bottom-up effort to give collective voice to cosmopolitan science had foundered on war-time enmities.

Paris, 1798–1799

The international conference to finalize the calculations of the meter had been scheduled to begin in Paris soon after the meeting in Gotha disbanded. But whereas the first Gotha meeting had been a self-organized and informal gathering (albeit one long meditated and designed in part to propagate the metric system), the Parisian meeting was a carefully planned, state-sponsored effort to give an international imprimatur to the metric system (albeit in ways that served the parochial interests of particular savants).

The meeting was the brainchild of Pierre-Simon Laplace, France's preeminent physicist and coauthor of the decision to define the meter as one ten-millionth of the distance from the North Pole to the equator. In 1797, as the mission to measure the meridian was drawing to a close, Laplace proposed that the Academy of Sciences ask the French government to invite a number of foreign savants to Paris to validate the results obtained by the two expedition leaders: Jean-Baptiste-Joseph Delambre and Pierre-François-André Méchain. At the same time, however, Laplace privately assured Delambre that the meeting would be "a mere formality," a way to rubber-stamp the decisions of its French creators while convincing the foreign savants "to consider the measure as belonging equally to them."[24]

Even so, the proposal was opposed by Jean-Charles de Borda, inventor of the instrument with which the measurements had been performed. Why, he wondered, should a measure based on nature need an international imprimatur? And what if the foreign savants proved less docile than Laplace supposed?[25] Borda's first concern proved profound, and his second prophetic, but in any case Laplace's proposal found two powerful backers. The first was France's foreign minister, Charles-Maurice de Talleyrand. In the early years of the revolution, Talleyrand had tried—and failed—to induce Britain to cooperate with France on common measures. Now that the European balance of power had shifted toward France, however, Talleyrand insisted on excluding the British, and inviting only representatives of those European client states he wanted to draw further into Paris's economic and administrative orbit. And though Laplace's second backer was the most junior member of the Academy, the young commander Napoleon Bonaparte had his own ambitions for European integration under French hegemony.

Indeed, soon after the representatives from Holland, Denmark, Switzerland, Spain, and the Italian states arrived in Paris for the September meeting, they began to suspect that they were little more than window-dressing for French scientific and imperial ambitions—and some resented it. Indeed, the French were stalling for time while Méchain completed his triangulations in southern France. And even after he returned to the capital in mid-November for the conference's "opening" gala—replete with grand speeches and copious wine—he refused to turn over his data to the commissioners and avoided their meetings altogether. Finally, in January 1799, the Danish commissioner, Thomas Bugge, announced he was fed up and returned to Copenhagen. He said he resented being "kept in the dark" and the "coldness and disdain with which the foreign savants had been treated."[26] His slurs soon found their way into the French press, irritating his hosts.[27] In Gotha, Zach heard rumors that the expedition's data were "worthless, poorly executed, inconclusive, and untrustworthy."[28] Then Borda died, and the delegates from Rome and Sardinia left Paris too.

With the conference on the verge of collapse, Laplace ordered Méchain to hand over his data in ten days.[29] Forty-four days later—and eight months late—Méchain finally presented his summary results. After all the suspense, the commissioners found them a marvel of precision. No wonder: his data had been massaged.

But at least the assembled savants could at last turn to their central task: boiling down those results into a single number—the meter. For the next few weeks, each commissioner calculated independently, using his own methods. The French mathematician, Adrien-Marie Legendre, deployed a new ellipsoid geometry; the Dutch astronomer, Jean-Henri Van Swinden, made use of traditional geodetic techniques; Johann Trallès, the Swiss delegate, used his own techniques; Lorenzo Mascheroni, a Milanese astronomer, tried a geometric method that did not find favor with his fellows; and Delambre employed methods he had recently published.[30] As each savant compared results, it became clear that the rumors were right: something was wrong. The meridian expedition had produced something shocking: genuine scientific novelty.

According to data gathered 50 years earlier at the equator and the arctic circle, the earth's eccentricity was 1/350. By contrast, Delambre and Méchain's data from France's meridian implied it was nearer to 1/150. Even more startling, the data from the intermediate latitude measures at Dunkirk, Paris, Evaux, Carcassonne, and Barcelona suggested that the curvature shifted with every segment of the arc. In fact, there is reason to believe that Laplace had originally urged that the meter be based on geodesy (rather than a pendulum, say) mainly to test this very proposition—at a cost of three times the Academy's annual budget.

There was only one problem. This discovery/confirmation invalidated the foundational premise of the mission. There was now no simple way to extrapolate from the idiosyncratic French sector of the meridian to the quarter-meridian of the world. This meant that the French and foreign delegates had real compromises to make. After heated discussion, the commissioners decided to combine the new data with results that were 50 years old in a way that several participants privately acknowledged was somewhat arbitrary.[31] In the end, a meter that was to have been based solely on nature was set by scientific consensus.

At a grand ceremony on June 22, 1799, the commissioners stood in strict (nonnational) alphabetical order before the French legislature to present that body with the platinum "Archive Meter" whose length had been set to match the calculated convention and which would henceforth serve as the legal standard. As the senior member of the expedition, Méchain had hoped to have the honor of formally presenting the meridian results; but he was asked to step aside in favor of a foreign delegate, the Dutch astronomer, Jean-Henri Van Swinden, so as to emphasize the international character of the metric system and the conference's deliberations.[32] On behalf of his fellow foreigners Van Swinden expressed gratitude for the iron facsimile that each delegate would carry back to his home country in the hope that the new measures might tie together the peoples of Europe "with fraternal bonds."[33] Then in his address—printed anonymously so as to not outshine his foreign colleagues—Laplace grandly touted the world-wide appeal of a measure derived from the size of the earth.[34] Of course, all the delegates were silent on the quasi-arbitrary compromises they had hit upon behind closed doors. Such adjustments might be necessary to make sense of complex data—and an open secret among colleagues—but even savants such as Zach agreed that the messy work of science was best kept out of public view.[35]

So the first state-sponsored international scientific convention had produced the first international metrical convention, but only because the delegates had been able to work out their differences in private meetings, face-to-face, behind closed doors. And even then, the metric system spread through western Europe, not on the say-so of an international assembly of delegates under French hegemony, but in the backpacks of Napoleon's armies, whose imperial administrators attempted to impose the meter on vassal states from Batavia to Sicily. (Several of the metric commissioners actually had to delay their return home because France's soldiers had again overrun their homelands.) And so, when the empire failed, so did the meter. Ordinary citizens, it turned out, preferred local measures, expressing local values, to universal rulers. Even within France, Napoleon abandoned most of the metric reform and the nation did not return to the metric system until the 1840s. But if the metric conference of 1798–1799 had failed in the short run, it had succeeded in imbuing the meter with an aura of internationalism and naturalness—two qualities that made it the sole plausible candidate for a transnational metrical standard in the decades to come.

Jérôme Lalande, however, had never given up on his dream of transnational scientific fellowship. One month after the metric commission disbanded, he decided to return to Gotha to spark another international assembly— even though he had no formal invitation from the duke. And this time he resolved to travel by the most newfangled means available: aerostatic balloon. Unfortunately, contrary winds forced his party down in the Jardin de Bagatelle, seven kilometers in the wrong direction. The newspapers had a field-day with his fiasco, which only further infuriated Lalande's would-be hosts. The duke arranged with the French foreign minister to forbid Lalande to ever return to Gotha.[36] There would be no further international scientific meetings for several decades.

Frankfurt-am-main, someday . . .

Both the Gotha and the Paris meetings paid homage to transnational scientific fellowship, well lubricated with drink and dance—plus the inevitable banquets and speechifying. And they demonstrated that such conferences were both technically feasible and scientifically productive. But Lalande and Zach's vision of a revitalized scientific cosmopolitanism, as expressed in a self-organized disciplinary congress, could not transcend the violent rivalries of the Revolutionary Wars fought with appeals to patriotic zeal and national autonomy. And Talleyrand and Napoleon's vision of Gallo-centric universalist science, as expressed via an imperial assembly of scientific delegates, likewise foundered on the shoals of rival nationalisms.

There would be one last-ditch effort to preserve the Republic of Letters by transforming it into a functioning republic—though this too came to naught. In 1816, soon after the Congress of Vienna reestablished peace in Europe, an anonymous pamphlet proposed that all the world's savants assemble once a year to discuss their common interests at an international "Congress of Savants" —to be located initially in Frankfurt-am-Main. The pamphlet noted that the town was centrally situated and able to accommodate many visitors, thanks to its moderately priced hotels. Sites for future meetings could, of course, be chosen by the participants themselves. This banal yet radical plan was the brainchild of Abbé Henri Grégoire, the revolutionary French cleric, linguist, and legislator, who had helped reestablish the Academy of Sciences after its dissolution during the revolution and then defended its autonomy from the interference of despots—first Napoleon, then the restored Louis XVIII. For his effrontery, Grégoire had been forced into exile in Germany, where he relied on the hospitality of foreign colleagues.

Grégoire cited the Greek Olympiads as one model for such gatherings. He also cited the only two modern precedents—the 1798 meetings in Gotha and Paris. Yet his ambition differed in important ways from the aspirations behind those meetings. Grégoire had a universalist assembly in mind, and he expected the delegates to come from every country without regard to "birth, sex, status, color or faith."[37] Among those he hoped would attend were learned men of Africa, notable women, and Persians. Practitioners of every scholarly discipline would be welcome. Even those without an official position would be invited. And the Congress would especially welcome students.

In a recent reedition of the pamphlet—the first notice taken of the proposal since its publication in 1816—its editor touts Grégoire's Congress as a precursor of UNESCO. This seems mistaken. Instead, this grand proposal, which was universally ignored, marks the apotheosis—and demise—of the Republic of Letters.[38] Grégoire's radical plan was to re-create the Republic in bodily form, not organize an assembly of elite delegates who represented their homelands. Unlike UNESCO (since 1954), Grégoire's Congress was to operate completely independently of the nation-state and without regard to discipline. But amid the era's intensification of nationalism and scientific professionalization— and a concommitant growth in the scale of science—such a gathering suddenly began to seem fantastical. The world's total number of savants at the time may well have been less than Frankfurt's population of 40,000, but that was not to last. How poignant that a fully materialized Republic of Letters only became imaginable when such a gathering became impractical.

To be sure, even as Grégoire's utopian dream of a universal assembly foundered, cross-border scientific exchange resumed, now that the Revolutionary Wars had given way to the industrial obsessions of the early nineteenth century. But Europe's newly nationalist regimes proved even less tolerant of their citizens' divided allegiances than the sovereign powers of the old regime. There was no scope for Lalande and Zach's vision of self-organized disciplinary assemblies, or Talleyrand's assembly of delegate-savants. Not until the middle decades of the nineteenth century did these two models reemerge, and then with telling differences. For instance, the state-sponsored model now took a form more attuned to the new liberal legal internationalism of the era, one in which autonomous sovereign powers dealt with one another on the grounds of ostensible equality, even while grouped into blocs allied with rival great powers.

Paris, 1869–1889

When the vogue for international scientific conventions finally took off in the second half of the nineteenth century, it was again the call for conventions/standards that brought the world's scientists (and nation-states) to the conference table. There were compelling reasons for this.

The late nineteenth century interwove shrill nationalism with international commerce. In Paris in 1863, an assembly of national delegations agreed on a world-wide postal treaty that defined transnational parcels in metric grams. As railroads and telegraphs bound the world together, national governments signed international treaties that wrapped the globe in time-zones. In the 1850s and 1860s, at a sequence of international meetings in Brussels, Paris, London, Berlin, Florence, and the Hague, statisticians employed in the bureaus of the various states of western Europe sought to bring their social and administrative data into alignment, and urged the adoption of the metric system to facilitate comparisons.[39] And in central Europe, where a new great power was ascendant, Prussia first sabotaged Austria's efforts to use the metric system to coordinate a trading zone among the western German states, then agreed in 1868 to adopt the metric system within its own free-trade zone of northern German states.[40] The metric system's appeal proved to be just what its French creators had hoped: as a measure (supposedly) taken from nature, the meter was acceptable to everyone because it favored no one. And just as in revolutionary France, this neutrality promised to facilitate national unification as much as international coordination.

Except this measurement norm was not based on nature, but on a platinum bar housed in the Parisian Archives, one whose length had been determined 70 years earlier by an international commission. And it was an open secret among geodesists that that this Archive Meter bar fell "short" of one ten-millionth of the quarter meridian. This discrepancy put the French in a painful position. They badly wanted the rest of the world to adopt their system. But they feared being hoist by their own rhetoric of nature's universality. Perhaps, they reasoned, an international scientific convention would lure other nations to accede to their metrical convention, The French government initially tried to summon these nations as if it were still Europe's hegemon. But in the end, as we will see, the scientists met in the spirit of the new age of liberal legal internationalism.

The technical challenge to the French Archive Meter had emerged in the 1860s during a series of international geodetic conferences held in Berlin under Prussian auspices, and organized by General Johann Baeyer, a Prussian army staff officer and cartographer.[41] For the past several decades, European geodesists had been calibrating their maps using reference bars that referred back to French originals. But these cartographers had increasingly noted incongruities between these reference bars, the Archive Meter, and its nominal definition.[42] At the second assembly of the *Europäischen Gradmessung* in 1867—which notably did not include a French delegate—the geodesists decided that to knit together their maps of central Europe they needed to remeasure the figure of the earth and forge a new standard meter bar—and they proposed creating an International Bureau of Measures to supervise this process.[43]

The French panicked. Would German precision supplant French mensuration, as German military prowess threatened to supplant the French army? French scientists bitterly debated the proper response. Some agreed that the length of the meter needed to be recalculated by remeasuring the earth. These nature foundationalists were quickly silenced, however, since such a move would have invalidated every ruler in France.[44] But even those hostile to remeasuring the meridian admitted that the current bar was beset by "humiliating" deficiencies.[45] As an "end standard," the Archive Meter had been used to calibrate measures by direct contact, meaning that 70 years of comparisons had worn down and pitted the bar. Moreover, the bar had not been composed of "pure platinum" as initially claimed, but of an alloy adulterated with iridium and other trace metals. These Frenchmen suggested inviting their foreign colleagues to Paris to discuss the forging of a new meter, but with the presumption that the new bar would be matched to the old.

Paradoxically, their position was aided by a report of eminent Russian geodisists who visited France in 1869 to evaluate the Archive Meter. Decades of progress in geodesy, the Russians noted, had revealed that the Archive Meter was not one ten-millionth of the earth's meridian; indeed, no such unit could be rigorously defined, given the differences among the earth's meridians, which were themselves subject to change over (geologic) time. Their conclusion? The Archive Meter was necessarily an "arbitrary" standard that was a matter of "convention."[46]

The French were so happy to seize this fig-leaf of convention that they even denied that their revolutionary predecessors had ever presumed that the lengths of meridians were equal or "absolute."[47] And on this basis, the Emperor Napoleon III in November 1869 invited all the major European and American nation-states to once again send scientific delegates to a metric conference in Paris, where once again (as the invitation promised) they would work with their French counterparts on a basis of "complete equality."[48]

Unfortunately, two weeks before the conference convened, France and Prussia went to war and the Prussian delegates stayed home. Instead, scientists from 15 other nations arrived on schedule—presumably by train—while the Prussian army mobilized with unprecedented speed, also by train. The assembled delegates agreed to postpone any decisions until *all* their colleagues were present. And the multilingual Swiss delegate—a geodesist who had helped draft the resolutions of the *Gradmessung* conferences—reassured his French colleagues that "no serious scientist in our day" would contemplate rederiving the length of the meter from the size of the earth.[49] The French relief was palpable and at the final session the leading French delegate, General Arthur-Jules Morin, toasted the spirit of "scientific co-fraternity" that had governed their friendly, but rigorous discussions: a demonstration, he said, of how scientists from different lands could labor together for the progress of world civilization, even under the most hostile conditions.[50]

Two weeks later, the French Emperor was captured at Sedan, and the French army collapsed. Two years later, in 1872—with the German empire now the preeminent power in Europe and having formally embraced the metric system—the French government again convened a metric conference in Paris. This time, to jump-start the negotiations, the French agreed in principle to consider something they had previously resisted: the possibility of an International Bureau of Weights and Measures. At the conference itself, the tone was collegial. Wilhelm Foerster, the head German delegate, was an enthusiast for metrical harmony; in his view, it was not the meter which was natural, but the idea that international mensuration should be governed by an international accord.[51] The scientific delegates unanimously resolved to forge a new standard to be as similar to the Archive Meter as possible, right down to the mix of the alloy, with each nation to receive a coequal, cocreated, and rigorously interrelated standard bar, with the principal reference bar to be held by the new International Bureau to be but a first among equals. It was a solution fit for the new era of liberal legal internationalism, in which autonomous nations, each with unitary sovereignty, negotiated within a framework of formal equality to create an international order that would benefit capital-intensive commerce.[52] But it was not yet law.

All that year the French scientists pressed their own government to accede to this internationalization, warning that refusal would lead foreign powers to look to Berlin for their standards.[53] But not until the "definitive" 1875 Metric Convention in Paris—where the scientific delegates were accompanied by plenipotentiary delegates authorized to legally assent on behalf of their governments—did the French (after some internal wrangling) vote with the majority in favor of creating a permanent International Bureau (leaving only the Dutch and British opposed). The French did, however, win the concession that the Bureau be located near Paris in the Pavillon de Breteuil, with the building, which had been badly damaged during the Prussian siege of Paris, to be rebuilt at international expense.[54]

Even then, it took 15 years of additional international meetings among commissioners duly elected under the provisions of the Metric Convention to determine exactly what everyone had agreed to in 1875. The "definitive" metrical standards were not forged until 1888, and not until 1889 did a final conference ratify them. The question, however, is not why consensus took so long, but how it was managed at all, given the international rivalries that fueled skepticism born of rival disciplines and personal predictions. The answer, of course, is by committee—that, and elaborate protocols, working documents, material artifacts, laboratory labor, and the ever-shifting amalgam of goodwill and mutual suspicion that spurred and constrained the efforts of the various assemblies.[55]

So what features of late-nineteenth-century life assured the success of these international metric conferences (and of international science), despite the era's bitter nationalism? Trains? Sure. But not because they carried the delegates—and the German army—to Paris. Rather because they broke down local markets and increased the value of interregional and transnational commerce to the point where nation-states found it worthwhile to adopt uniform standards that were those of their neighbors—even though their neighbors were also their military rivals. And even then, the telling motive in each case was that the state would thereby bring uniformity to its own *national* measures. This had been the case in revolutionary France in the 1790s, as it was in breakaway Belgium in 1830, reunified Italy in 1863, and now, the new imperial Germany. The historical record shows that states have only adopted the metric system at times of sharp ruptures in *national* sovereignty: revolution, conquest, national unification, or (de)colonization. And the exception here proves the rule: among the major nations, only the United States has yet to adopt the metric system, and only the United States has had no break in sovereignty since the founding of the metric system in 1799.

Conclusion: Global science and the return to natural standards

Several historians have documented the expansion of international science in the latter decades of the nineteenth century.[56] This expansion, they have argued, occurred in part because the era's fierce nationalism rekindled a countervailing ethos of transnational cooperation among scientists, an

apparent contradiction that many scientists resolved by presenting themselves at home as competitors in the scientific arena abroad for the greater
glory of their nation. This chapter's argument is not at odds with this conclusion. But we must also account for the fact that much of international science
was devoted to finding consensus amid this competition, which could not
in any event proceed without common standards. Indeed, a sizeable percent
of the international scientific meetings in the later nineteenth century were
devoted to setting standards for various scientific/disciplinary communities
and nation-states.[57] So long as reaching agreement required intense negotia-
tion on substantive matters, international conventions (aka junkets) were one
of the preferred venues for thrashing out such standards.

In his comments at a symposium devoted to the history of international
science, John Heilbron once remarked that it was not so much the spirit of
scientific universalism as money that made the world of modern international science go around—and that this fact, as much as anything, separated
the utopian aspirations of Francis Bacon's House of Solomon from the practices of institutionalized science.[58] Certainly it is money that has made such
meetings possible—as have trains, and more recently, planes. Even more, it is
the money to be made from large-scale commerce that spurred the need for
international conventions—in both senses of the term—and which thereby
distinguished the rise of international science from the older cosmopolitan
form known as the Republic of Letters.

Comparing these two eras of transnational science enables us to understand how each produced the sort of standards they did—and suggests how
we might understand our own contemporary form of transnational science.
Over the course of the early modern period, the savant-to-savant networks
of the cosmopolitan Republic of Letters increasingly operated as an adjunct
to formal academies with a national (or municipal) imprimatur. During this
period transnational norms in science were set by imitation and emulation,
not formal agreement. Not until this cosmopolitanism was in tatters during
the wars of the French Revolution did some savants see the need for assemblies
of colleagues from diverse nations. Three distinct models for such assemblies
were proposed, each with a different vision of how standards might be set.
The first model (pioneered in Gotha in 1798) was organized by the practitioners themselves around disciplinary affiliation; it showed how savants might
successfully adjudicate standards within a discipline (here, astronomy), but
it failed to promulgate standards for public use. In any case, this workshop
model ran afoul of nationalist sentiment and would not resurface until the
late nineteenth century when disciplinary practitioners assembled under the
aegis of their professional societies and delegated the authority to set standards for their fields to elected sub-committees. The second model (pioneered
in Paris in 1798–1799) was sponsored by a hegemonic nation-state to advance
its imperial interests and was attended by scientific delegates from client
states. This model did not find imitators in the nationalist and conservative
ethos of the early nineteenth century, and when the French tried to revive it
in the middle of the nineteenth century, they discovered that the European

balance of power had shifted and that any international convention had to mirror the new liberal legal internationalism. This new international system found expression in the Metric Convention of 1875. The third model (proposed by the Abbé Grégoire) fully embodied the cosmopolitan ethos of the Republic of Letters, but did so in a radical form that asked savants to set aside their national affiliations. It has never been realized.

What then can we say about the transnational conventions of more recent times? And in what sense can we still refer to the norms that they have established as conventions?

The international metric assemblies of the late nineteenth century openly justified their choice of standard as a convention, confidently asserting that it was based on nothing more than the assent of each nation (with voting power assigned in proportion to national prowess). The late nineteenth century, of course, was also the era in which some philosophically minded scientists acknowledged that scientific laws were themselves a kind of convention expressing nothing more than the scientific community's agreement on how to most efficiently express the results of reliable measuring instruments. In retrospect, of course, it is clear that the most potent "convention" of the nineteenth century was the sovereign nation-state itself.

Yet this positivist account of nature was even then in tension with the longstanding counter-claim that the regularity expressed in scientific laws derived from a constancy in the properties of nature, which might thereby serve as a reliable source for technical norms (and perhaps even social and political ones). This nature-foundationalism was the basis on which the original metric expedition had been launched in the 1790s, and it continued to confer legitimacy on the Archive Meter well into the nineteenth century—even after it was proven to be erroneous. In the twentieth century, when this nature-foundationalism returned with a vengeance, the physical meter bar was finally replaced with a standard based on nature—this, after decades of debate and growing pressure from users who found the current standard insufficiently exact for their purposes. In 1960 the meter was defined as an integer multiple of the wavelength of a frequency of a particular kind of light. And in 1983, it was again redefined as the distance traveled by light (whose speed was a natural constant) during a time interval defined by an integer multiple of the periodicity of an atomic clock (again, a natural constant). Of course, in both cases, these integer multiples were carefully chosen so that the new meter would match, as nearly as possible, the 1889 bar built in accordance with the Metric Convention of 1875. It is worth noting that this twentieth-century "return" to a natural standard occurred in an era that placed renewed faith in assertions of natural and universal rights, on the one hand, while claims of national sovereignty were being increasingly eroded by supranational organizations, multinational corporations, and global finance, on the other. We have entered an era of global science in which scientists increasingly assemble under the aegis of transnational professional societies, supranational agencies, and NGOs, and seek to coordinate standards on a global scale.[59]

Again, the exception here may prove the rule. To date, there is only one standard of the International System of Units that is still defined by an artifact, and in that sense, by convention. The kilogram is still embodied by the platinum-iridium mass that was created by Johnson Matthey of England in 1878, ratified as the "International Prototype Kilogram" at the first Conference Générale des Poids et Mesures in 1889, and declared to be the legal standard of weight at the third Conference in 1901. Yet tiny divergences between its mass and that of its cocreated replicas have led leading metrologists to pine for a natural standard of weight. At the twenty-first Conference in 1999 they urged a re-definition based on an electric balance calibrated by the Planck constant; they reaffirmed this ambition at the twenty-third Conference in 2007, and they hoped to see it resolved at the twenty-fourth Conference held in 2011. By then, however, they had to contend with a rival proposal to define the kilogram as a fixed number of silicon atoms. So in the end the delegates agreed to postpone their decision until the twenty-fifth conference in 2014; or so they hoped.[60] So, should the kilogram standard ultimately be redefined in terms of its natural properties, the general pattern will still prevail: even when they're based on nature, scientific conventions are still the products of scientific conventions.

Notes

The earliest sections of this chapter debuted as a species of simultaneous discovery. It was some 20 years ago that I first learned, some 20 minutes before I was to give my first professional talk as a grad student—a conference paper on the origins of the metric system—that the illustrious historian of science, John Heilbron, had recently addressed just this topic in his 1989 Sarton Memorial Lecture, and was about to publish an extended version in a forthcoming collected volume. As I stood at the conference book display, frantically leafing through the pages, I further learned, to my relief and horror, that our interpretations agreed at various points and diverged at others. I recall being more relieved about the divergences than the overlaps, more anxious at the time to be thought original than horrified at being thought wrong. Afterward, I sent my paper to Vice Chancellor Heilbron and arranged to meet him. He could not have been more cordial. I first presented a version of this essay at the 2009 Sarton Memorial Lecture, which I dedicate to John Heilbron, whose work has preceded, dogged, and inspired mine.

1. Our newfangled computer-generated maps of point-to-point international correspondents implicitly remind us that these correspondents never assembled as an international group. Anthony Grafton, "Sketch Map of a Lost Continent: The Republic of Letters," *Republic of Letters* 1 (2009), http://www.stanford.edu/group/arcade/cgi-bin/rofl/issues/volume-1/issue-1.
2. See also Claude Tapia and Jacques Taieb, "Conférences et congrès internationaux de 1815 à 1913," *Relations internationales* 5 (1976): 11-35.
3. Simon Schaffer, "Newton on the Beach: The Information Order of Isaac Newton's *Principia Mathematica*," *History of Science* 47 (2009): 23–76.
4. A central theme of the history of the early Royal Society is how it might be made into an acceptable form of (corporate/regulated) public assembly; Steven Shapin and Simon Schaffer, *Leviathan and the Air-Pump: Hobbes, Boyle, and the Experimental Life* (Princeton: Princeton University Press, 1985).

5. Quoted in James E. McClelland, *Science Reorganized: Scientific Societies in the Eighteenth Century* (New York: Columbia University Press, 1985), 5.
6. Peter Collinson to Cadwallader Colden, August 23, 1744, in *The Letters and Papers of Cadwallader Colden* (New York: New York Historical Society, 1918), 3: 69.
7. Michael Guenther, "Enlightened Pursuits: Science and Civic Culture in Anglo-America, 1730–1760" (PhD dissertation, Northwestern University, 2008).
8. Elise Lipkowitz, "'The Sciences Are Never at War?': The Republic of Science in the Era of the French Revolution, 1789–1815" (PhD dissertation, Northwestern University, 2009).
9. Deidre Dawson, "In Search of the Real Pangloss: The Correspondence of Voltaire with the Duchess of Saxe-Gotha," *Yale French Studies* 71 (1986): 93–112.
10. Peter Brosche, *Der Astronom der Herzogin: Leben und Werk von Franz Xaver von Zach, 1754–1832* (Frankfurt: Deutsch, 2001).
11. Archives de l'Observatoire de Paris (hereafter AOP) MS1090: Zach to Lalande, May 25, 1798; March 12, 1792.
12. AOP MS1090: Zach to Lalande, February 24, 1798.
13. AOP MS1090: Zach to Lalande, June 30, 1796; May 25, 1798.
14. Johann Horner to family, August 1798, in Rudolff Wolf, *Astronomiche Mittheilungen* 61 (1884): 25–26.
15. Johann Bode, "Ueber meine Reise nach Gotha," in Bode, ed., *Astronomisches Jahrbuch für das Jahr 1801* (Berlin: Dümmler, 1798), 235–239.
16. AOP MS1090: Zach to Lalande, December 31, 1798.
17. *Review, or Annals of Literature* (London) 33 (December 1801): 547. AOP MS1090: Zach to Lalande, May 25, 1798. Zach to Karl Ludwig Edler von Lecoq, October 2, 1798, in Brosche, *Zach*, 99.
18. *Anti-Jacobin Review Magazine* (London) 5 (April 1800): 515.
19. AOP MS1090: Zach to Lalande, October 17, 1798.
20. Georg Christoph Lichtenberg, *Lichtenbergs Briefe*, ed. Albert Leitzmann and Carl Schüdderkopf (Leipzig: Dieterich'sche Verlagsbuchhandlung, 1904), 3: 206, quoted in Dieter B. Herrmann, *The History of Astronomy from Herschel to Hertzsprung*, tr Kevin Krisciunas (New York: Cambridge University Press, 1984), 186.
21. Jérôme Lalande, *Bibliographie astronomique* (Paris: Imprimerie de la République, year XI [1803]), 798.
22. Zach to Karl Ludwig Edler von Lecoq, October 2, 1798, in Brosche, *Zach*, 99.
23. Bode, "Reise nach Gotha," 235–239.
24. Laplace to Delambre, 10 pluviôse VI [January 29, 1798] in Yves Laissus, "Deux lettres de Laplace," *Revue d'histoire des sciences* 14 (1961): 288. See Maurice Crosland, "The Congress on Definitive Metric Standards, 1798–1799: The First International Scientific Conference?" *Isis* 60 (1969): 226–231; Ken Alder, *The Measure of All Things: The Seven-Year Odyssey and Hidden Error That Transformed the World* (New York: The Free Press, 2002).
25. See Académie des Sciences, Paris, *Procès-Verbaux* (Hendaye: Imprimerie de l'Observatoire d'Abbadia, 1910–1922) 1 (1, 5 pluviôse VI [January 20, 24, 1798]): 334–335; plus the alterations to the minutes for those same days, in Archives de l'Académie des Sciences, Paris.
26. Kongelige Bibliotek, Denmark (hereafter KBD) NKS1304: Bugge to Bernstoff, Minister of Foreign Affairs, November 17, 1798.
27. *Décade philosophique* 15 (February 18, 1799): 372. Thomas Bugge, *Reise nach Paris in den Jahren 1798 und 1799* (Copenhagen: Beummer, 1801), 711–718.
28. AOP MS1090: Zach to Lalande, May 28, 1799.
29. AOP E2–19: Méchain to Delambre, 18 pluviôse VII [February 6, 1799].
30. Jean-Baptiste-Joseph Delambre, ed., *Base du système metrique* (Paris: Badouin, 1806–1810), 1: 93.

31. Because the commission did not keep any minutes of their meetings, we cannot assess how they handled these discrepancies; but for some self-criticism of how they (mis)handled the arc of Peru and various "artifices of calculation," see New York Public Library, Ward Papers: Delambre to Adrien Legendre, 12 floréal VII [May 1, 1799].

32. KBD NKS1304: Méchain to Bugge, September 1, 1799; 10 brumaire VIII [1 November 1799].

33. Van Swinden, in Delambre, *Base*, 3: 648.

34. [Laplace], "Discours," 4 messidor VII [June 22, 1799], in Delambre, *Base*, 3: 581–589.

35. AOP MS1090: Zach to Lalande, April 29, 1799; July 10, 1799.

36. Simone Dumont, *Un Astronome des lumières, Jérôme Lalande* (Paris: Vuibert, 2007), 268.

37. Henri Grégoire, *Plan d'association générale entre les savans, gens de lettres et artists* (n.p., 1816), 48, quoted in *L'abbé Grégoire et la République des Savants* ed. and trans. Bernard Plongeron (Paris: CTHS, 2001), 250.

38. For the development of Grégoire's ideas, see Grégoire, "[L]es moyens de perfectionner les sciences politiques," *Mémoires de l'Institut national, Classe des sciences morales et politiques* 1 (year VI [1798]): (7 germinal IV [March 27, 1796]), 564; and Grégoire, "Essai sur la solidarité littéraire entre les savans de tous les pays," in Plongeron, *République* (Paris: n.p., 1824), 267–289.

39. M. Engel, ed., *Compte-rendu général des travaux du congrès international de statistique dans ses séances tenues à Bruxelles 1853, Paris 1855, Vienne 1857, et Londres 1860* (Berlin: Imprimerie Royale, 1863), xx, 56, 192–193. On the international movement for the metric system, see Edward Franklin Cox, "The Metric System: A Quarter-Century of Acceptance, 1851–1876," *Osiris* 13 (1959): 358–379.

40. "Maas und Gewichtsordnung für den Norddeutschen Bund," August 27, 1868, in *Bundes-Gesessblatt des Norddeutschen Bundes*, 28 (1868): 473–480.

41. J. J. Levallois, "The History of the International Association of Geodesy," *Bulletin géodesique* 54 (1980): 248–271.

42. Michael Kershaw, "The *'nec plus ultra'* of Precision Measurement: Geodesy and the Forgotten Purpose of the Metre Convention," *Studies in History and Philosophy of Science* [2012], in press.

43. C. Bruhns, W. Foerster, and A. Hirsch, eds., *Bericht über die Verhandlungen...der Europäischen Gradmessung* (Berlin: Reimer, 1868), 126.

44. Pontécoulant, "Observations," *Comptes Rendus de l'Académie des Sciences* (hereafter *CR*) 69 (1869): 728–730; Faye, *CR* 69 (1869): 737–741; Mathieu, *CR* 69 (1869): 741–742.

45. AOP MS1060-II-C-1: [Le Verrier] to Morin, May 13, 1862. See the debate between Faye (speaking for the Bureau of Longitudes and pro-cooperation) versus Le Verrier (speaking for the Paris Observatory and anti-cooperation), in *CR* 56 (1863): 28–37, 66–72.

46. Otto Struve, Heinrich Wild, H. M. Jacobi, "Rapport de la Commission," May 20, 1869, *Rapports adressés à l'Académie Impériale des sciences de St. Petersberg concernant la confection des étalons prototypes* (St. Petersberg: Imprimerie de l'Académie Impériale, 1870), 1–3.

47. Dumas et al., "Rapport sur les prototypes," *CR* 69 (1869): 514–518.

48. Ministre des Affaires Etrangères to French diplomatic agents, November 16, 1869, in Guillaume Bigourdan, *Le système métrique des poids et mesures* (Paris: Gauthier-Villars, 1901), 272–273.

49. Adolph Hirsch, quoted in Commission Internationale du Mètre, *Session de 1870, Procès-verbaux des séances* (Paris: Baudry, 1871), 16.

50. Morin and Otto Struve, in Commission Internationale du Mètre, *Session de 1870*, 41–42.

51. Commission Internationale du Mètre, *Procès-verbaux des séances du Comité des Recherches Préparatoires*, April 1872 (Paris: Viéville, 1872), 59.

52. Martti Koskenniemi, *The Gentle Civilizer of Nations: The Rise and Fall of International Law, 1870–1960* (Cambridge: Cambridge University Press, 2002).

53. Archives Nationales, Paris F17/3715: Pres. of Bureau of Longitudes to Min. of Instruction Publique, October 1873.

54. Charles-Edmond Guillaume, *La création du Bureau International des Poids et Mesures* (Paris: Gauthier-Villars, 1927).

55. The most complete account remains Bigourdan, *Système métrique*. For an interesting case where international scientific diplomacy only papered over ambiguities in the standards—here, electrical norms—see Michael Kershaw, "The International Electrical Units: A Failure in Standardization?" *Studies in History and Philosophy of Science* 28 (2007): 108-131.

56. Elisabeth Crawford, "The Universe of International Science, 1880–1939," in Tore Frängsmyr, ed., *Solomon's House Revisited: The Organization and Institutionalization of Science* (Canton, Mass.: Science History Publications, 1990), 251–269. Elisabeth Crawford, Terry Shinn, and Sverker Sörlin, eds., *Denationalizing Science: The Contexts of International Scientific Practice* (Dordrecht: Kluwer, 1993), 1–42.

57. For a typology of the agendas of international scientific conferences, see Anne Rasmussen, "Jalons pour une histoire des congrès internationaux au XIXe siècle: Régulation scientifique et propagande intellectuelle," *Relations internationales* 62 (1990): 115-133; and B. Schroeder-Gudehas, "Les congrès scientifiques et la politique de coopération internationale des académies des sciences," *Relations internationales* 62 (1990): 135-148.

58. J. L. Heilbron, "Why Revisit Solomon's House," *Solomon's House Revisited*, 331–342.

59. Paul Edwards, *A Vast Machine: Computer Models, Climate Data, and the Politics of Global Warming* (Cambridge, Mass.: MIT Press, 2010).

60. Jonathan Keats, "The Search for a More Perfect Kilogram," *Wired Magazine* (October 2011). Rachel Courtland, "The Kilogram, Reinvented," *IEEE Spectrum* (May 2012).

2

Practicing Eighteenth-Century Science Today

Hasok Chang

Introduction

What is the use of the history of science? Or more broadly, of the history, the philosophy, and the sociology of science, technology, and medicine? Over 20 years ago John Heilbron formally exhorted the members and the leadership of four major professional societies to take this question seriously, to consider what "applied history of science" could contribute in the areas of general education, science education, and science policy.[1] Heilbron's call has only been heeded to a small extent. Although there are surely a good number of scholars in science studies who have engaged with the pressing problems concerning science education or the social impact of science, at least for historians such "applied" work tends to remain very separate from their scholarly work. I believe that to a large extent this disconnection stems from an increasingly entrenched tendency of historians to shy away from making value judgments on science in their professional work. Against that tendency toward neutrality and detachment, I find inspiration in Paul Forman's exhortation for historians of science to embrace "the obligation to decide for ourselves what is the good of science, and by our historical research and writing to advance that good."[2] A restoration of epistemic and political judgment in our work is a necessary step for our scholarly work to become truly applicable.

As I work mostly from a philosophical angle, I am sensitive to most historians' wariness of ahistorical judgments passed on the basis of some supposedly eternal and universal epistemic criteria. In this chapter I wish to demonstrate that epistemic judgment can be historically situated, taking place in a dialogue between the past and the present, rather than by the present simply imposing its standards on the past. My main concern, when I do history, is with what we can learn from the past, or "how the past can improve our future," to borrow Neil Postman's evocative phrase.[3] My notion of learning from the past is not based on the assumption that the future will be like the past so that we can draw useful lessons for the future by generalizing from the past. I am also not primarily concerned with the enterprise of understanding the present better by seeing in the past where and how it originated. In the latter context Heilbron expressed a worry that the teaching of premodern history may be quite useless for helping citizens understand modern science.[4] I do not have that worry, because my focus is more on learning from the past by finding out what valuable things we have lost and recovering them where possible.

My contention is that the history of science can help us improve present and future science. Professional historians can serve a uniquely important role here, by dredging up exactly those parts of past science that scientists themselves tend not to notice or remember because they do not fit nicely into current conceptions or customs. What matters most in this context is to notice important differences between the present and the past, not the continuity and similarity between them. There are several dimensions to these differences that we can learn from. As the section "Learning from past science" illustrates, there are phenomena that past scientists used to know that have been forgotten by many present scientists. As the section "Being humbled by past science" illustrates, there are past scientific questions that present scientists have abandoned for no convincing reason. As hinted in the section "Why study 1800 science?" and elaborated further in the section "Practicing 1800 science today," there are also valuable past manners of organizing and promoting inquiry that present science does not countenance.

Why study 1800 science?

For all three of those dimensions of learning from the unfamiliar past, I believe that the most fruitful place in the past of science to look to is late eighteenth-century Europe, for those of us living in the "modern" scientific–technological civilization that originates from the European and American domination of the world. When I say "the late eighteenth century" what I really mean is the period covering a few decades on either side of the year 1800, so I should rather say "*circa* 1800." Since there is no standard designation for this period, I will be saying things like "the 1800 period" and "1800 science" as a shorthand.

Historians of science will need no persuading that the 1800 period was a very important phase in science. Emblematically, the year 1800 itself was quite an eventful one. In that year Alessandro Volta announced his invention of the battery, which led immediately to the first electrolysis of water by William Nicholson and Anthony Carlisle. William Herschel discovered infrared rays in solar radiation (followed within two years by the discovery of ultraviolet rays by William Hyde Wollaston and Johann Wilhelm Ritter), and Thomas Young began his advocacy of the wave theory of light. Meanwhile Count Rumford (born Benjamin Thompson) founded the Royal Institution of Great Britain and would shortly employ Young to lecture there, and later Humphry Davy, too. This remarkable year in science ended well, too, with the discovery of Ceres, the first asteroid, by Giuseppe Piazzi on January 1,

1801. More generally, the rapid progress in European science *circa* 1800 was impressive indeed. Within just a decade on either side of 1800, scientists established electrochemistry, atomic chemistry, crystallography, comparative anatomy, and the metric system. We also remember Rumford's and Fourier's works on heat, Jenner's smallpox vaccination, Lamarck's ideas on evolution, and Laplace's perfection of Newtonian celestial mechanics. It must have been an exciting time for the "men of science," as they called themselves, including the women.

The scientific achievements of this period are made even more interesting by their links with broader social developments, which laid so many aspects of the foundations of our modern life, with the flourishing of the Enlightenment, romanticism, liberal economics, and democratic revolutions. There was a remarkably free traffic between science and other walks of life. This was an age when Richard Watson (1737–1816) could be appointed Professor of Chemistry at Cambridge while freely admitting that he didn't actually know any chemistry. He did become a rather good chemist, and a Fellow of the Royal Society by 1769, but soon moved on to become Regius Professor of Divinity, and by 1782 the Bishop of Llandaff. The educational entry-barrier to science was also very low, perhaps more than in any era before or after it. For example John Dalton (1766–1844), the originator of the chemical atomic theory, never even went to secondary school, not to mention university; the situation was similar with many other greats including Humphry Davy (1778–1829) and Michael Faraday (1791–1867).

Joseph Priestley (1733–1804) traveled from religion to science, in the opposite direction from Watson. Priestley's remarkable achievements in chemistry began when he went to Leeds to preach at the Mill Hill Chapel there. He happened to move into a house next to a brewery, got curious about the fixed air (carbon dioxide in modern terms) collecting in the fermenting vats there, and began making experiments. Much of Priestley's pioneering work on the dozen or so new gases he made, including oxygen (or "dephlogisticated air" as he conceived it), was done at home, in the "constantly warm mice-ridden Yorkshire cottage kitchens," in J. G. Crowther's phrase.[5] His apparatus was simplicity itself, including mice as his best indicator of air quality. Priestley's method of making artificially carbonated water became a sensation all over Europe and brought him to the attention of Lord Shelburne, later to be prime minister, who employed Priestley as his political advisor in residence.[6]

An even more extraordinary example of the free traffic between science and other areas of life is Rumford, the American soldier-of-fortune who became a count of the Holy Roman Empire. Rumford's great insight that heat was a form of motion rather than a material substance was reached while he supervised the manufacture of cannons. That was in Munich, where he ran the army and the police, rounded up beggars into workhouses, invented the soup kitchen to feed the poor, and created the wonderful English Gardens. After that he came to London and became rich and famous by remodeling the homes of the rich and famous, putting in efficient fireplaces and kitchens of his own invention. He also advocated the use of coffee as a healthy drink for the masses instead of gin and invented the drip coffeemaker. To promote useful applications of science (including his own inventions) he founded the Royal Institution. Rumford's colorful life continued with a move to Paris, marriage to the widow of the great chemist Antoine-Laurent Lavoisier, followed by divorce, and a lonely death.[7]

Another notable polymath was Jean-André De Luc (1727–1817), a native of Geneva who eventually settled in England as tutor to Queen Charlotte. De Luc was a businessman with a passion for meteorology, geology, mountain-eering, and theology. A meticulous metrologist, he attained his initial fame by improving barometers and using them to measure the heights of mountains. In his long life he also made major contributions to chemistry and electricity.[8] De Luc's work brings me to the particular pieces of science that I want to discuss in some depth.

Learning from past science: The boiling point

De Luc was reputed to be a very boring man, but he gave me the most exciting moment of my academic life by allowing me to make a genuine scientific discovery—or I should say, recovery. This was my first vivid experience of learning from past science. All this transpired in the summer of 2004, which I spent boiling water.[9] While studying the early history of thermometry for my book *Inventing Temperature*, I had come across many reports of unruly variations in the boiling point of water.[10] I do not mean the well-known effects of pressure variations, or impurities—I am talking about the boiling of pure, distilled water under standard pressure. De Luc and many others observed that the boiling temperature depended greatly on the material of the vessel employed, on the exact manner of heating, and on the amount of dissolved air present in the water. I reported these observations in my book, but like a good historian I did not get into the business of saying whether they were correct or not. After the book went off to press, however, curiosity got the better of me, and I had to see for myself.

Now, we have all boiled water. But how many people have really observed water boiling? Sat and stared at it for hours on end, taking notes? After a careful observational experience, the historian will be able to make much better sense of certain puzzling things in the record of past science. For example, take the thermometer frame attributed to George Adams (the Elder), dating from around 1860, held at the Science Museum in London.[11] There are *two* boiling points marked on this scale: "water boyles vehemently" at 212° Fahrenheit, and "begins to boyle" at about 204°F. Was Adams simply incompetent? That is not so likely as he was the official instrument-maker to George III. Equipped with a simple thermometer and a Bunsen burner or a kitchen stove, we can try to see for ourselves whether Adams was hallucinating or basing his instrument on something real.[12]

Put distilled water in an ordinary glass beaker over a Bunsen burner flame, monitoring the temperature with an ordinary mercury thermometer or a now-standard digital thermometer.[13] It is worth your while to observe carefully the number, size, shape, and frequency of bubbles, and how they change as the heating progresses. Tiny bubbles begin forming very early on, which are most likely dissolved air being released, since the solubility of air in water decreases with temperature. Vapor bubbles also start forming at quite low temperatures, but for some time, these bubbles do not make it through to the surface of the water. In my experience, something that looks vaguely like boiling usually begins around 97°C (not so far from where that point is marked on the Adams thermometer), and there is quite active boiling from around 98°C. After full boiling begins the temperature stabilizes around 100°C, then creeps up slowly, eventually reaching nearly 101°C. By that point, the vapor bubbles arise from only a few spots at the bottom of the beaker. Watching this makes one realize that boiling is a complicated phenomenon and not likely to take place at a precisely fixed temperature.

The end of that experiment led naturally to the next thing I wanted to check out, namely, the claim by Joseph-Louis Gay-Lussac, endorsed by Jean-Baptiste Biot, that the temperature of boiling water was 101.232°C in a glass container, while it was exactly 100°C in a metallic container.[14] What I observed after boiling water for a prolonged period in a glass vessel was not too far from Gay-Lussac's claim, especially considering that he was not using Pyrex glass and a Bunsen burner. And what happens in other kinds of containers? It is very easy to show by experiment that the boiling temperature of water is indeed lower in a metallic container than in a glass container. The difference is not only in the temperatures but also in the shape, size, and number of bubbles forming in the two different vessels. The boiling behavior is affected not only by the material of the vessel but also by the exact state of its inner surface. For example, putting in fine scratches on the inside bottom surface of a glass vessel helps the formation of bubbles;[15] in a scratched-up beaker, the water temperature is clearly lower than 100°C even at full boiling. In most ceramic mugs the boiling temperatures are very high, easily reaching 102°C; bubbles form and detach themselves with great difficulty and a characteristic noise. With bubbles not forming fast, the water cannot lose heat quickly enough and ends up in a "superheated" state. In a stainless steel pot, the temperature is much lower. The variability of boiling behavior and temperature is illustrated most clearly in Teflon-coated pots: bubbles form very eagerly on this surface from a very low water temperature, and the temperature of both the onset and the peak of boiling is very low, reaching the maximum of only about 99°C. All of these variations are largely forgotten by modern physicists, though they are known to various engineers and chemists.

Past science had even bigger surprises in store for me. These are effects relating to the presence of dissolved gases. For this I come back to De Luc, who noted in his book of 1772 that the bubbles formed during normal boiling come only from the layer of water that is immediately in contact with the heated surface, which must be much hotter than the main body of the water in which we insert the thermometer. To find out the temperature of what he called "true ebullition," he tried various experiments.[16] In a notable experiment that I have replicated, De Luc tried to bring the whole body of water to the same temperature by heating the water slowly while minimizing the loss of heat at the surface. He took a round flask with a long, thin neck and heated it by immersing it in a bath of hot oil. After trying a more or less exact replication, I later found that essentially the same phenomena can be reproduced by a simpler arrangement—heating water in a volumetric flask (a glass vessel with a long, thin neck) on a hotplate, very hot but still much gentler than a naked flame.

The behavior of the water in this setup is very different from boiling driven by a more intense heat source in a wide-mouthed vessel. As the temperature approaches 100°C, the water starts to boil in a normal way. As boiling continues, however, the temperature continues to rise, while the bubbles get bigger but less frequent; they also rise more irregularly, often in bursts. The temperature goes over 100°C, easily reaching 101–102°C while the boiling is reasonably steady. This is what nineteenth-century observers termed "bumping." Later in the process we can observe the "puffing" behavior: with continued heating, the bubbles can become even less frequent, while temperature creeps up further; often there are long quiet periods punctuated by isolated large bubbles. Sometimes the puffs are explosive, throwing some water out of the flask (this is one practical reason to avoid using a hot oil bath). It is very easy to produce temperatures up to 104°C during puffing, which is entirely consistent with what De Luc had reported in 1772.

It may seem puzzling that the boiling behavior changes as it goes on, becoming more and more irregular. De Luc worked out that the formation of vapor bubbles was facilitated by the presence of dissolved air in the water. As the process of boiling has the effect of sweeping air out, boiling becomes more difficult as it goes on. For De Luc, boiling facilitated by dissolved air was not true boiling; he wanted to study boiling in truly pure water. To remove the last bit of air that still remains even after prolonged boiling, De Luc used a kinetic method. If you have ever made the mistake of shaking a bottle of fizzy drink before opening it, you know that mechanical agitation tends to dislodge dissolved gases. So, shaking is what De Luc did. He reported: "This operation lasted four weeks, during which I hardly ever put down my flask, except to sleep, to do business in town, and to do things that required both hands. I ate, I read, I wrote, I saw my friends, I took my walks, all the while shaking my water."[17]

De Luc reported that his degassed water reached 112°C without boiling, and then exploded. That, too, I was able to reproduce many times during the summer of 2004. Only I must confess that I did not have De Luc's dedication to spend four weeks shaking a bottle of water. In the end I found another method that is almost as good, and takes less than an hour. For those readers curious enough to try it for themselves, here is my method: start with water that has been boiled in a loosely covered pot for some time (10–20 minutes will do), will have had most of the air removed already. Then pour this water carefully into a long-necked flask and place it on a hotplate; boiling in this water is very bumpy, and the temperature goes well beyond 100°C, resulting in further degassing. After a while remove the flask from the hotplate and allow it to cool slightly.

Having thus prepared a flask of degassed water, one must heat it again gently to reproduce the explosive effect. This is done most safely and conveniently in a bath of graphite (instead of oil), keeping the temperature of the graphite below 250°C. The temperature of the water should be monitored by inserting a thermometer occasionally. I recommend "occasionally," because at high degrees of superheating the insertion of a thermometer excites violent boiling, as the roughness at the tip of the thermometer serves as a site for bubble-formation, or "nucleation." Undisturbed, the water will be absolutely still most of the time, although its temperature is very high, easily reaching 107–108°C. Inserting the thermometer prompts very active boiling, bringing the temperature down. When there are higher degrees of superheating, the water will explode on contact with the thermometer, or sometimes spontaneously. (It is highly advisable to wear goggles, and one must not look directly into the opening of the flask, though that will be tempting.) If the surface area of the water is relatively large, the fast evaporation that happens at the surface of the superheated water can cause heat-loss that matches the rate of heat input from the graphite, so we can easily have water superheated to 105–106°C sitting there indefinitely with no bubbling.

The immediate lesson from these experiments on boiling is that we can learn fresh things about nature from past science. It seemed unbelievable and wonderful to me that a 230-year-old text could teach me something basic that I had never heard of in my years of studying physics at today's elite universities. I have presented this material to many audiences, and received a whole range of reactions. Some people do not believe me at all, until they see the experiments (or even afterward). There is also the other end of the spectrum. Once I spoke at the Royal Academy of Engineering in London and faced some real anger from an eminent professor who had written a whole book about bubbles. He said that everyone, even his barman, knew about the effects I was talking about—so why was I making a big fuss? Now, many engineers who work on heat transfer, as well as some physical chemists, do know a great deal about the intricacies of boiling.[18] But even these specialists do not know everything; especially, the effect of dissolved gases does not seem to be fully understood.

But what exactly the specialists know and don't know is not quite the issue here. Why should something like the boiling of water, at least in its basic phenomenology, be consigned to the realm of specialists? Most of us boil water on a daily basis. It is not right that we go around repeating that pure water under standard pressure always boils at 100°C, scolding children and marking down students if they don't agree with that piece of untruth. If boiling is only for the specialists, what is left for the poor nonspecialists?

Being humbled by past science: The Voltaic cell[19]

Before coming to more general points, I would like to present one more detailed case in which a look back at very old science has generated some fresh scientific thinking.[20] The subject this time is electrochemistry, and it starts with the replication of a very simple and most intriguing experiment published in 1801 by William Hyde Wollaston (1766–1828)—the London-based physician-turned-chemist, master of platinum, codiscoverer of ultra-violet radiation, and an early advocate of Dalton's atomism. Wollaston began with the well-known observation that certain metals dissolve in acids, releasing hydrogen gas. For instance, one can easily dissolve a zinc wire in fairly weak hydrochloric acid (HCl), producing fine bubbles of hydrogen in the process. Insert a copper wire to the same pot of acid, and no reaction takes place, since HCl does not react with copper readily. But just make the two wires touch, and hydrogen bubbles immediately start issuing from the copper as well the zinc wire.[21]

This experiment is extremely easy to do. *Understanding* it is surprisingly difficult. Wollaston thought that the acid attacked zinc and released the electric fluid from it, which was then conducted over to the copper wire. Modern textbook accounts say that hydrogen ions in the acid take electrons from zinc, turning themselves into hydrogen gas; this transfer of electrons also ionizes zinc, which dissolves in the aqueous acid. But if that is what happens, how does the reaction generate any excess electrons that travel over to the copper side to make hydrogen gas there? In my humble opinion, this is an incomplete account of what acids do to metals. According to the common Brønsted–Lowry theory it is hydrogen ions that define acidity, and H^+ concentration is indeed what pH meters measure. But it seems to me that a crucial role is also played by the anion (that is, the negative ion), which is specific to each acid. This would also help make sense of the fact that hydrochloric acid is quite powerless to attack copper but nitric acid dissolves it readily, while both acids should provide an abundance of H^+ ions. Also note that the nitric-acid reaction produces not hydrogen gas but nitrogen oxide, which promptly reacts with oxygen in the air to create the red fumes of nitrogen dioxide. I learn from T. M. Lowry's own textbook that there is no simple story about what happens in this reaction.[22]

I am not entirely alone in having these unorthodox thoughts about the role of anions.[23] For example, the article on "Battery and Fuel Cell" in the *Encyclopedia Britannica Kids* explains the flow of electrons in a Wollaston-type setup by reference to anion action.[24] With copper and zinc immersed in sulphuric acid, the proposed mechanism is precisely the action of sulphate ions (SO_4^{2-}) in removing zinc ions (Zn^{2+}) from the zinc wire, producing loose electrons, which go over to the copper side and combine with the hydrogen ions (H^+) in that vicinity. Now, this is billed by *Britannica* as an explanation of the Voltaic cell. Why? Well, because it actually is. The topology of Wollaston's experiment is the same as that of Volta's cell: namely, two different metals with an electrolyte between them. In my version of the Wollaston experiment there is typically a voltage of 0.6–0.7V between the copper and the zinc wires. When multiple such cells are connected, one literally has a *battery* of them, which is the origin of that term.

Volta himself had such an arrangement, which he called "the crown of cups" (though it is less famous than his so-called pile, which has pairs of metallic disks separated by layers of wet paper).[25] This is also easily recreated. By connecting up six cups of hydrochloric acid in series with joined-up pairs of copper and zinc wires, I produced a potential of 4.3V; the voltage decreases in steady steps if the cups are removed one by one from the circuit.

Now, if Wollaston's setup is the Voltaic cell, then we should be able to understand what is going on in Wollaston's experiment simply by referring to the modern explanation of the Voltaic cell. So, what is the standard modern explanation of the Voltaic cell? Surprisingly, there isn't one. What we do have, almost everywhere we turn, is an explanation of the Daniell cell in which the electrolyte consists of two different solutions, connected by a salt bridge, or a porous barrier. In this setup each metal is dipped in its own solution, and the electrical action is easily explained in terms of the imbalance of the redox potentials on the two sides. Volta's original cell cannot be explained in this way, and consequently, it has disappeared from basic electrochemical thinking, so much so that people now commonly refer to the Daniell cell as the "Voltaic cell." Volta's original theory, which had attributed the electrical action to the contact between two different metals, has also disappeared.

There was a long and complex debate that raged throughout the nineteenth century between Volta (and his followers) and those who believed (with Wollaston) that the electricity originated from chemical reactions. The details of this debate are now lost to everyone except for a handful of expert historians of science. Helge Kragh has given an overview, which concludes that the dispute was never really resolved, even in the twentieth century.[26] Sungook Hong has given a detailed account of one curious phase of this history in which Kelvin revived Volta's contact theory in the 1860s.[27] There is also much to be found in the older secondary literature, such as J. R. Partington's history of chemistry and most of all Wilhelm Ostwald's text on electrochemistry.[28] But all of this is now at the risk of becoming lost even among historians of science.

This history includes many intriguing phenomena that I wish to replicate. De Luc made a "dry pile" that involved no electrolytes at all. Davy confounded Volta by making a cell with no metals but a piece of charcoal and two different liquids. Volta himself, fascinated by the thought that his "pile" was a realistic model of the torpedo (electric fish), made a battery using pieces of bone instead of metal.[29] Priestley claimed that the electrolysis of water stopped after a while if he covered the surface of the water with oil, preventing the entry of atmospheric oxygen.[30] And so on. All of these phenomena are lying buried in the historical record. Most historians who come across them don't know what to do with them; most philosophers don't know where to look; and most scientists don't care.

I have been discovering my own intriguing phenomena in the lab, too. For instance, I can generate a decent voltage (up to 0.6V) using my thumb as the layer between copper and zinc, with no wetness. In the wet cell, the voltage actually decreases when the concentration of acid is increased beyond a fairly low threshold. The maximum voltage I can reach with copper and zinc in hydrochloric acid is 0.99V, and that happens at quite a low concentration of acid (pH 2.5, which is only as strong as vinegar).

I am beginning to put together my own way of understanding these phenomena, with a modernized version of the eighteenth-century one-fluid theory of electricity, treating the old electric fluid as a collection of free electrons. In understanding how the electric fluid is pushed around, it is necessary to bring in Volta's contact potential (recognized by modern physicists as an expression of the different values of the work function in different metals). Electrons can be liberated for a variety of reasons but in batteries the cause seems to be mostly chemical. However, the bubbling up of hydrogen on the zinc side of the Voltaic cell is mostly an irrelevant sideshow as far as the production of electric current is concerned. A series of simple experiments I carried out demonstrates this point briefly. First, in the original Wollaston setup, a current of about 13mA flowed through the circuit. Next, increasing the amount of zinc dipped in the acid by about 20 times made the zinc-side reaction quite excitingly vigorous, but there was no appreciable increase in the current. Instead, increasing the amount of copper dipped in the acid resulted in a marked increase of current, to about 100mA. Even with a minimal amount of zinc dipped in the acid, we can produce a very good amount of current as long as there is a lot of copper.

Also consider the fact that Volta himself actually used salt water instead of acids, which squares with my notion that the active species here is not H^+ but the anion, in this case Cl^-. The current is small in this setup, in the order of 0.1mA. But the voltage is very good, in fact higher than in the setup using acids. I think that the particular effectiveness of acids as electrolytes in the Voltaic cell is due to the provision of H^+ ions in the vicinity of the copper, to receive electrons there and to allow an easy flow of current. This is also confirmed by an early nineteenth-century experiment by William Sturgeon (1783–1850), who made a Voltaic cell using zinc–mercury amalgam instead of plain zinc in acid, which produced a very good amount of electricity with no production of hydrogen on the zinc side. This experiment was replicated successfully by my student Alexandra Sinclair.[31]

Complementary science

What I have presented in the last two sections are clearly very different kinds of investigations from what historians or philosophers of science usually engage in. They are examples of the mode of study that I have dubbed "complementary science."[32] Complementary science begins with a recognition that the cutting edge is not all there is to science. There are valuable scientific questions that current specialist science does not address. Specialists do not and cannot work with complete freedom. Their line of thinking is severely constrained by particular traditions, which is also what enables them to focus so effectively on detailed and esoteric topics of research. This is the truly lasting part of Thomas Kuhn's insights about the nature of paradigms.

The neglect of certain basic questions is not detrimental to specialist science. How water boils is no longer fundamental to thermodynamics; nor is how the Voltaic cell works to electrochemistry. As science develops, nothing of importance may rest any longer on its historical origins. Metaphorically speaking: the upper layers of a tower can be supported by structures other than what they first rested on in the process of construction. We know that the Eiffel tower stands very well with a great empty space at the bottom[33]— likewise for specialist science. However, that does not mean that the no-longer-fundamental questions are now unimportant in an absolute sense. Someone should still be investigating them.

The discipline of history and philosophy of science (or HPS, to use the common abbreviation in our business) can serve as a refuge for these and other excluded scientific questions. In that way, HPS becomes an enterprise that complements specialist science, neither hostile nor subservient to it. HPS in this complementary mode is not *about* science; rather, it *is* science, only not as we know it. Nature, rather than science, is its primary object of study. HPS in this mode of operation can serve as a useful shadow discipline to specialist science, like the shadow cabinet in British politics picking up on what the real cabinet neglects. To use another metaphor, we need complementary science like we need philanthropy or a welfare system, to help us meet social needs neglected by the capitalist economy, efficient as it is in what it does.

There are several types of investigations one can make in complementary science:

1. The most obvious one is the project of recovering natural phenomena that have become lost to modern science, such as those discussed in the last two sections. When I read old science, I actively go hunting for things that sound wrong; the more wrong they sound, the better, especially if they come from great scientists.[34] These bizarre reports are everywhere, and they should not be disregarded; many of them open doors to a fascinating store of lost scientific knowledge.
2. Textual recovery of lost phenomena is not enough. We need to check out these reports in the lab.
3. Once in the lab, things won't always go as we expect; nature is not that boring. This is why scientists celebrate so heartily the rare moments when experiments do go as expected. So the experimental work intended for checking past reports can easily lead to the discovery of genuinely new phenomena.
4. New and recovered phenomena also stimulate fresh theorizing. I actually find it exhilarating to encounter natural phenomena that I cannot quite understand, either in terms of modern science or according to the thinking of those who first stumbled upon them.
5. Theoretical work is also needed on some very familiar phenomena, such as the Voltaic cell. Another great example is frictional electricity. We all know that rubbing certain objects together generates static electricity. The common facile explanation of this is that some materials have greater

attraction for electrons than others. But we have to ask how it is that the electrons are disengaged in the first place. Except in metals, electrons should be all securely locked away in molecules; it seems unlikely that one can set them free by crude mechanical agitation, and moreover without changing the chemical properties of the materials.[35]

6. Complementary theorizing can start for theoretical reasons, too. My standard starting-point is a sympathetic understanding of past theories that are now discarded. Often it turns out that apparently crazy ideas—such as phlogiston, caloric, and ether—were held for very sensible reasons, and discarded for less than convincing reasons. In other places I have argued, for example, that the concept of phlogiston that Priestley defended so valiantly not only had very good uses at the time but also could easily have been maintained and developed into the concept of chemical potential energy on the one hand, and free electrons on the other.[36] Cultivating such lost ideas raises our critical awareness, and may lead to useful new ideas, too.

Engaging in these complementary-scientific investigations has given me a unique perspective on the nature of HPS.

In philosophy of science, I think we need to lose the habit of simply deferring to scientists on scientific questions. The balance to maintain is to be critical while respecting specialists in their areas of specialization. Many philosophers have actually been doing this type of respectful critical work on topics such as the interpretation of quantum mechanics, or the reduction of psychology and medicine to genetics. We need to develop this kind of work further. The great temptation to resist is the recent tendency toward so-called naturalism, which would reduce philosophy to a branch of cognitive science and give the ultimate authority on all epistemic questions over to neuroscientists. Proper naturalism in philosophy, if I may be allowed to redefine the term, ought to mean that philosophers engage in their own independent considerations of nature, not merely serve as the scientists' mouthpiece. Some philosophers may feel that digging into scientific details does not constitute proper philosophy. On the contrary, I think complementary science is very much in keeping with the most fundamental mission of philosophy: to ask questions without restrictions. So it makes perfect sense that philosophers should pick up important questions that scientists neglect, whether they be questions of methodology, purpose, ethics, or actual scientific content.

Interestingly, something very similar can be said about the remit of the history of science, too. When the history of science asserts its independence from science itself, its domain is apt to be defined negatively, to encompass whatever elements of past science that current science cares not to retain in its institutional memory. A new vision for the history of science arises from complementary work, reaching beyond both the antiquarianism of learning the details of past science without asking whether it was a good way of understanding nature and the current fashion of treating science purely as a social phenomenon with no judgment on its content. Both of these types of history treat past science as something dead, merely an object of study. That is not the way to learn from the past.

Even more important is a reform of science education, based on a recognition that only a tiny fraction of students will grow up to become research scientists. So, what is the purpose of teaching science to all students, and how can that purpose be best served? My proposal is that it will be beneficial, for the students themselves and for society at large, to incorporate complementary science into science education at all levels. Teachers of science often behave like overprotective parents, guiding students carefully on a strict and narrow path toward current specialist knowledge. This is how we have lost the Voltaic cell in electrochemistry, superheating in thermodynamics, and so many other things like that. In making the learning of science safe, we also make it devoid of original thinking and independent inquiry.[37] In any case most students are not able to keep on the prescribed narrow path, or even interested in doing so. This is how we end up treating the majority of science students like failures, who are shown the narrow path and soon enough told that they are not good enough to walk it. And why should students bother with acquiring a system of knowledge that they have no stake in and that they feel they will never use in any real sense?

I am just beginning a dialogue with interested science educators on these issues. Meanwhile, I am not inclined to sit around and wait for science teaching to change. The spirit of complementary science dictates the breaking down of many boundaries. Breaking down the boundary between science and HPS implies that what I do in my job is science education, whether I am teaching students taking degrees in HPS or science students taking optional courses. And my claim that nonexperts can make valuable contributions to knowledge will be mere lip service if I cannot even find ways of getting university students to participate in the production of it. That is why the integration of research and teaching has been so important in my work. Similarly, the pursuit of complementary science also blurs the distinction between research and popularization; this I see as a fundamental way of going beyond the so-called deficit model in science communication.

Practicing 1800 science today

From the viewpoint of complementary science, I would like to take another look at 1800 science. This period of science embodied a way of scholarship that is very congenial to the aims of complementary science. This makes sense because complementary science attempts to provide what is lost when we only have specialist science, and it was shortly after 1800 that specialist (or professional) science became really established. This is why I propose the quaint enterprise of *practicing 1800 science today*. In this regard, I have had some inspiration from the radical educationist Neil Postman, who argues that many answers to our modern problems can be found by a creative look back at the eighteenth century.[38] I fashion my own aim as building bridges to

1800 science: to recover and develop valuable knowledge from past science, while cultivating what was best in the culture of science from that period.

Earlier I mentioned the low entry-barrier to science in the 1800 period. This is quite important for complementary science. In April 1797 in London, a little-known event of capital significance happened: William Nicholson founded *A Journal of Natural Philosophy, Chemistry, and the Arts*. This journal had no institutional basis of any kind, and Nicholson invited contributions from anybody at all. He judged the submissions himself for their interest and significance, rather than sending them out for peer review by experts. The journal was published every month and distributed widely. All this was quite a contrast to the slow and exclusive process of publishing in the *Philosophical Transactions of the Royal Society*, for example. There was a wonderful diversity of authors and topics represented in the pages of *Nicholson's Journal*. Nicholson's editorial work created a broad scientific community including many nonspecialists who actually contributed to the progress of science.

As historian Samuel Lilley put it, this was "popular research"[39] (not "popular science" in which experts tell the ignorant masses about science in simple terms), meaning that original work was actually carried out by the broader public, sometimes in patient empirical steps, sometimes in great leaps of imagination. In fact, following the publication of Nicholson's own paper on the electrolysis of water, *Nicholson's Journal* for a time became the premier venue in Britain for the publication of new research in electrochemistry, some of it by the likes of Davy and the rest by a multitude of now-forgotten people. I have been so impressed by the story of this journal that for six years I have had my undergraduate students at University College London simulate it; they wrote on a particular topic each year as amateur scientists from around 1800; I played editor, publishing the best contributions in our *Virtual Nicholson's Journal*.[40]

The broadening of scientific community effected by *Nicholson's Journal* and other similar initiatives in this period had significant consequences. Allowing nonspecialists to set scientific questions meant that science was obliged to seek knowledge of things that mattered to these people, either for curiosity or for practical benefits. There is an interesting contrast here with modern specialist science, which has a tendency to focus only on what it is good at—that is, to address questions that are most amenable to attack by today's standard methods.

An important consequence of the willingness to ask awkward questions is humility. It was quite common for scientists then to confess that they did not have the final story about the universe. Priestley had a particularly instructive notion of humility, which was dynamic: "Every discovery brings to our view many things of which we had no intimation before." He had a wonderful image for this : "The greater is the circle of light, the greater is the boundary of the darkness by which it is confined." As knowledge grows, so does ignorance.[41] "But," Priestley continued, "notwithstanding this, the more light we get, the more thankful we ought to be. For by this means we have the greater range for satisfactory contemplation. In time the bounds of light will be still farther extended; and from the infinity of the divine nature and the divine works, we may promise ourselves an endless progress in our investigation of them: a prospect truly sublime and glorious."[42] For Priestley all this was based on the infinity of God, but nonbelievers can simply think in terms of a basic plenitude of nature.

From such humility follows pluralism, based on the recognition that one's own attempts at understanding nature are so limited and so uncertain that other attempts ought to be given a chance, too. Priestley is often remembered as a dogmatic defender of the phlogiston theory, but nothing could be farther from the truth. His whole life was spent in advocacy of tolerance—religious, political, and scientific. His last major defense of phlogiston was published in 1796, by which time he was living in exile in America, having been hounded out of England for supporting the French Revolution; meanwhile Lavoisier had met his end in that revolution, guillotined in 1794 for his involvement in privatized tax-collecting. Priestley's preface is addressed to Lavoisier's surviving colleagues, and contains a plea for a tolerant pluralism in science: "But you will agree with me, that no man ought to surrender his own judgment to any mere *authority*, however respectable. Otherwise, your own system would never have been advanced. As you would not, I am persuaded, have your reign resemble that of *Robespierre*, few as we are who remain disaffected, we hope you had rather gain us by persuasion, than silence us by power."[43]

This is not the place for my full-blown argument for scientific pluralism,[44] so I am just going to leave you with a joke.[45] It is a brief exchange between a teacher and a pupil.

> "*Teacher*: Clyde, your composition on 'My Dog' is exactly the same as your brother's. Did you copy his?"
> "*Clyde*: No, Sir. It's the same dog."

We laugh about this, but when it comes to science we tend to get very ceremonious in dismissing any suggestion that there might be two different and equally valid and good stories about the same object.

To conclude: a careful look back at 1800 science helps us realize that modern specialist science only deals with a restricted range of things in a restricted range of ways. The brilliant successes of today's science may make it seem that we have securely worked out the basic story about nature, with only some details left to be determined. But just scratch the surface, and you begin to see so much more, even in very simple and mundane phenomena. All sorts of things that may seem as boring as watching paint dry will turn out to be fascinating subjects for further investigation. (Come to think of it, the drying of paint is probably an interesting and complex chemical and physical process worthy of a PhD dissertation!) The *wonder* of nature all around us is something we often neglect in our view of science today—so heavily colored by concerns about exams, peer review, and research grants. What I have proposed here are some concrete steps toward a far-reaching aim—that of enabling the educated public to participate, once again, in the wonderful enterprise of building the knowledge of our universe. I cannot think of a better use for the history of science.

Notes

This chapter is an updated and extended version of the inaugural lecture delivered at University College London on May 13, 2009.

1. J. L. Heilbron, "HSS Lecture: Applied History of Science," *Isis* 78 (1987): 552–563. This address was delivered at the 1986 joint meeting of the History of Science Society, the Philosophy of Science Association, the Society for History of Technology, and the Society for the Social Studies of Science.
2. Paul Forman, "Independence, Not Transcendence, for the Historian of Science," *Isis* 82 (1991): 71–86, on p. 86.
3. Neil Postman, *Building a Bridge to the 18th Century: How the Past Can Improve Our Future* (New York: Alfred A. Knopf, 1999).
4. Heilbron, *Isis*, 78: 554.
5. J. G. Crowther, *Scientists of the Industrial Revolution* (London: The Cresset Press), 218.
6. On Priestley see Hasok Chang, "Priestley, Joseph (1733–1804)," in Andrea I. Woody, Robin Findlay Hendry and Paul Needham, eds., *Philosophy of Chemistry*, Handbook of the Philosophy of Science, vol. 6 (Amsterdam: Elsevier, 2012), 56–62, and references therein.
7. On Rumford see Sanborn C. Brown, *Benjamin Thompson, Count Rumford* (Cambridge, Mass.: The MIT Press, 1979).
8. On De Luc see John Heilbron, "Citoyen de Genève and Philosopher to the Queen of England," *Archives des Sciences* 58 (2005): 75–92; Paul A. Tunbridge, "Jean André De Luc, F. R. S.," *Notes and Records of the Royal Society of London* 26 (1971): 15–33.
9. For enabling this work I thank Andrea Sella, Crosby Medley, Mike Ewing, and other members of the Chemistry Department at University College London, who literally gave me the keys to one of their teaching labs that summer, and the Leverhulme Trust and the ESRC for a research grant that easily met the modest expenses in apparatus and materials.
10. Hasok Chang, *Inventing Temperature: Measurement and Scientific Progress* (New York: Oxford University Press, 2004), Chapter 1.
11. See Chang, *Inventing Temperature*, cover illustration.
12. This, of course, is assuming that water essentially behaves today like it did in the eighteenth century. That level of inductive faith, though admitted to be fallible, is necessary for any historical investigation.
13. This and other experiments described here are discussed in more detail, with video clips, in Hasok Chang, "The Myth of the Boiling Point," http://www.hps.cam.ac.uk/people/chang/boiling/. The videos were made expertly by Matt Aucott and his colleagues in the UCL Multimedia Section.
14. Jean-Baptiste Biot, *Traité de physique expérimentale et mathématique*, 4 vols. (Paris: Deterville, 1816), vol. 1, 41–43.
15. This is apparently a technique used by expert bartenders to get better bubbling in champagne.
16. Jean-André De Luc, *Recherches sur les modifications de l'atmosphère* (Geneva, 1772), vol. 2, supplement, Chapters 8–10 (pp. 356–398, §§ 980–1073).
17. Ibid., 387 (§ 1048).
18. See, for instance, Frank P. Incropera and David P. DeWitt, *Fundamentals of Heat and Mass Transfer*, 4th ed. (New York: John Wiley and Sons, 1996).

19. The content of this section is largely a reflection of the state of my research at the time when this chapter was first presented as a lecture in 2009. A more detailed and up-to-date report can be found in Hasok Chang, "How Historical Experiments Can Improve Scientific Knowledge and Science Education: The Cases of Boiling Water and Electrochemistry," *Science and Education* 20 (2011): 317–341.

20. I was able to start this project thanks to the generosity of Daren Caruana, who has given me lab space and expert advice; I also thank everyone in the Caruana lab, and Rosie Coates and Georgette Taylor for their collaboration, and the Leverhulme Trust for financial support.

21. William Hyde Wollaston, "Experiments on the Chemical Production and Agency of Electricity," *Philosophical Transaction of the Royal Society of London* 91 (1801): 427–434.

22. T. M. Lowry, *Historical Introduction to Chemistry* (London: Macmillan, 1936), 91. I think the question of causality in the metal–acid reaction is philosophically just as interesting as the question of causality in gene expression or in the EPR experiment.

23. I have located some discussion from the first half of the twentieth century about the action of anions in the solution of metals. See, for example, L. Whitby, "The Dissolution of Magnesium in Aqueous Salt Solutions," *Transactions of the Faraday Society* 29 (1933): 415–425, 853–861; U. R. Evans, "Behaviour of Metals in Nitric Acid," *Transactions of the Faraday Society* 40 (1944): 120–130.

24. Voltaic cell [Art], *Britannica Online for Kids*, http://kids.britannica.com/comptons/art-106622, accessed on July 24, 2012.

25. Alessandro Volta, "On the Electricity Excited by the Mere Contact of Conducting Substances of Different Kinds," *Philosophical Transaction of the Royal Society of London* 90 (1800): 403–431, Fig. 1, on the plate between p. 430 and p. 431.

26. Helge Kragh, "Confusion and Controversy: Nineteenth-Century Theories of the Voltaic Pile," in Fabio Bevilacqua and Lucio Fregonese, eds., *Nuova Voltiana: Studies in Volta and His Times*, vol. 1 (Milan: Hoepli, 2000), 121–151.

27. Sungook Hong, "Controversy Over Voltaic Contact Phenomena, 1862–1900," *Archive for History of Exact Sciences* 47 (1994): 233–289.

28. J. R. Partington, *A History of Chemistry*, vol. 4 (London: Macmillan, 1964); Wilhelm Ostwald, *Electrochemistry: History and Theory*, 2 vols., translated from the German by N. P. Date (New Delhi: Amerind Publishing Co., 1980); originally published in 1895.

29. Giuliano Pancaldi, *Volta: Science and Culture in the Age of Enlightenment* (Princeton: Princeton University Press, 2005), 205.

30. Joseph Priestley, "Observations and Experiments relating to the Pile of Volta," *A Journal of Natural Philosophy, Chemistry and the Arts (Nicholson's Journal)*, new series 1 (1802): 198–204, on p. 201. No one seems to have taken notice of that observation, but Davy later made a possibly related observation that the presence of dissolved oxygen in the electrolyte enhanced the action of the battery; see Humphry Davy, "The Bakerian Lecture [for 1806]: On Some Chemical Agencies of Electricity," *Philosophical Transactions of the Royal Society* 97 (1807): 46–47.

31. How do we decide whether my ideas have any merit? What exactly is the nature of explanations we are looking for here? That is another whole story, of course. For now I will just say that I have an instrumentalist and pluralist take on this: I will use any concepts that make sense of observations for me, and if there are multiple explanations for the same phenomena, all the better. If I think that my ideas have merit, it is not at all to suggest any devaluing of the standard specialist work in electrochemistry.

32. See Chang, *Inventing Temperature*, Chapter 6.

33. For photos showing the stages of the construction of the Eiffel Tower, see "Onarchitects", http://onarchitects.blogspot.co.uk/2012/06/eiffel-tower-how-is-made.html, accessed on 1 August 2012. Initially the lower platform of the tower had to be propped up by a thick column underneath it, which was later removed.

34. One fascinating example is Rumford's work on "frigorific rays"; see Hasok Chang, "Rumford and the Reflection of Radiant Cold: Historical Reflections and Metaphysical Reflexes," *Physics in Perspective* 4 (2002): 127–169.

35. I think all this can be explained by reviving Kelvin's old vision of atmospheric electricity, according to which everything in touch with the earth is negatively electrified, that is to say, soaked in free electrons. But I really need to leave this subject for another occasion, except to thank my student Mat Paskins for teaching me about it.

36. Hasok Chang, "We Have Never Been Whiggish (About Phlogiston)," *Centaurus* 51 (2009): 239–264. See also Hasok Chang, *Is Water H₂O? Evidence, Realism and Pluralism* (Dordrecht: Springer, 2012), Chapter 1.

37. I am reminded of my own early education in the Confucian tradition, which instills a beneficial respect for the past but on the other hand stifles innovation by teaching that we should only try to have original thoughts after achieving a complete mastery of inherited wisdom. Trying to reinvent the wheel would be the worst sin in Confucian scholarship. Even in the West in the twenty-first century, science education is somewhat Confucian in this way—how dare you have an original thought, before you have earned your right to do so by suffering through years of problem sets untangling the Schrödinger equation and the like, and long sessions in the lab doing titrations, standard syntheses, damped harmonic oscillation, etc. As long as we are operating within that basic framework, no amount of enquiry-based or problem-based learning will encourage true originality. There is no use in telling students they should go and find out the answer to a question themselves if you also tell them that there is one right answer and you know that answer; then naturally students will figure out it is better and easier to get the answer from the teacher rather than risk getting it wrong by thinking for themselves. This futility can be avoided by means of the pluralism inherent in complementary science.

38. Postman, *Building a Bridge*. For example, after a cartoon in *The Los Angeles Times*, he asks us (pp. 90–91) to imagine a wonderful new technology for information-transfer: "It requires no batteries or wires; no maintenance contract is needed; it is lightweight, recyclable, and biodegradable; it is absolutely portable...[and] completely quiet...; no secret numbers, access codes, or modems are needed...; one has unlimited use of it for about twenty dollars a month; it comes pre-edited for pornography, fraud, and typos...; and, last but not least, the product does not...contribute to the bank account of Bill Gates." Yes, you guessed it — this amazing new product is called the daily newspaper.

39. S. Lilley, "'Nicholson's Journal' (1797–1813)," *Annals of Science* 6 (1948–1950): 78–101, on p. 93, section heading.

40. http://www.ucl.ac.uk/sts/staff/jackson/nicholsons, accessed on 24 July 2012. After my departure from UCL, Catherine Jackson continued the Virtual Journal for two further years.

41. For an illustration of this image, see figure 5.1 in Chang, *Is Water H₂O?*, p. 256.

42. Joseph Priestley, *Experiments and Observations on Different Kinds of Air, and Other Branches of Natural Philosophy, Connected with the Subject*, 3 vols. (Birmingham: Thomas Pearson, 1790), vol. 1, xviii-xix.

43. Joseph Priestley, *Considerations on the Doctrine of Phlogiston, and the Decomposition of Water* (Philadelphia: Thomas Dobson, 1796; reprinted in 1969 by Kraus Reprint co., New York), 17. The content of this publication was reiterated in a further publication of 1803, published shortly before his death: *The Doctrine of Phlogiston Established, and That of the Composition of Water Refuted*.

44. For such an argument, see Chang, *Is Water H₂O?*, Chapter 5.

45. I thank my mother-in-law, Elva Siglar, for sharing this joke.

3

The Textbook Case of a Priority Dispute: D. I. Mendeleev, Lothar Meyer, and the Periodic System

Michael D. Gordin

Introduction

I have no idea who discovered the periodic system of chemical elements, and I am going to tell you why. When you open a chemistry textbook today, you can often find, next to its periodic table, a sidebar with a grizzled bearded man who is depicted as "the discoverer" of the periodic law, the formula tor of the table whose checkered countenance greets you from the wall of every chemistry laboratory in the world. Almost always, that bearded man is Dmitrii Ivanovich Mendeleev (1834–1907), a chemist from St. Petersburg who published his version of this system in 1869—or maybe in 1871, depending on how you figure it. Sometimes he shares the space with the grizzled beard of Julius Lothar Meyer (1830–1895), who published his version in 1864, or 1868,[1] or 1870.[2] A hundred years ago, German textbooks might simply have presented Meyer, and some esoteric texts would have also depicted John Newlands, or Gustav Hinrichs, or one or two others—grizzled beards all. The textbooks are endowed with a certainty I do not have; they know what the periodic table *is*, and therefore they know who discovered it first.

Their framework rests on a preconceived notion of what "the discovery" is, what the fact or theory consists of *in essence*. The difficulty with this approach, however, can be illustrated by drawing a lesson from philosopher Ludwig Wittgenstein's conception of language as a game. That is, there are no specific ostensive meanings to certain words, or given grammar rules written in stone, but rather simply guidelines that only make sense within the framework of a specific set of circumstances. Disagreement can stem from stressing either too few of the similarities or too many of the differences between two concepts:

> If someone were to draw a sharp boundary [around a term such as "game"] I could not acknowledge it as the one that I too always wanted to draw, or

had drawn in my mind. For I did not want to draw one at all. His concept can then be said to be not the same as mine, but akin to it. The kinship is that of two pictures, one of which consists of colour patches with vague contours, and the other of patches similarly shaped and distributed, but with clear contours. The kinship is just as undeniable as the difference.[3]

Kicking a ball might be a solitary exercise, or a move in a game of soccer, or an illegal action in basketball. I could select one of these as the meaning of "kick" and exclude all the others as "not kicks," depending on how I am *using the concept at that moment*. The surrounding context gives meaning to that word.

This might not seem to be too much of a problem in science if you are talking about something like a solar eclipse. We might all agree that the observation of the eclipse happened and give credit to the person who saw it first—assuming our watches were synchronized and that we agreed on whether credit should go to the first person who *saw* it, or who *wrote* it down in a notebook, or who *published* it, or who *explained* it, or who *predicted* it. So, even here, in a case of an ostensibly simple observation of the natural world, we encounter an almost irreducible problem of how to assign credit if credit is to be apportioned with respect to *being first*.[4]

The worries get much worse when we talk about the periodic system of chemical elements. Just about every individual who has had even the most cursory science education can recognize a periodic table on sight; it may be, in fact, the most widely recognized icon of science in the world. It would be really nice to be able to give credit to the person who "discovered it." Here we encounter conceptual difficulties, both in terms of what it means to "discover" the system and then concerning what "it" is.

What *is* the periodic system of chemical elements? Is it the abstract idea of a system? Is it recognition of a periodic law undergirding the ordering of chemical elements? Is it representation of that law and system in a tabular format? Which tabular format? (There are roughly one hundred topologically distinct representations of the periodic system.)[5] We find in the scholarly literature a number of competing definitions by chemists, philosophers, and historians of science as to the essence of the table and therefore who should get credit for having arrived at it first. Candidates for the crucial feature include:[6]

1. Recognition that properties of elements repeat periodically with increase of atomic weight.
2. Arranging a subset of the elements in a two-dimensional grid to present this relation.
3. Using this system to classify *all* known elements.
4. Leaving gaps in the system for elements that have not yet been discovered but whose existence can be inferred from the properties of known elements.
5. Correcting measured properties of known elements using the system (also known as retrodiction).
6. Predicting detailed properties of new elements to fill the gaps.

Depending on which of these claims you take to be the essence of the periodic system of chemical elements, you will end up with a different discoverer who is assigned priority for being first.

I have two problems with this picture: the first is with the notion that there is one law and therefore only one discoverer, and the second is with how we as present-day observers of history detect who came first.[7] First to the problem of essentializing discovery with respect to the periodic table. The periodic table is one of the classic cases of so-called "simultaneous discovery," with six individuals vying for credit in the 1860s alone (bracketing supposed "precursors"). Depending on your commitments to the six points above, you will give the credit to Alexandre-Émile Beguyer de Chancourtois (#1, #2),[8] John Newlands (#2),[9] William Odling (#3),[10] Gustav Hinrichs (#3),[11] Lothar Meyer (#4 and arguably #5),[12] and Dmitrii Mendeleev (#6).[13] Tell me who you think discovered the periodic system in the 1860s, and I will tell you what you think the periodic system *is*. This may be an amusing philosophical parlor game, but it is rather dubious history, because it forces us to project back our conception of what the correct system is and look for its antecedents among this plethora of discoverers/codiscoverers.

Now to the problem of how historians measure "Firstness." *Why* were there so many different systems emerging in the 1860s? The 1860s proved a tumultuous period in the history of chemistry—when almost every concept and theory was up for redefinition, rearticulation, or rejection.[14] In September 1860, attendees of the International Congress of Chemists at Karlsruhe witnessed a seminal speech by the Italian chemist Stanislao Cannizzaro, who argued for a revitalization of Amedeo Avogadro's (or Charles Gerhardt's—another priority mess!) hypothesis to provide for standardized atomic weights. By applying Avogadro's rules consistently, it was possible to reconcile many seeming anomalies among atomic-weight determinations (from $C = 6$ to $C = 12$, for example) and thus be in a position to compare the corrected weights to each other and seek relationships among them. Two attendees at this Congress, Meyer and Mendeleev, later cited Cannizzaro's influence as crucial in their individual paths to the periodic system.[15] By the late 1860s, only 63 elements had been discovered (very few of them rare earths), so classification of the substances in a two-dimensional grid was simpler than it might have appeared later. Six periodic systems within the decade; none earlier.

So how do we know who came first? Because most scholars who have examined this question are in thrall to a pre-Wittgensteinian notion of essences of theories, they have searched among scientific articles published in the specialized chemical press. If you believe in individualized nuggets of discovery, this is the perfect place to stalk your quarry, since scientific articles focus on specific claims and they cite predecessors. In this way, you can make a claim that someone did not (or did) know about someone else's work and look for which of our six features was affirmed by the author.

My approach is different. I contend that the genre of the scientific article has often structured how we look at the history of science, a bias that is particularly harmful to understanding episodes in the middle of the nineteenth century when that genre was just beginning to congeal. Instead, I take Wittgenstein's concept of a game seriously. In many of the claims to discovering periodicity, one finds that the periodic system emerged in the context of the writing of a chemistry textbook. Yet the histories of periodicity are written mostly or entirely from journal articles, with scant attention to the textbooks. Here I consider Mendeleev's *Principles of Chemistry* and Meyer's *Modern Theories of Chemistry* as loci of the creation of each individual's periodic system.[16] By exploring how the periodic system fits in the composition and then revision of each of their textbooks, I hope to reorient the discussion a smidgen away from who-found-what-first to what-did-each-want-to-do-with-it. In the context of the systems' deployment in the textbooks, we see that both Mendeleev's and Meyer's systems encoded a picture of what chemistry as a whole was about, and as a result we grasp a crucial difference between these two major claimants—specifically, why Lothar Meyer did not predict the properties of any new elements to fill the gaps in his system, while Mendeleev did. I defer here the interesting history of how these two systems got ripped out of their textbooks and placed in the agonistic field of journal disputation, or the importance of scientific priority over Germans for Russian nationalist politics in this period, as well as an extension of this analysis to the other four contenders for priority. The priority dispute proper took place among the scientific community writ large; the systems, however, were born with the classroom in mind.

Mendeleev's *Principles of Chemistry*

If you recognize the name Dmitrii Ivanovich Mendeleev, you probably heard of him in school—for it is in current chemistry textbooks that he is introduced as *the* discoverer of the periodic law, full stop. I will not adjudicate claims of priority here; I only wish to demonstrate what it means when someone gives sole credit to Mendeleev—which features of the periodic system are emphasized and which features are elided. This section will summarize the process by which Mendeleev came to his formulation of the periodic law in the course of writing his textbook, *Principles of Chemistry* (*Osnovy khimii*) in 1869–1871, and then point to how the pedagogical origins of "Mendeleev's periodic law" stresses particular features as the essence of the periodic system.

Mendeleev was born in Tobol'sk, Siberia, in 1834, the last child of a school inspector and the daughter of a factory owner who had fallen on hard times.[17] After his strong (but not exceptional) performance in school, his recently-widowed mother decided to enroll her son in university and conveyed him first to Moscow (where he was turned down by Russia's oldest university) and then to St. Petersburg (where he failed to gain admission to St. Petersburg University but eventually matriculated in 1851 at his father's *alma mater*, the Chief Pedagogical Institute).

Mendeleev graduated with an emphasis in the natural sciences, especially physics and chemistry, and then undertook study for a master's degree in chemistry at St. Petersburg University. After a number of travails—including a stint teaching at a high school in the Crimea, which he detested—he was sent abroad to Heidelberg University for additional postgraduate study.[18] He returned to St. Petersburg in early 1861, two weeks before Tsar Alexander II abolished serfdom, and took on several adjunct positions—including one at St. Petersburg University for a few months before it was closed for two years due to student unrest—until settling into an extraordinary professorship at the St. Petersburg Technological Institute. In this period of relative penury, he first tried his hand at textbook composition to earn some extra money, penned *Organic Chemistry* very rapidly, and received the additional boon of the Demidov Prize of the Academy of Sciences for the final product in 1862.[19] This textbook, composed around the central concept of Charles Gerhardt's and Auguste Laurent's type theory, was soon eclipsed by the structural framework of Aleksandr M. Butlerov, chemistry professor at Kazan (and soon St. Petersburg), whose textbook, *Introduction to the Complete Study of Organic Chemistry* (*Vvedenie k polnomu izucheniiu organicheskoi khimii*), soon became a classic of Russian chemical pedagogy.[20]

Mendeleev was promoted to professor of chemistry at St. Petersburg University in October 1867. This new position demanded that he teach the introductory inorganic chemistry lecture course, a requirement for all students in the rapidly expanding natural sciences faculty. To do this, he needed to assign a textbook. Unfortunately, Russian-language chemistry textbooks did not exactly grow on trees, especially in the late 1860s, when all prior textbooks quickly became superannuated by the rapid developments in contemporary chemistry. A Russian professor had two choices: pick an up-to-date textbook in French, German, or English and translate it (amending it in the process); or write one from scratch.[21] Mendeleev, concluding that scientific developments would likely eclipse the first option by the time the translation was completed and that he was more likely to turn a self-composed textbook into a lucrative financial venture, opted for the second. The idea to write *Principles of Chemistry* was born.

This was a fortunate decision for us, since Mendeleev's formulation of the periodic system of elements grew directly out of the process of composition of this text.[22] *Principles of Chemistry* consisted of two volumes. Volume 1 was largely written in 1868 and concluded in the first month of 1869. The idea for a periodic arrangement of elements was introduced as Mendeleev attempted to map out an outline for volume 2. Volume 1 consisted of a largely empirical introduction to the practices of being a chemist—providing multilayered and detailed introductions to hydrogen, carbon, oxygen, and nitrogen, as well as the halogen family. This left just under seven-eighths of the 63 known elements for volume 2. Mendeleev needed to come up with an organizational system that would compress them into the same span with which he had dealt with only eight elements. What began as an outline for grouping elements together to ease their exposition soon developed, by late February 1869, into a suggestion for an underlying pattern that united *all* elements into a natural system (figures 3.1 and 3.2).

```
                                  Ti = 50    Zr = 90     ? = 180.
                                  V = 51     Nb = 94    Ta = 182.
                                  Cr = 52    Mo = 96     W = 186.
                                  Mn = 55    Rh = 104,4  Pt = 197,4
                                  Fe = 56    Ru = 104,4  Ir = 198.
                          Ni = Co = 59   Pl = 106,6     Os = 199.
        H = 1                     Cu = 63,4  Ag = 108   Hg = 200.
              Be = 9,4   Mg = 24  Zn = 65,2  Cd = 112
              B = 11     Al = 27,4  ? = 68   Ur = 116   Au = 197?
              C = 12     Si = 28    ? = 70   Sn = 118
              N = 14     P = 31    As = 75   Sb = 122   Bi = 210?
              O = 16     S = 32    Se = 79,4 Te = 128?
              F = 19     Cl = 35,5 Br = 80   J = 127
        Li = 7 Na = 23   K = 39    Rb = 85,4 Cs = 133   Tl = 204.
                         Ca = 40   Sr = 87,6 Ba = 137   Pb = 207.
                          ? = 45   Ce = 92
                        ? Er = 56  La = 94
                        ? Yt = 60  Di = 95
                        ? In = 75,6 Th = 118?
```

Figure 3.1   The first published version of Mendeleev's periodic system, dated February 17, 1869, produced while composing *Principles of Chemistry*.

Source: D. I. Mendeleev, *Periodicheskii Zakon. Klassiki Nauki*, ed. B. M. Kedrov (Moscow: Izd. AN SSSR, 1958), 9.

[31]	Группа I	Группа II	Группа III	Группа IV	Группа V	Группа VI	Группа VII	Группа VIII. Переход к группе I
	H = 1							
Типические элементы	Li = 7	Be = 9,4	B = 11	C = 12	N = 14	O = 16	F = 19	
Первый период — Ряд 1-й	Na = 23	Mg = 24	Al = 27,3	Si = 28	P = 31	S = 32	Cl = 35,5	
— 2-й	K = 39	Ca = 40	— = 44	Ti = 50?	V = 51	Cr = 52	Mn = 55	Fe = 56, Co = 59, Ni = 59, Cu = 63
Второй период — 3-й	(Cu = 63)	Zn = 65	— = 68	— = 72	As = 75	Se = 78	Br = 80	
— 4-й	Rb = 85	Sr = 87	(?Yt = 88?)	Zr = 90	Nb = 94	Mo = 96	— = 100	Ru = 104, Rh = 104, Pd = 104, Ag = 108
Третий период — 5-й	(Ag = 108)	Cd = 112	In = 113	Sn = 118	Sb = 122	Te = 128?	J = 127	
— 6-й	Cs = 133	Ba = 137	— = 137	Ce = 138?	—	—	—	— —
Четвертый период — 7-й	—	—	—	—	—	—	—	
— 8-й	—	—	—	—	Ta = 182	W = 184	—	Os = 199?, Ir = 198? Pt = 197?, Au = 197
Пятый период — 9-й	(Au = 197)	Hg = 200	Tl = 204	Pb = 207	Bi = 208	—	—	
— 10-й	—	—	—	Th = 232	—	Ur = 240	—	
Высшая солянная окись	R²O	R²O² или RO	R²O³	R²O⁴ или RO²	R²O⁵	R²O⁶ или RO³	R²O⁷	R²O⁸ или RO⁴
Высшее водородное соединение			(RH⁵?)	RH⁴	RH³	RH²	RH	—

Figure 3.2   Short-form periodic system. This version, taken from a November 1870 article by Mendeleev, is virtually identical to one which appeared in the first edition of the *Principles*.

Source: D. I. Mendeleev, *Periodicheskii Zakon. Klassiki Nauki*, ed. B. M. Kedrov (Moscow: Izd. AN SSSR, 1958), 76.

Understandably, Mendeleev did not fully grasp in February 1869 the implications of the periodic system, but certain features of the incomplete first system (such as the question marks embedded in figure 3.1) indicate that he was well on the way to thinking them through. He continued to develop the system for the next two years, during which time he revised the second volume of his textbook, and he completed both the research cycle and the textbook in late 1871. Although Mendeleev would of course tinker with the system throughout his life—even adding a whole group of noble gases for the seventh and eighth editions—he insisted that the essence of the law could be found in the first edition. For example, consider this statement from Mendeleev's fifth (1889) edition: "I would like to show in an elementary exposition of chemistry the tangible utility of the application of the periodic law, which appeared before me in its entirety precisely in 1869, when I wrote this composition . . . In this, 5th, edition I did not change a single essential feature of the original work, but only supplemented it."[23] (This notwithstanding the fact that Mendeleev continued to work on the periodic system, and in each of the eight editions of the *Principles of Chemistry* he elevated its significance and its status to a periodic *law*.)[24] Shortly after the publication of the first edition, Mendeleev claimed in a letter to Emil Erlenmeyer—at that moment editor of *Liebigs Annalen*—that even the 95-page research article he had submitted was inferior in detail to the textbook itself: "Despite its size, the present article does not go over the course of my ideas in all the details, which are developed more completely and fully in my Russian articles and in my 'Principles of Chemistry,' and which I would happily acquaint the German public with."[25]

Mendeleev always stressed not only the periodic system's pedagogic origins but also its continued pedagogic utility (a feature of the system appreciated by chemistry teachers to the present day). Statements on this were so important that he preserved them in numerous translations of his original Russian articles: "I will add still another remark. it is that the use of the periodic law facilitates the learning of chemical facts by beginners. I have come to this conclusion during the courses of lectures that I have given for two years, and during the preparation of my 'Traité de Chemie Inorganic,' now published (in Russian), which treatise is based on the periodic law."[26] For however flighty and mercurial Mendeleev might have been as a natural scientist and a professional colleague, he was deeply committed to undergraduate pedagogy and left a lasting impression on generations of students (he retired from the university, although not from lecturing at various other institutions, in 1890).[27] For Mendeleev, the periodic system was pedagogically inflected into its core because it represented a hypothesis-free (to his lights) means of conveying chemistry. He emphasized this in the same letter to Erlenmeyer quoted above: "I want only that you will pay attention to the fact that I do not set up any hypotheses, because in my view these often seduce students as false keys and thus tend to slow down the free development [of] science."[28]

The pedagogic core of the periodic law reflected Mendeleev's deep commitments as to what were admissible and inadmissible hypotheses in chemistry, such as his skepticism about both atomism and valency. It may appear somewhat counterintuitive that Mendeleev remained for most of his life (he recanted somewhat in his final decade) hostile to the very two concepts—the existence of atoms and the integral units of chemical bonding—that seem to many today to be the central features of the periodic system. In 1877, British chemist William Crookes, in an evaluation of the periodic system, observed that "M. Mendeleeff himself declares that the Periodic Law cannot be harmonised with the Atomic theory without inverting known facts."[29] Mendeleev insisted that the periodic system did not provide any evidence either way on the existence or nonexistence of atoms, and he professed himself happier as an agnostic about their ultimate reality. He was deeply suspicious of Prout's hypothesis, which in its earliest form proposed that all atoms were glommed-together compounds of hydrogen atoms; since this original formulation was clearly ruled out by fractional atomic weights, such as chlorine's 35.5, it was later modified as an umbrella term for any belief that atoms were composite in nature. For Mendeleev, Prout's hypothesis was an instance of unwarranted hypothesizing along the same lines as traditional atomism.[30] His suspicion of valency deepened his general hostility toward overly microscopic interpretations of atomic behavior with a competitive defense of his older type-theoretic organic chemistry in juxtaposition to the Kekulé-Butlerov structure theory.[31]

To today's chemists, Mendeleev's views seem rather bewildering—and they seemed so to his contemporaries as well. While he was not the only chemist who resisted atomism and valency, he was one of a dwindling number, and most of his coskeptics were theoretical reactionaries who resisted even the periodic system. With one exception, on every major theoretical speculation in late nineteenth-century chemistry—atomism, substructure to atoms, the existence of the electron, the existence of noble gases, valency, radioactivity—Mendeleev was on the conservative, *incorrect* side.[32] The exception, of course, was the use of the periodic system to predict the properties of unknown elements. Mendeleev was almost alone in advocating this as a feasible use of the system in the early 1870s, and he was spectacularly right three times—correctly foreseeing the properties of elements eventually discovered as gallium (1875), scandium (1879), and germanium (1886). And these successful predictions are the sole reason we now see Mendeleev as a chemical visionary instead of a chemical reactionary. In the textbook context, we very clearly observe Mendeleev's essential conservatism on the chemical-theoretic issues of the day and notice how the periodic system fits this frame beautifully—an organization of the elements that does not require presumptions about Proutian "primary matter" ("protyles"), or adherence to a specific theory of valency. It was supposed to teach students how to reason chemically with a knowledge of the substances and a resistance to fancy speculation.

In the context of scientific journal articles, however, prediction was quickly elevated not only as the major differentiating point between his claim to priority and Lothar Meyer's (which is true enough), but also as the essential feature of the periodic system. The fact that most historians have assiduously analyzed only these journal articles has resulted in an overweening emphasis on prediction in accounts of Mendeleev's formulation of the system.

Mendeleev's system was announced in foreign chemical journals in basically two ways. First, it was reported in the proceedings of the Russian Chemical Society's meetings, a standard informational bulletin.[33] Second, it emerged in Mendeleev's own translated articles. The first of these pieces, in the *Zeitschrift für Chemie* in 1869, contained a translation error that in itself was the source of much dispute between Mendeleev and Meyer.[34] It is fairly clear from archival sources that Mendeleev had previously been unaware of alternative periodic systems that had appeared either in textbooks or in journals. Now that others were laying claim to having provided the foundation for Mendeleev's obviously more comprehensive and refined system, he became both more defensive and aggressive in his priority claims. He soon declared himself "an enemy of all questions of priority," which is a good indication that the speaker is anything but.[35] But how could he defend himself when he was manifestly the last person to publish a periodic system in the 1860s?

He opted for two main points of attack: independence of his system, and its greater completeness. Both came together under a theory of credit-distribution in the sciences. First, independence:

> I consider it necessary to impart, that during the formulation of the periodic system of elements I used the earlier works of Dumas, Gladstone, Pettenkofer, Kremers, and Lenssen on the atomic weight of similar elements, but that I was unaware of the apparently preceding works of de Chancourtois in France (Vis tellurique or the spiral of elements based on their properties and equivalences) and of J. Newlands in England (Law of octaves, according to which e.g. H, F, Cl, Cr, Br, Pd, J, Pt form the first and O, S, Fe, Se, Ru, Fe, Au, Th form the second octave), in which some embryos of the periodic law are to be seen.[36]

Leaving Lothar Meyer, of course, unmentioned, the man he accused of having stolen periodicity. He only ceded Meyer some credit after the Royal Society awarded the Davy Medal for the periodic system jointly to both men in 1882.[37] (After Meyer's death, Mendeleev started to be positively cordial to the man—but only as a precursor, not as the initiator of a full-fledged competing system.)[38]

Once he had established his independence, Mendeleev made a virtue of coming last, arguing that even though others had found germs of the idea, historical exemplars indicated that true credit should only go to the one who fully realized all the system's implications (in analogy to oxygen being attributed to Antoine Lavoisier as opposed to Joseph Priestley): "It is right to consider as the creator (*Schöpfer*) of a scientific idea he who not only recognized the philosophical concern but also the real side of a matter, who knows how to so illuminate the issue that anyone could be convinced of its truth and it becomes general. Only then would the idea, like matter, become indestructible."[39] This naturally implied that the correct parameter to judge credit was who drew out the furthest correct implications. Once one frames the field in this way, the answer becomes obvious: he who correctly predicted the properties of unknown elements. And we all know who that was—not Lothar Meyer.

Meyer's *Modern Theories of Chemistry*

Based on his background, it is somewhat odd that Lothar Meyer became a chemist at all.[40] He was born in Varel, Oldenburg, on August 19, 1830, the fourth of seven children of a local physician and the daughter of a physician. With this pedigree, his father wanted his sons to become doctors, and Meyer was happy to acquiesce, even more definitively so after his father's death in 1850. Although Meyer was five years older than Mendeleev, the two were exact contemporaries in terms of their careers, since Meyer's father was forced to withdraw his son from school at the age of 14 because of the boy's intense headaches. Meyer was apprenticed for a few years to a gardener (which apparently helped with the migraines), and he reenrolled in school and graduated from the gymnasium in Oldenburg in 1851 (a year after Mendeleev). He matriculated from Zürich University in medicine in May 1851, studied under Carl Jakob Löwig and Carl Friedrich Wilhelm Ludwig, moved to Würzburg (and Rudolf Virchow) after two years, and completed his training on February 25, 1854.

That year he moved to Heidelberg—yet another parallel with Mendeleev—to study with Robert Wilhelm Bunsen, whom he adored. Here the divergences with Mendeleev become clearer, for Meyer loved his time in Heidelberg and continually referred back to it. As one of his obituaries put it: "The years spent at Heidelberg were times of great moment, and their influence is to be distinctly traced in the subsequent work of his life."[41] The work performed there went into his dissertation concerning gases in the blood, published in 1857 in Königsberg, which included the first correct analysis of the mechanism of carbon monoxide poisoning: the displacement of oxygen molecule for molecule in the blood. To develop his growing interest in physical chemistry as he moved further away from medicine, in 1856 Meyer moved to Königsberg to study physics with Franz Ernst Neumann, joining his elder brother Oskar Emil Meyer. He left to take a *Privatdozent* position in physics and chemistry at Breslau in February 1859. There he displayed a sharp talent for chemical theory in his critical work: "On the Chemical Doctrines of Berthollet and Berzelius." He also attended the Karlsruhe Congress.

He was called to his first independent position at the School of Forestry at Neustadt-Eberswalde in 1866. In 1868 he succeeded Carl Weltzien as a professor of chemistry and the director of the chemical laboratory at the Karlsruhe Polytechnic Institute, and settled in 1876 in Tübingen, where he taught until his death on April 11, 1895. His biographers always point to his commitment to pedagogy—he trained over 60 doctoral candidates in chemistry at Tübingen (another contrast to Mendeleev, who trained very few). He taught inorganic chemistry during the winter semester and organic chemistry during the summer, and supplemented the latter with a special lecture course on an advanced topic, often having to do with chemical theory. He served twice as dean and was rector the year before his death. (His last documented official action as rector was awarding Otto von Bismarck an honorary doctorate from the Natural Sciences Faculty in honor of his eightieth birthday.)[42] Running like a scarlet thread through this biography, from Virchow to Bunsen to Neumann to Tübingen, is the importance of pedagogy.

As committed as Meyer was to teaching, he was even more passionate about the proper construction of textbooks so that they included a prominent role for chemical theory, which he felt was underemphasized in most classrooms of the day.[43] Like Mendeleev's, Meyer's periodic system emerged during the composition (and revision) of his textbook, *Modern Theories of Chemistry and Their Significance for Chemical Statics* (*Die modernen Theorien der Chemie und ihre Bedeutung für die chemische Statik*), and throughout his life he continued to develop methods by which the system could be used in the classroom.[44] Unlike Mendeleev, however, Meyer drew a direct line from Cannizarro's development of the theory of atomic-weight determination to his own system, thus placing himself within a continuous development: "After Cannizzaro had established the correct principles for the determination of atomic weights, the regularities which had been observed up to that time took shape in the first edition of my 'Modern Theories,' in 1864."[45]

This book was published while Meyer was still in Breslau and comprised a slim 147 pages. It occupied a liminal space between theoretical treatise and textbook, and was intended as a survey of relevant theories in chemistry, especially atomism and valency. Both of these, he emphasized early in the text, were *chemical* theories, and the purpose of this book was to differentiate the domains of theory in chemistry from those theories that were proper to physics:

It is undeniable that through the adoption and development of the atomic theory chemistry becomes more and more alienated from its near relation physics. The areas became more sharply differentiated; each discipline went on its own path; the common border districts remained in many cases undeveloped when chemistry has not alone seized them, as more often seems to be the case. Yet almost daily new relations were being discovered between chemical and physical phenomena; but even the greatest discoveries produced by the application of physical methods to the area of chemistry could not establish stronger ties across the loose rift between both disciplines, because the goals of both had become different.

Chemists were concerned, first and foremost, with the countless compounds whose possibility atomic theory allowed one to predict, to produce the largest possible number of them, to study them and to order them systematically. Thus chemistry became more and more a descriptive natural science, in which general theoretical speculations, such as those Berthollet had set in the foreground, only occupied a background significance. This change was necessary . . . [A] theoretical chemistry was demanded for an exact knowledge of an extraordinarily large number of chemical compounds, without which there was a very near danger that it would run aground . . . [P]erhaps only in the coming century can one build a theory of chemistry that, as now the theory of light or electricity [in physics], can teach us to calculate the phenomena from given conditions in advance.

From this goal that Berthollet had in mind, chemistry is even today still endlessly far away . . . Today's chemistry resembles a plant which has its roots spread out in the soil and gathers nutrients for the later sudden flourishing of stalks, flowers, and fruits. The rich material that the rapid development of atomic theory has enabled guarantees for chemistry its lasting autonomy; it will never again be a dependence, a subdivision of physics.[46]

This lengthy extract highlights several crucial points: that chemistry and physics occupied very different domains, and that this difference stemmed from the different role of theory in each; that chemistry was not yet endowed with overarching predictive theories like those in physics; and that the purpose of theoretical developments was to order empirical data into broad schemes. Yet Meyer noted that chemists tended to be skeptical of overhasty generalizations based on theory: "There thus emerged a feeling of uncertainty or doubt about the value of theoretical efforts in general, that speculations about causes and the essence of phenomena were usually hurried and suggestive, often even not directly stated, leaving the reader to abstract them himself."[47] If Meyer wanted to defend the utility of chemical theory in this textbook—and particularly the importance of atomism—he would have to calm this concern of his peers and show how theory could be useful without necessitating leaps to unfounded conclusions.

	4 werthig	3 werthig	2 werthig	1 werthig	1 werthig	2 werthig
	—	—	—	—	Li $=7{,}03$	(Be $= 9{,}3$?)
Differenz $=$	—	—	—	—	16,02	(14,7)
	C $= 12{,}0$	N $= 14{,}04$	O $= 16{,}00$	Fl $= 19{,}0$	Na $= 23{,}05$	Mg $= 24{,}0$
Differenz $=$	16,5	16,96	16,07	16,46	16,08	16,0
	Si $= 28{,}5$	P $= 31{,}0$	S $= 32{,}07$	Cl $= 35{,}46$	K $= 39{,}13$	Ca $= 40{,}0$
Differenz $=$	$\frac{89{,}1}{2} = 44{,}55$	44,0	46,7	44,51	46,3	47,6
	—	As $= 75{,}0$	Se $= 78{,}8$	Br $= 79{,}97$	Rb $= 85{,}4$	Sr $= 87{,}6$
Differenz $=$	$\frac{89{,}1}{2} = 44{,}55$	45,6	49,5	46,8	47,6	49,5
	Sn $= 117{,}6$	Sb $= 120{,}6$	Te $= 128{,}3$	J $= 126{,}8$	Cs $= 133{,}0$	Ba $= 137{,}1$
Differenz $=$	$89{,}4 = 2.44{,}7$	$87{,}4{,} = 2.43{,}7$	—	—	(71 $= 2.35{,}5$)	—
	Pb $= 207{,}0$	Bi $= 208{,}0$	—	—	(Tl $= 204$?)	—

Figure 3.3 Lothar Meyer's table of elements from the first edition of *Modern Theories of Chemistry* (1864).

Source: Lothar Meyer, *Die modernen Theorien der Chemie und ihre Bedeutung für die chemische Statik* (Breslau: Maruschke & Berendt, 1864), 137.

An excellent illustration of this point was his system for organizing the elements on the twin axes of atomism and valency, often called (anachronistically) his first periodic table (figure 3.3). The image appears late in the book and is meant to show the regularities of the amount of increase of atomic weight within groups of similar valency (*Werthigkeit*), the differences being indicated by the calculations between the rows. The point here was to solidify and emphasize the conceptual utility of both atomic weights and valency theories by showing that they, heretofore treated independently in the book, seemed connected by deeper regularities. This link was mostly implicit in Meyer's account. He introduced the table thus: "The following table gives such relations [between the atomic weights] for six related well characterized groups of elements."[48]

This partial table is pretty impressive; one might think that one could use it as a springboard for evaluating empirical results. But Meyer was very careful to exclude precisely this use of the system:

It is surely not to be doubted, that a definite regularity (*Gesetzmässigkeit*) prevails in the numerical values of atomic weights. It is rather improbable that it is as simple as it appears, if one leaves aside the relatively small deviations in the values of the evident differences. In part indeed these deviations can justifiably be seen as brought about through incorrectly determined values of atomic weights. But this can hardly be the case for all of them; and entirely certainly one is not justified, as is seen all too often, to want to arbitrarily correct and change the empirically determined atomic weights due to a suspected regularity, before experiment has set a more exact determined value in its place.[49]

Thus, immediately after introducing a system of elements, Meyer turned its suggestiveness into an object lesson in theoretical humility. The purpose of this system, and the whole book, was to provide a middle ground in defending the restrained utility of theory as opposed to unrestrained empiricism. As he commented in his conclusion: "The more science progresses, the more it will be possible to keep in abeyance the damaging influence of hypotheses and theories. Also in chemistry one will more and more be in the position, as is now the case in physics, to always keep in view the dependence between each hypothesis and the results of observation compared with theoretical consequences."[50]

The second edition of *Modern Theories*, published in 1872, and at 364 pages now ballooned to over double its original size, further developed his table of elements into a "true" periodic system and insisted even more forcefully on restraint in using it for prediction. Meyer expanded the work to make it more useful as a textbook: "Through this expansion of observational material the book has come to approach more closely the form of a textbook or handbook."[51] Emphasizing his pedagogical intent, he dedicated it to his mentor, Bunsen. There were many interesting features in this new edition—including a mention of Prout's hypothesis, absent from the first edition—but perhaps none more striking than how he now treated the periodic system.[52]

The two most salient aspects of Meyer's development of the periodic system in this second edition were his handling of priority claims and his attitude to prediction. As a rule, both in the textbook and in his journal articles about periodicity, Lothar Meyer was scrupulous about acknowledging both "precursors" and giving lavish attention to Mendeleev (although the latter felt the attention was not lavish enough).[53] Given Meyer's goal of enhancing the status of theoretical developments in chemistry, this distribution of credit made a great deal of sense; by showing a continuous development of atomism through Cannizzaro and to the periodic system, he could demonstrate the utility of continuous attention to theory.

With respect to the possibility—and the *desirability*—of prediction, he was much more circumspect. After displaying a modified periodic table (figure 3.4) and his famous curve of increasing atomic volumes (figure 3.5), Meyer noted:

> As one runs through the row of elements by magnitude of atomic weight, one sees the periodicity of properties in their dependence upon the magnitude of atomic weight very clearly. While the differences of the atomic weights that immediately follow each other seem to pertain to no simple law, one sees between the atomic weights of members of one and the same family entirely regular relations.[54]

This is hardly a ringing endorsement for reformulating chemistry around the periodic system. (Recall, however, that Mendeleev also proposed nothing of the sort; he did not enhance the *structural* centrality of the system as a

	1	2	3	4	5	6	7	8	H	9	10
I.									H 1	Li 7,01	Be 9,3
II.	B 11,0	C 11,97	N 14,01	O 15,96	F 19,1						
III.	Al 27,3	Si 28	P 30,96	S 31,98	Cl 35,37					Na 22,99	Mg 23,94
IV.	? 47?	Ti 48	V 51,2	Cr 52,4	Mn 54,8	Fe 55,9	Co 58,6	Ni 58,6		K 39,04	Ca 39,90
V.	? 70?	? 72?	As 74,9	Se 78	Br 79,75					Cu 63,3	Zn 64,9
VI.	? 88?	Zr 90	Nb 94	Mo 95,6	? 98?	Ru 103,5	Rh 104,1	Pd 106,2		Rb 85,2	Sr 87,2
VII.	In 113,4	Sn 117,8	Sb 122	Te 128	J 126,53					Ag 107,66	Cd 111,6
VIII.	? 173?	? 178?	Ta 182	W 184,0	? 186?	Os 198,6	Ir 196,7	Pt 196,7		Cs 132,7	Ba 136,8
IX.	Tl 202,7	Pb 206,4	Bi 207,5							Au 196,2	Hg 199,8

Figure 3.4   One of Lothar Meyer's 1872 periodic tables.

Source: Lothar Meyer, *Die modernen Theorien der Chemie und ihre Bedeutung für die chemische Statik*, 2d. ed. (Breslau: Maruschke & Berendt, 1872), 301.

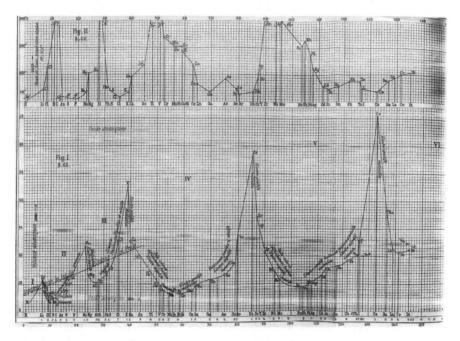

Figure 3.5 Lothar Meyer's atomic-volume curve, also published in the 1872 textbook on the inside cover. This more complete version comes from his 1870 journal article.

Source: Lothar Meyer, "Die Natur der chemischen Elemente als Function ihrer Atomgewichte," *Annalen der Chemie und Pharmacie,* Supp. VII (1870): 354–364, insert.

pedagogical tool in later editions of his *Principles*, although he did accentuate the powers of the periodic law for conceptual understanding.) The importance of Meyer's claim that the periodic system reflected "no simple law" was to exclude the possibility of making predictions based on the system—in a textbook published the year after Mendeleev had done just that. "We are however conscious of the weakness of our weapons," he continued, "so it is as always allowed for us to test our powers through this, that we can predict the properties of still undiscovered elements with the greatest possible probability, in order to later perhaps compare them with the actually observed ones and then be able to judge the value or lack of value of our theoretical speculations."[55] To clinch the point:

> If chemistry is to be spared new deeply distressing catastrophes, one must before all else strive for a correct valuation of hypotheses and theories, that, as we hope, will soon become a general resource for all researchers. As we have happily overcome the false disdain for hypotheses and theories and the overweening fear of their dangerousness, so we must also take care to avoid the opposite extreme in chemistry, the simplistic erection, overvaluation and dogmatization of hypothetical assumptions.[56]

Thus, from the textbook context, we can clearly see that Meyer refrained from making detailed predictions of undiscovered elements—although he left gaps in his table and engaged in some interpolations—not because of timidity or fear of hypotheses but to prove a point about the conjunction of observation and theory.[57] This was a pedagogical point, a point to be drilled into students. (Meyer's extensive experimental work on the accurate determination of atomic weights offered a complementary research agenda to exemplify his pedagogical stance.)[58]

Each further edition of *Modern Theories*, even after he expanded it from chemical "statics" to chemical "mechanics" (the change took place in the fourth edition), continued to downplay prediction, denying that it formed any part of chemistry's domain (at least at present) and assigning it to physics. This was true even in his third lightly-revised edition of 1876, published after the discovery of gallium and the first successful confirmation of Mendeleev's predictions.[59] The fourth edition expanded to 607 pages and included a great deal about atomic dynamics (derived from innovations from organic chemistry), and began ever more to resemble a textbook organized around the twin principles of atomism and valency.[60] His revisions continued to be minor and in the direction of comprehensiveness rather than transformation until the sixth edition, which was published posthumously by his brother. While preparing this version (it had, after all, been thirty years since the first), Meyer decided to split the book into three separate volumes—as it had indeed been split internally since the fourth edition. He had finished the first third and sent it off to the publisher on the morning of April 11, 1895; that afternoon he suffered the stroke that killed him by evening.[61]

Modern Theories was not Meyer's only textbook venture. He also published a more traditional textbook, *Essentials of Theoretical Chemistry* (*Grundzüge der theoretischen Chemie*) in 1890, dedicated to his other pedagogical idol, Franz Neumann.[62] Even though this textbook came after all three of Mendeleev's successful predictions, and was even more emphatically in favor of Prout's hypothesis and other controversial theories, Meyer still urged caution to students in thinking about the reliability of theory: "Never however are we allowed to take even the best established theory for absolute truth; high probability is the highest that we attain."[63] This avowal of a healthy skepticism continued into the multiple revised editions produced posthumously by his students.[64]

Thus we find a striking divergence between Mendeleev and Meyer in terms of their stances on the controversial issues of the day. On every contemporary theoretical issue of consequence—the existence of atoms, their substructure, the validity of Prout's hypothesis, the centrality of valency—Meyer not only stood clearly in their favor but also gave answers that are now considered by chemists to be right on each point, while Mendeleev's are not.[65] We are thus faced with an intriguing contrast: Mendeleev was hostile to most forms of speculative chemistry, was fundamentally conservative on theory, and still made astonishingly accurate predictions of the properties of yet-undiscovered elements; on the other hand, Lothar Meyer felt a strong affinity to theories in chemistry and asserted their validity but refused to consider the periodic system a stable enough platform from which to speculate. What are we to make of this difference?

Conclusion: A question of "boldness"?

Russian (and especially Soviet) authors typically surmounted the impasse by endorsing Mendeleev's scheme for giving credit to the most "developed" system and systematically ignoring his puzzling theoretical myopia. Western scholars have mostly shied away from this approach, but several efforts have harnessed prediction to resolve the strange antisymmetry between Meyer and Mendeleev. In this framework, what is to be explained is not why Mendeleev was inconsistent on the issue of theoretical speculation but assumed that prediction was the natural end result of the periodic system, and frame the question instead as: Why didn't Meyer make any predictions? The answers boil down to an issue of personality—declaring Mendeleev a more "bold" (*kühn* in the German) chemist in hazarding predictions and faulting Meyer for an implied timidity:

> But it is especially in the deductive application of the system, that we find the Russian scientist much in advance of the German; the scope of the phenomena encompassed, the definiteness and lucidity of the reasons adduced for the conclusions arrived at, the number and importance of the predictions made together the marvelous way in which these have been verified, have combined to make this part of Mendeleeff's work one of the greatest scientific achievements of the century, one of the most striking confirmations of the modern method.[66]

Even Meyer ascribed "boldness" to Mendeleev in the third edition of *Modern Theories*.[67]

There is some justification in the historical record for this emphasis on prediction as the relevant axis for differentiating the two chemists. The idea of prediction excited quite a few chemists from the beginning, however skeptical they were toward the correctness of Mendeleev's claims. In one of the first characterizations of Mendeleev's predictions to the German Chemical Society on December 18, 1870, for example, V. von Richter atypically waxed emphatic about the possibility of predicting the properties of yet-undiscovered elements: "Interesting predictions, if some of these elements are eventually really discovered!"[68] In the fifth edition of *Principles of Chemistry*, Mendeleev himself mocked Meyer for not "rushing" to make predictions.[69]

Yet this explanation is unsatisfying, for several reasons. First, it fails to explain why Mendeleev refused to be bold about other "speculations" in chemistry that were rather less radical than his predictions—such as, say, the existence of atoms. Mendeleev's clearly conservative stance on many political and social matters seems to indicate that his caution was more typical than his "boldness," which should suggest that his willingness to predict needs to be *explained*, not presumed.[70] Further, this interpretation ignores the clear evidence of Meyer's enthusiasm for theoretical elaborations in many instances (which, to be Whiggish again for a moment, one might reiterate happened to be correct). Finally, this metric of audacity naturalizes and fixes certain features of chemistry—that it is supposed to be a predictive natural science—that were openly disputed at the time.[71] Reduction to a matter of personal courage obscures much more than it reveals in what should be, at least in part, a story about chemistry's disciplinary boundaries.

Much more appropriate is a consideration of the pedagogical motivations for each chemist and the context of textbook-writing in the development of each of their systems. Both systems emerged as solutions to problems of textbook composition (Mendeleev) and pedagogical presentation of theories (Meyer). In the textbook context, *both* scientists refused to draw extensive implications from their systems: Meyer quite explicitly and Mendeleev by leaving extensive discussion of predictions out of his *Principles*. The difference stems from what happened once the periodic system moved into the journal literature: *there*, Mendeleev began to expand on speculative predictions, while Meyer held his system much closer to its original pedagogic context. Recall that there are two questions that need to be explained: why did Meyer refuse to predict, and why did Mendeleev feel comfortable predicting? The textbook origins of the periodic system provide an answer to the first question. The second question still remains to be answered—indeed, remains to be asked—by philosophers and historians of chemistry.

The purpose of this essay was to clarify and reframe some assumptions of present-day observers as they think about the periodic-table priority dispute. My goal is not to allocate credit differently—or to attribute credit at all, for that matter. Late-nineteenth-century contemporaries already solved that problem to their satisfaction by assigning both men the 1882 Davy Medal for their work on the periodic system, solomonically splitting credit down the middle. Yet even this compromise did not last very long. At a meeting of the British Association for the Advancement of Science in Manchester in 1887, both Mendeleev and Meyer were in attendance at an awards banquet, and already then one could observe Meyer being eclipsed by Mendeleev's shadow. According to an eyewitness:

> [W]hen, at the conclusion of Dr. Schunk's address, there was a call for a speech from Mendeléef[f], he declined to make an attempt to address the section in English, and simply rose in his place to bow his acknowledgments, an action followed by the rising of Meyer from his seat next to Mendeléeff, and who, as if to prevent any misconception, prefaced his speech with the declaration, "I am not Mendeléeff," a statement which may, perhaps, have disappointed some of his hearers, but the round of applause which greeted his further remark, "I am Lothar Meyer," proved that the feeling, if it existed at all, was more than counterbalanced by the anticipation of the pleasure of listening to the words of one whose name

will ever in the annals of our science be justly associated with that of the great Russian chemist.[72]

The audience that day knew something that the textbooks relating the discovery of the periodic system have forgotten—that Meyer was not a usurper, a false claimant to the title of discoverer. He was not simply "not Mendeléeff"; he was a chemist with his own approach to the periodic system, a different but related system that was enmeshed in a complex of other pedagogical goals. Yet simultaneously, that audience signaled something else—that after the dust settled, Mendeleev structured the storyline of the periodic law, and Meyer's importance, such as it was, came from being "justly associated" with his Russian counterpart.

Mendeleev's shadow in the story of chemistry has swallowed up any number of others. In 1974, at the beginning of his first published book, *H. G. J. Moseley: The Life and Letters of an English Physicist, 1887–1915*, historian of science John Heilbron found the same effect. Moseley was a striking character for a number of reasons—not least his death at Gallipoli, a sizable blow to British science—but his scientific reputation rests primarily on his use of x-rays to establish that the elements in the periodic system were arranged not by increasing atomic weight (for there were exceptions, such as heavier tellurium preceding lighter iodine) but by the rising quantity of nuclear charge, what came to be known as atomic number. If we were playing the "who discovered the periodic table" parlor game, we could add a seventh point to our earlier list: "Explained the ordering of the elements and the repetition of their properties." Credit under that definition would probably fall to Harry Moseley. Heilbron, as one might expect, knew better than to embark down that path. The closest he came was in his second epigraph, quoted in French from the noted experimental physicist Maurice de Broglie: "Moseley's law justifies Mendeleev's classification; it justifies even the little tweaks that one has been obliged to give to this classification."[73] He, too, was not Mendeleev, and his law mostly survives as an adjunct to a discovery that had been credited to the Russian before Moseley was born. Ask not who discovered the periodic system; ask why you want to know the answer.

Notes

1. This is the odd date out. It comes from examining drafts of Meyer's textbook dated *before* Mendeleev's 1869 publications. See Karl Seubert, "Zur Geschichte des periodischen Systems," *Zeitschrift für anorganische Chemie* 9 (1895): 334–338.
2. Sharing credit between Meyer and Mendeleev used to be more common than it is today. See, for example, Curt Schmidt, *Das periodische System der chemischen Elemente* (Leipzig: Johann Ambrosius Barth, 1917), 22 and Karl Seubert, ed., *Das natürliche System der chemischen Elemente: Abhandlungen von Lothar Meyer und D. Mendelejeff* (Leipzig: Wilhelm Engelmann, 1895), 122.
3. Ludwig Wittgenstein, *Philosophical Investigations*, tr. G. E. M. Anscombe (Oxford: Basil Blackwell, 1953), § 76, 36e.

4. For the classic and still relevant discussion of this problem, see Robert K. Merton, "Priorities in Scientific Discovery: A Chapter in the Sociology of Science," *American Sociological Review* 22 (1957): 635–659; and Merton, "Singletons and Multiples in Scientific Discovery: A Chapter in the Sociology of Science," *Proceedings of the American Philosophical Society* 105, no. 5 (1961): 470–486.
5. Edward G. Mazurs, *Graphic Representations of the Periodic System during One Hundred Years* (University: University of Alabama Press, 1974 [1957]).
6. This list is drawn from a synthesis of: Eric R. Scerri, *The Periodic Table: Its Story and Its Significance* (New York: Oxford University Press, 2007); J. W. Van Spronsen, *The Periodic System of Chemical Elements: A History of the First Hundred Years* (Amsterdam: Elsevier, 1969); Hinne Hettema and Theo A. F. Kuipers, "The Periodic Table—Its Formalization, Status, and Relation to Atomic Theory," *Erkenntnis* 28 (1988): 387–408; and Heinz Cassebaum and George B. Kauffman, "The Periodic System of the Chemical Elements: The Search for Its Discoverer," *Isis* 62 (1971): 314–327.
7. For thoughtful critiques of the notion of "discovery" in the sciences, see Theodore Arabatzis, *Representing Electrons: A Biographical Approach to Theoretical Entities* (Chicago: University of Chicago Press, 2006), 19–26 and Augustine Brannigan, *The Social Basis of Scientific Discoveries* (Cambridge: Cambridge University Press, 1981).
8. P. Lecoq de Boisbaudran and A. de Lapparent, "Sur une réclamation de priorité en faveur de M. de Chancourtois, relativement aux relations numériques des poids atomiques," *Comptes rendus* 112, no. 2 (1891): 77–81 and P. J. Hartog, "A First Foreshadowing of the Periodic Law," *Nature* 41 (1889): 186–188.
9. The most strenuous advocate of Newlands's priority was Newlands himself; see especially J. A. R. Newlands, *On the Discovery of the Periodic Law, and on Relations Among the Atomic Weights* (London: E. & F. N. Spon, 1884).
10. See Gerstl's correspondence from London, January 29, 1871, printed in *Berichte der Deutschen Chemischen Gesellschaft* 4 (1871): 132 and Cassebaum and Kauffman, "The Periodic System of Chemical Elements," 320.
11. Carl A. Zapffe, "Gustavus Hinrichs, Precursor of Mendeleev," *Isis* 60 (1969): 461–476 and J. W. Van Spronsen, "Gustavus Detlef Hinrichs Discovered, One Century Ago, the Periodic System of the Chemical Elements," *Janus* 56 (1969): 46–62.
12. Tentatively argued in Scerri, *The Periodic Table*, 93 and more vigorously in Friedemann Rex, "Zur Erinnerung an Felix Hoppe-Seyler, Lothar Meyer und Walter Hückel: Berufungsgeschichten und Periodensystem," *Bausteine zur Tübinger Universitätsgeschichte* 8 (1997): 103–130, on p. 130.
13. François Dagognet, *Tableaux et Langages de la Chimie* (Paris: Éditions du Seuil, 1969), 97; Don C. Rawson, "The Process of Discovery: Mendeleev and the Periodic Law," *Annals of Science* 31 (1974): 181–204; Masanori Kaji, "Mendeleev's Discovery of the Periodic Law of the Chemical Elements: The Scientific and Social Context of His Discovery (english [sic] summary)," in *Mendeleev's Discovery of the Periodic Law of the Chemical Elements—The Scientific and Social Context of His Discovery* [in Japanese] (Sapporo: Hokkaido University Press, 1997), 365–380; George Gorin, "Mendeleev and Moseley: The Principal Discoverers of the Periodic Law," *Journal of Chemical Education* 73 (1996): 490–493; and F. P. Venable, *The Development of the Periodic Law* (Easton, PA: Chemical Publishing Co., 1896), 94.
14. See the excellent discussion in Alan J. Rocke, *The Quiet Revolution: Hermann Kolbe and the Science of Organic Chemistry* (Berkeley: University of California Press, 1993).
15. For Meyer's acknowledgement of Cannizzaro's influence, see Lothar Meyer, ed., *Abriss eines Lehrganges der Theoretischen Chemie vorgetragen an der Universität Genua von Prof. S. Cannizzaro*, tr. Arthur Moliati (Leipzig: Wilhelm Engelmann, 1891) and Gerhard Fritz, "Lothar Meyer in Karlsruhe," *Bausteine zur Tübinger*

Universitätsgeschichte 8 (1997): 75–78. For Mendeleev's, see his contemporary report on the Congress, D. I. Mendeleev to A. A. Voskresenskii, September 7, 1860 (O.S.), published as "Khimicheskii kongress v Karslrue," *S.-Peterburgskie Vedomosti*, November 2, 1860, #238 and his later Faraday Lecture—after having established his claim on the periodic law—in D. I. Mendeleev, "The Periodic Law of Chemical Elements," *Journal of the Chemical Society* 55 (1889): 634–656.

16. I draw liberally on a series of recent studies on using textbooks to analyze the development of nineteenth-century chemistry, especially Anders Lundgren and Bernadette Bensaude-Vincent, eds., *Communicating Chemistry: Textbooks and Their Audiences, 1789–1939* (Canton, Mass.: Science History Publications/USA, 2000).

17. The biographical details are drawn from Michael D. Gordin, *A Well-Ordered Thing: Dmitrii Mendeleev and the Shadow of the Periodic Table* (New York: Basic Books, 2004).

18. For details, see Michael D. Gordin, "The Heidelberg Circle: German Inflections on the Professionalization of Russian Chemistry in the 1860s," *Osiris* 23 (2008): 23–49.

19. D. I. Mendeleev, *Organicheskaia khimiia* (St. Petersburg: Obshchestvennaia pol'za, 1861). On the importance of this text for Mendeleev's later thought in inorganic chemistry, see Michael D. Gordin, "The Organic Roots of Mendeleev's Periodic Law," *Historical Studies in the Physical and Biological Sciences* 32 (2002): 263–290.

20. The textbook is reproduced as volume 2 of A. M. Butlerov, *Sochineniia*, 3 vols. (Moscow: Izd. AN SSSR, 1953–1958). On the three editions (two Russian, one German) of this textbook, see G. V. Bykov, "Materialy k istorii trekh pervykh izdanii 'Vvedeniia k polnomu izucheniiu organicheskoi khimii' A. M. Butlerova," *Trudy Instituta istorii estestvoznaniia i tekhniki* 6 (1955): 243–291.

21. K. Ia. Parmenov, *Khimiia kak uchebnyi predmet v dorevoliutsionnoi i sovetskoi shkole* (Moscow: Akademiia pedagogicheskikh nauk RSFSR, 1963). Chapter 3 of this work discusses the lasting impact of Mendeleev's *Principles of Chemistry*.

22. Gordin, *A Well-Ordered Thing*, Chapter 2, and references therein.

23. D. I. Mendeleev, from the introduction to the fifth edition of *Osnovy khimii* (1889), as reproduced in D. I. Mendeleev, *Periodicheskii zakon: Dopolnitel'nye materialy. Klassiki nauki*, ed. B. M. Kedrov (Moscow: Izd. AN SSSR, 1960), 381. Ellipses added.

24. Gordin, *A Well-Ordered Thing*, 183–187.

25. Mendeleev to Erlenmeyer, [August 1871?], repr. in Otto Krätz, "Zwei Briefe Dmitri Iwanowitsch Mendelejeffs an Emil Erlenmeyer," *Physis* 12 (1970): 347–352, on p. 351. The article in question is Mendeleev's famous "Die periodische Gesetzmässigkeit der chemischen Elemente," *Liebigs Annalen der Chemie und Pharmacie*, Supp. VIII (1872): 133–229.

26. D. Mendeleeff, "The Periodic Law of Chemical Elements," *Chemical News* 41 (1881): 2–3, on p. 3.

27. V. P. Veinberg, *Iz vospominanii o Dmitrii Ivanoviche Mendeleeve kak lektor* (Tomsk: Tip. gubernskago upravleniia, 1910); V. A. Krotikov and I. N. Filimonova, "Ocherk pedagogicheskoi deiatel'nosti D. I. Mendeleeva v Peterburgskom universitete (1856–1867 gg.)," *Vestnik Leningradskogo universiteta* 10 (1958): 126–132; Krotikov and Filimonova, "Ocherk pedagogicheskoi deiatel'nosti D. I. Mendeleeva v Peterburgskom universitete (1867–1881 gg.)," *Vestnik Leningradskogo universiteta* 16 (1958): 140–148; and Krotikov and Filimonova, "Ocherk pedagogicheskoi deiatel'nosti D. I. Mendeleeva v Peterburgskom universitete (1881–1890 gg.)," *Vestnik Leningradskogo universiteta*, no. 4 (1959): 112–119.

28. Mendeleev to Erlenmeyer, [August 1871?], repr. in Krätz, "Zwei Briefe Dmitri Iwanowitsch Mendelejeffs an Emil Erlenmeyer," 351.

29. [William Crookes], "The Chemistry of the Future," *Quarterly Journal of Science* 7 (1877): 289–306, on p. 306.

30. Don Rawson has suggested that Mendeleev's hostility to Prout liberated him from the numerological tendency one often observes in earlier claimants to discovery of the periodic law, particularly Newlands: Rawson, "The Process of Discovery." On Prout and his hypothesis, see W. H. Brock, *From Protyle to Proton: William Prout and the Nature of Matter, 1785–1985* (Bristol: Adam Hilger Ltd., 1985). On skepticism toward atomism and the possibilities for subatomic structure in this period, see:: Alan J. Rocke, *Chemical Atomism in the Nineteenth Century: From Dalton to Cannizzaro* (Columbus: Ohio State University Press, 1984).

31. A. A. Makarenya, "Development of the Valency Concept in the Aspect of the Theory of Periodicity," in V. I. Kuznetsov, ed., *Theory of Valency in Progress*, tr. Alexander Rosinkin (Moscow: Mir Publishers, 1980): 75–84.

32. Gordin, *A Well-Ordered Thing*, Chapter 8. To be clear: these positions are considered incorrect *today*. I flirt with Whiggish language here because Mendeleev's views were also regarded as incorrect to most practicing chemists in his own day.

33. V. von Richter, "[Correspondence from St. Petersburg]," *Berichte der Deutschen Chemischen Gesellschaft zu Berlin* 2 (1869): 552–554 and von Richter, "[Correspondence from St. Petersburg]," *Berichte der Deutschen Chemischen Gesellschaft zu Berlin* 3 (1870): 988–992.

34. D. Mendelejeff, "Ueber die Beziehungen der Eigenschaften zu den Atomgewichten der Elemente." *Zeitschrift für Chemie*, n.s. 5 (1869): 405–406. Mendeleev was aware of the translation error (*stufenweise* vs. *periodisch*; "gradual" vs. "periodic") and took Meyer to task for not checking the Russian original. (Mendelejeff, "Zur Frage über das System der Elemente," *Berichte der Detuschen Chemischen Gesellschaft* 4 (1871): 348–352, on p. 351.) Meyer's response reflected his exasperation: "It seems to me too strong a demand that we German chemists should read, not merely the memoirs appearing in the Germanic and Romantic languages, but those also which are produced in the Slavic tongues and should test the German abstracts for their accuracy." Lothar Meyer, "The History of Atomic Periodicity," *Chemical News* 43 (1881): 15. The dynamics of these linguistic disputes lie beyond the scope of this essay.

35. Mendelejeff, "Zur Frage über das System der Elemente," 352.

36. D. Mendelejeff, *Grundlagen der Chemie*, tr. L. Jawein and A. Thillot (St. Petersburg: Carl Ricker, 1890), 683–684n8.

37. Mendelejeff, *Grundlagen der Chemie*, 684n8. He did give (690n12) Meyer some credit for the 1864 table although he pointed out its incompleteness. For a more detailed, point-by-point, and almost *ad hominem* attack on Meyer, see Mendelejeff, "Zur Geschichte des periodischen Gesetzes," *Berichte der Deutschen Chemischen Gesellschaft* 13 (1880): 1796–1804, on p. 1801.

38. D. Mendeléeff, "Comment j'ai trouvé le système périodique des éléments," *Revue générale de chemie pure et appliquée* 4 (1901): 533–546, on p. 538.

39. Mendelejeff, "Zur Geschichte des periodischen Gesetzes," 1802.

40. Biographical details are drawn from Otto Theodor Benfey, "Meyer, Lothar," in Charles Coulston Gillespie, ed., *Dictionary of Scientific Biography* (New York: Scribner, 1970), IX and X, 347–353; P. Phillips Bedson, "Lothar Meyer Memorial Lecture," *Journal of the Chemical Society* 69 (1896): 1403–1439; Klaus Danzer, *Dmitri I. Mendelejew und Lothar Meyer: Die Schöpfer des Periodensystems der chemischen Elemente*, 2nd ed. (Leipzig: B. G. Teubner Verlagsgesellschaft, 1974) and Friedemann Rex, "Lothar Meyer im Spiegel seiner Veröffentlichungen," *Bausteine zur Tübinger Universitätsgeschichte* 8 (1997): 89–102.

41. Bedson, "Lothar Meyer Memorial Lecture," 1405.

42. Bernd Stutte, "Lothar Meyer in Tübingen," *Bausteine zur Tübinger Universitätsgeschichte* 8 (1997): 79–88 and Danzer, *Dmitri I. Mendelejew und Lothar Meyer*, 62.

43. See the dispute he had with Erlenmeyer concerning the latter's pedagogical practices, which were heavily empirical and, Meyer thought, counterproductively excluded theory. Lothar Meyer to Emil Erlenmeyer, November 28, 1882, HS-242/17, Archive of the Deutsches Museum, Munich, Germany.

44. Lothar Meyer, "Ueber den Vortrag der anorganischen Chemie nach dem natürlichen Systeme der Elemente," *Berichte der Deutschen Chemischen Gesellschaft* 26 (1893): 1230–1250.

45. Meyer, "On the History of Atomistic Periodicity," *Chemical News* 41 (April 30, 1880): 203.

46. Lothar Meyer, *Die modernen Theorien der Chemie und ihre Bedeutung für die chemische Statik* (Breslau: Maruschke & Berendt, 1864), 7–8. Ellipses added.

47. Meyer, *Die modernen Theorien der Chemie* (1864), 1st ed., 12

48. Meyer, *Die modernen Theorien der Chemie* (1864), 1st ed., 137.

49. Meyer, *Die modernen Theorien der Chemie* (1864), 1st ed., 139.

50. Meyer, *Die modernen Theorien der Chemie* (1864), 1st ed., 144.

51. Lothar Meyer, *Die modernen Theorien der Chemie und ihre Bedeutung für die chemische Statik*, 2nd ed. (Breslau: Maruschke & Berendt, 1872), viii–ix.

52. For Prout, see Meyer, *Die modernen Theorien der Chemie* (1872), 2nd ed., 292.

53. Meyer, *Die modernen Theorien der Chemie* (1872), 2nd ed., 294–300. On his distress at Mendeleev's "violent reply" to his articles on periodicity, see Meyer, "The History of Atomic Periodicity," 15. It should be said that while Meyer would specifically credit Mendeleev for his predictions, he also pointed out the places where the Russian chemist was inexact: Lothar Meyer, *Grundzüge der theoretischen Chemie* (Leipzig: Breitkopf & Härtel, 1890), 60–61.

54. Meyer, *Die modernen Theorien der Chemie* (1872), 2nd ed., 302–303. For Meyer's analysis of his curve, see ibid., 307.

55. Meyer, *Die modernen Theorien der Chemie* (1872), 2nd ed., 344.

56. Meyer, *Die modernen Theorien der Chemie* (1872), 2nd ed., 362.

57. One can observe some of these interpolations and gaps through minute inspection of the atomic-volumes curve. Lothar Meyer, "Die Natur der chemischen Elemente als Function ihrer Atomgewichte," *Annalen der Chemie und Pharmacie*, Supp. VII (1870): 354–364, on p. 360.

58. Lothar Meyer and Karl Seubert, *Die Atomgewichte der Elemente aus den Originalzahlen neu berechnet* (Leipzig: Breitkopf & Härtel, 1883).

59. Meyer continued to give Mendeleev a great deal of credit even here: Lothar Meyer, *Die modernen Theorien der Chemie und ihre Bedeutung für die chemische Statik*, 3rd ed. (Breslau: Maruschke & Berendt, 1876), xvii. On the other hand, his patience wore thin with Mendeleev's tone about priority: "His [1869] scheme then still contained much arbitrariness and irregularities that were later eradicated." Ibid., 291n.

60. Lothar Meyer, *Die modernen Theorien der Chemie und ihre Bedeutung für die chemische Mechanik*, 4th ed. (Breslau: Maruschke & Berendt, 1883).

61. Lothar Meyer, *Die modernen Theorein der Chemie und ihre Bedeutung für die chemische Mechanik*, 6th ed., vol. 1: *Die Atome und ihre Eigenschaften* (Breslau: Maruschke & Berendt, 1896), viii.

62. Meyer, *Grundzüge der theoretischen Chemie*, v.

63. Meyer, *Grundzüge der theoretischen Chemie*, 4. On Prout, see ibid., 49.

64. E. Rimbach, *Lothar Meyers Grundzüge der theoretischen Chemie*, 4th ed. (Leipzig: Breitkopf & Härtel, 1907).

65. See Lothar Meyer, Review of Benjamin Brodie's *The Calculus of Chemical Operations*, *Zeitschrift für Chemie*, n.s. 3 (1867): 478–480; Meyer, "Die Natur der chemischen Elemente als Function ihrer Atomgewichte," 354–355; Van Spronsen, *The Periodic*

System of Chemical Elements, 131; and Britta Görs, *Chemischer Atomismus: Anwendung, Veränderung, Alternativen im deutschsprachigen Raum in der zweiten Hälfte des 19. Jahrhunderts* (Berlin: ERS, 1999), 109.

66. Ida Freund, *The Study of Chemical Composition: An Account of its Method and Historical Development* (Cambridge: Cambridge University Press, 1904), 474. For two further examples (among many), see Schmidt, *Das periodische System der chemischen Elemente*, 23 and Stephen G. Brush, "The Reception of Mendeleev's Periodic Law in America and Britain," *Isis* 87 (1996): 595–628, on p. 618.

67. Meyer, *Die modernen Theorien der Chemie* (1876), 3rd ed., 291n.

68. Von Richter, "[Correspondence from St. Petersburg (1870)]," 991.

69. Mendelejeff, *Grundlagen der Chemie*, 692–693n13.

70. Mendeleev's conservatism is discussed at length in Gordin, *A Well-Ordered Thing*, especially Chapters 1, 6, and 8.

71. See the helpful analysis in Mary Jo Nye, *From Chemical Philosophy to Theoretical Chemistry: Dynamics of Matter and Dynamics of Disciplines, 1800–1950* (Berkeley: University of California Press, 1993).

72. Bedson, "Lothar Meyer Memorial Lecture," 1409.

73. "*La loi de Moseley justifie la classification de Mendéleeff; elle justifie mêmes de pouce que l'on avait été obligé de donner à cette classification.*" Quoted in J. L. Heilbron, *H. G. J. Moseley: The Life and Letters of an English Physicist, 1887–1915* (Berkeley: University of California Press, 1974), vii.

4
Complex Systems and Total War: British Operational Research and the PM Statistical Branch at the Beginning of World War II

Dominique Pestre

The purpose of this chapter is to document and give historical meaning to the changes that occurred in the analysis, design, and monitoring of military and economic action in the United Kingdom at the onset of World War II. Optimization and practical efficiency here and now were the goals of these approaches based on the careful study of past and present actions, on field studies, on the creation of, and constant comparison between statistical data and analyses that led to proposals for immediate reforms and actions.[1]

My central thesis is that the emergence of such concerns and practices in the United Kingdom—*and not in Germany or the United States*—and the intensity with which they manifested themselves in the Blackett circus and the Prime Minister (PM) Statistical Branch during the Battle of Britain and the Blitz are mainly the result of the exceptional situation then facing the United Kingdom. Germany was directly threatening the integrity of British territory, thus producing a feeling of absolute urgency and the need to mobilize anything available to survive and resist better. The point might seem obvious but it has not been explicitly made up to now for several reasons.

First because the early British Operational Research as practiced by Blackett and his friends between August 1940 and the end of 1941 has often been read as derived from radar work—as an extension of it, as another instantiation. It is true that the expression "Operational Research" (OR) has first been coined in that context but the meaning it takes when revived by Blackett is profoundly different. Another reason that might explain the confusion is that the early British OR is often identified with the practices that became common in the second part of the war in the American context, and with what it became after the war in business and the RanD Corporation contexts. Among historians, the invention of OR has also often been identified with the scientific elite desire to get "a more prominent role in British military policy", as a way to realize part of their ongoing "political agenda"—so masking the importance of the exceptional situation Britain was facing in 1940 and 1941.[2] To me, this frame of interpretation is too simple, it creates too much continuity with, and gives too much weight to, the narratives that the leftist part of British scientific academia developed before and after the war. It forgets the novelties that the situation allowed, *the emergent nature* of what was concretely invented and done in 1940 and 1941. And that explains my strategy to combine in this chapter a precise description of OR beginnings with the parallel invention of the Statistical Branch around quite different kinds of people—Churchill and Lindemann. As will be shown, both are motivated by the same situation and develop parallel strategies.

To get at what was proper to the early phase of OR in the United Kingdom, a quick comparison with the task entrusted by the Office for Scientific Research and Development (OSRD) to Warren Weaver in the fall of 1940 in the United States could be of use.[3] Weaver was asked, for example, to consider the question of automatic firing control, notably of antiaircraft firing—which he did by bringing together academics, companies, and military engineers, and by establishing numerous research groups. The project was essentially a medium-term R&D project, and it became operational only in 1944. This shows a striking contrast with the UK, where a drastic effort was also made at the beginning of the war to address the inefficiency of antiaircraft fire. However, the imminent threat, and soon enough the daily bombings, did not leave time for detours or speculations on scientific and technical excellence. Contrary to what happened in the United States, British scientists and engineers were assigned to batteries and operational commands to improve their efficiency by any means at hand. And it is mainly in that context that OR started.

This chapter focuses on two groups. The first one consists of scientists and engineers who worked on antiaircraft and antisubmarine warfare. Their work involved technical tasks but mainly required field work and design of varied information-processing systems. Members of this group included some memorable figures of the academic left who made it their mission to save the country from the incompetence and defeatism of "the Establishment". Let me mention World War I veterans Henry Tizard and Archibald V. Hill—as well as Patrick Blackett who would join Frederick Alfred Pile, the chief commander of Anti-Aircraft Command, in August 1940.

The second group is smaller in number, more spatially contained and quite different in political outlook. This group, put together by Lindemann, later Lord Cherwell, was known as the first lord of the Admiralty (later PM) Statistical Branch (SB). Formed when Churchill became head of the navy, the group was maintained after Churchill's appointment as prime minister and worked in his immediate proximity. Lindemann had been a close friend of Churchill for many years and belonged to his intimate circles. He had occupied the Physics Chair at Oxford since 1919. He was a militant conservative, a rich aristocrat, a racist, virulently hostile to the academic left, but, like Hill or Blackett, he was obsessed by the grave nature of the situation. From the beginning, he played more of a political and global role at Churchill's side than simply that of a technical advisor. Nevertheless, like the academic left, he contributed to a profound reform of administration and statistical data collected in the UK.

To be clear, I have no intention, in this text, of giving another account of the myth of physicists who won the war. This standard discourse (whether on OR, on the SB, or on the development of the atomic bomb[4]) is known to be unsatisfactory. However, confronted with a new situation, these people were led to progressively define a new role for themselves—and thus to significantly change the ways in which war was conceived and practiced.[5]

A last word before proceeding. John Heilbron, to whom this chapter is dedicated, is known for having worked on nearly all topics connected with the history of physics—and to have visited nearly all archives in the world. In the frame of the courses he gave at Berkeley on the mobilization of science in World War II, he thus studied and presented the first World War II applications of OR, rightly stressing, among other things, how far a little mathematics directed by inspired common sense could be efficient. My chapter should thus be read as a modest tribute paid to him—as a way for me to express my admiration by offering John a study that, I know, interests him and comforts his position.

On OR as practiced and defined by Blackett in 1940 and 1941[6]

On research concerning the operationalization of the coastal radar chain in 1939–1940

OR as practiced by Blackett and his friends in 1940 and 1941 is not the mere extension of radar work. It was initially articulated on it, however, and I need to come back to the creation and operationalization of the early British radar system to show continuities and discontinuities. Neither do I want to recall the decision to create a defense structure based on the RDF system in the mid-1930s (RDF was the common name for radar techniques prior to World War II), nor do I intend to narrate the progressive implementation of hardware. That history has been told many times, and I only want to mention the Committee for the Scientific Survey of Air Defence chaired by Henry Tizard from its creation in 1934—a Committee that was politically decisive in the construction of the Home Chain and that counted among its members not only Hill, Blackett and Wimperis, the director of scientific research for the Air Ministry, but also, for a time, Lindemann.[7]

In 1938, questions arose about the operationality of the system. The large-scale exercises carried out during that year showed that radar outputs contained too much contradictory information. Airplanes coming at the radars were not always distinguished from airplanes that were behind the radars, interfering noise often made identification extremely complex, airplanes flying at low altitudes eluded the detection system (this would lead to the development of Home Chain Low radars), and interception guidance was totally ineffective (a recurring problem was the radar's poor estimation of the altitude of the spotted planes, which often placed pilots in a bad position for attacking bombers).[8]

The question of making the radar system an efficient system, able to detect incoming enemies and direct the planes of Fighter Command, was first tackled by a group of telecommunications engineers that included Eric C. Williams, Harold Lardner, and G. A. Roberts. They were placed under the responsibility of Raymund G. Hart, a Royal Air Force officer and squadron leader. This group was initially in charge of the radar stations, their administration, and operation. Based at the Bawdsey pilot station (the technical center of the Home Chain and the only radar-equipped station before 1937), it was probably the only group at that time in a position to understand how the detection system could be made operational. The group thus played a key role in perfecting the techniques and procedures used to interpret and filter radar information (not only separating noise from signal but also comparing information coming from different sources), and in facilitating plotting (the human part of the system where the radar information had to be translated in command rooms). At that time, it was estimated that no more than four minutes could elapse between the first sighting of a radar signal and the take off of an airplane unit—and managing that urgency was all that mattered.

The tasks of filtering and plotting, in particular, became central in 1939–1940. The analysis of radar data (the way to process information) was a matter of expert judgment (learning how to read screens without getting too distracted by echoes and interference for example). It was also a matter of materially organizing rooms that enabled a centralized view of all incoming information. This work was done in places where the floor was covered by a map of the battlefield, separated from an overhanging platform where officers took seat. Women with headphones connected to detection centers moved markers representing the forces on the map, like croupiers in a casino. A sophisticated system of colors and forms allowed to distinguish between enemies and allies, to discern the reliability of information, and to know if the information was up to date. The transmission of information by telephone was the object of intense study by linguists and acousticians, and procedure rules were published—for example on how to spell words to avoid misunderstanding without altering the speed of communication.

In February of 1940, Dowding, the commander in chief of Fighter Command, demanded that the whole of Hart's group be brought to his headquarters at Bentley Priory—and most operational procedures were revised again. This stemmed from the fact that things were accelerating— the system was then transformed from week to week by the constant arrival of new materials, it was drastically enlarged in geographical terms (in 1938 there were but five stations), and it was no longer a matter of exercise or training. The circulation of the "Operational System Research Memoranda," of tactical documents based on battle reports and of interception exercises, intensified accordingly. There were also regular conferences at Fighter Command for debating all such questions. In short, activity significantly intensified with the purpose of analyzing actions that could possibly turn an uneven system into an efficient system.

On Anti-Aircraft Command in the Battle of Britain and the Blitz, and Patrick Blackett's arrival by the side of General Pile

In the early days of August 1940, P. M. S. Blackett joined Pile, the officer in charge of Anti-Aircraft Command. In previous years, Blackett had worked on various instruments—a bombsight, which was a success, and photoelectric proximity fuses, which were never realized.[9]

On Hill's initiative, a conference on Anti-Aircraft gunnery was organized on August 9, 1940. The central issues were radar/battery coupling, no doubt, but more globally the overall inefficiency of Anti-Aircraft gunnery—and the distress of its commander, General Pile. Held under Pile's presidency, the conference was attended by six senior officers and nine scientists.[10] Hill concluded that the ballistic committee to be reinforced "would usefully extend its functions by getting more closely in touch with the 'user' and his needs." "All the factors on which the scientific instruments in connection with AA fire are based are proving illusory," he added, and he suggested that a special advisor be attached to the commander in chief. Hill was thinking of Blackett whom he introduced to Pile.

When Blackett arrived in August 1940, a key problem was the use of the radars associated with AA batteries (these were small radars independent from the Home Chain). Leaving technical issues to engineers, Blackett took on the task of studying their concrete use. The major problem was that these local radars could not be directly fixed onto the mechanical predictors that were used to orient the guns (it was not until 1944, thanks to Weaver's program, that the integration of radar data and predictors would become organic). Blackett and his group thus proposed ad hoc solutions that were locally effective for the existing equipment. These solutions ranged from training artillerymen to transform raw radar data into data that could be quickly used as input in the predictors, to self improved predictors (Blackett's group modified predictors from the Sperry company).

However, Blackett soon considered a different role for himself—that of reconsidering the entire functioning of the Anti-Aircraft system. His first move was to create a common and explicit metric for measuring battery efficiency. He did so by bringing to general use what his group identified as "the best practice." Reports were certainly filed before the arrival of Blackett (who himself used these reports) but it was he who initiated the systematic statistical treatment of data, as well as systematic field work to understand the nature of inconsistencies. For example, the comparison clearly showed that the average success rate for sea-facing batteries was two times higher than that of inland batteries. This intrigued Blackett who "proved" that it mainly was due to counting bias, as artillerymen at sea tended to overestimate the number of airplanes that they shot down—"facts" that were more easy to verify when airplanes crashed on land.

The novelty of Blackett's approach, and the aim he had in mind, are evidenced by another study, which considered the geographical distribution of batteries around London in light of the fact that attacks were then largely carried out at night by airplanes flying higher and faster, bombings were uniformly spread across the city, and light projectors were ineffective in most situations—and with an insufficient number of radar sets available to service all batteries (the radars were not too effective but there were no alternative anyway at night). After weighing advantages and disadvantages of grouping batteries, his conclusion was to group all batteries in only 15 locations, equipped with the 15 radar sets that were then operational.

What emerged under Blackett between August 1940 and March 1941 was thus a new way, under profound inefficiency of the system, of looking at Anti-Aircraft Command's activities. By not only reevaluating these activities through systematic collection of old and new data but also by defining objectives and checking if they were realized, Blackett was able to propose new ways to proceed that could be implemented instantly. Clear institutional support made his approach possible. Because Anti-Aircraft Command was aware of the very poor efficiency of the system, Blackett and his men were given full autonomy. Pile authorized them to access data of any kind, to investigate wherever they wished, and to tackle any issue—all of these under the condition that Blackett reported to Pile, and to Pile only. Blackett accepted the conditions, so that Pile had no fear that Blackett's critiques or suggestions could be turned against him.

On Coastal Command at the onset of the battle of the Atlantic

In March of 1941, Blackett joined Coastal Command (CC), a command that was part of the air force and that oversaw a fleet of airplanes engaged in antisubmarine warfare.[11] The primary reason for moving Blackett to CC was that the air force had found his interventions effective and that Anti-Aircraft Command was no longer central (the Blitz was under control). The prime urgency was now to target the seemingly unshakeable efficiency of German submarines—and Churchill himself decided to preside over the new committee in charge of organizing this effort.

Continuing with the assessment of Blackett's methods, I would like to describe one of the problems that his group decided to tackle upon arriving at CC—that of the exceedingly small number of submarines that were spotted by CC airplanes compared to the multitude of submarines known to navigate and the number of dispatched airplanes. The problem demanded consideration of a large number of assumptions: German submarines might float in the deep seas, but that was denied by German prisoners; these submarines might have better radar equipment than CC airplanes, which was disproven by captured vessels; the method used to survey oceans was perhaps ill-conceived— all issues were discussed but were ultimately regarded as lacking significance to account for the wide gap between reality and models.

The hypothesis that prevailed was that airplanes were sighted (or heard) by the submarines' lookout guards before the former detected the latter. Analysis of this hypothesis led to a reconsideration of the camouflage paint used on CC aircrafts. These airplanes had been painted a dark color because every aviator knows that black is the best color for eluding searchlights at night. Studies that were conducted at sea, however, revealed that the problem was more complex: in the context of antisubmarine warfare, the issue of search-lights was not the point, and at sea, including at night time, an airplane's silhouette stands out more as a dark object against the light background of the sky. CC aircrafts were therefore repainted off-white, after much debate on the levels of gray in the color. The efficiency of the measure, however, could not be established by Blackett, since it was put into place at the same time that new radars were installed on CC aircrafts.

Other problems were redefined over the course of this study, such as ways to survey a surface with a finite number of airplanes—a mathematical problem that accounts for endless documents in the archives; ways for human beings to maintain their attention while scrutinizing a mainly uniform surface for several hours (a physiological and psychological issue); and also ways of regulating detonators of explosive charges dropped on submarines.

On the "invention" of OR as a new set of tools and methods

It is in two memos dated October 1941 and October 1943[12] that Blackett begins to theorize "Operational Research" as a new modus operandi, which he defines to be a novelty, a set of general methods to be "unified" under one name. The expression had previously been in use around the Home Chain, but not in a consistent manner. For example, upon attachment to Downing in 1939, Hart's group, formerly known as the "Operational Research Group [of the HC system]," became the "RDF Research Section [of Fighter Command]." It neither had an unambiguous meaning nor notably the meaning that Blackett gave to the expression in 1941 and 1943. With his memos, Blackett gave autonomy to a practice that had progressively developed since August 1940 and for which he claimed authorship. And it is in this narrow sense that I will now use the expression.[13]

In both texts, Blackett defined OR as conducting four types of work.

First, *studying the techniques and new weapons systems proposed to various commands by technical services and industry*—not so much for the sake of improving them, Blackett always insisted that this was the responsibility of the technicians who developed them, but for the sake of studying how they perform in practice (on that point the heritage is direct with the work performed around radars in the late 1930s). According to Blackett, one has to first weigh the benefit of introducing new technical elements into existing systems. An old system could perform well, and it is not necessarily good policy to think that the introduction of new gadgets would necessarily improve things.[14] Blackett believed that experience and know-how should always take precedence over technical novelty. When a decision is made to integrate a new system, an analysis should be performed on how to do this, how to codify the use of the system, and how to train the people. In peacetime, such studies are carried out by a series of successive interventions.[15] New weapons are first acquired and tested by technical services; they are then validated by operational services, first on particular ships or aircrafts, then in a few combat units; if the result is satisfactory, the system is brought to general use and systematic training is put into place. During a war, time is short, technical

turnover rates are smaller and these stages tend to collapse. It is in such context that OR could prove useful—as it is the means of managing overlaps in
a state of emergency.

Second, *OR involves studying the functioning and optimization of local systems*—a plotting room, for example, or the use of radar sets near batteries.
The solution in this case is well known to managers. It consists of closely
examining the entire chain of actions, carrying out in-depth field investigations, locating bottlenecks, and proposing new organizational charts.
Williams accomplished this for Blackett while trying out means to accelerate
CC airplane rotations in maintenance services in 1941. This work was a result
of long visits to airfields and company managers could have done it. This
situation was more common in the United States—and around Churchill
and Lindemann: many industrial leaders were sent to reorganize production,
install new sites, rethink the organization of labor, and so on. In this sense,
OR does not present specificity.

Third, *on a larger scale, OR demands the consideration of a system in its entirety.*
I would like to begin by taking as an example the study carried out by Blackett
in 1943 on *all* convoy attacks in the North Atlantic since 1940. This analysis
considered the relationship between the number of escort vessels (from 1 to
15), the size of the group of attacking submarines (from 1 to 20), the size of
convoys, their itinerary, direction, speed, and so on. The number of sunk
ships was studied as a function of the size of the submarine group, survey
tactics, modes of attack, size of escort vessels, air escort support, and so on.
This was complemented by studies on spatial positions of ships in convoys:
number of lines and columns, position of escort vessels, differences between
daylight and nighttime conditions, and so on. The study resulted in recom-
mendations—including, among many things, an optimal size for all convoys
and escort vessels.

As I have already suggested, modern war is very much a paper war—a war of
office, a war of reports. Every military engagement, every ship's route, every
combat between two airplanes, is made the subject of a report. Archives hold
dozens of thousands of reports, and the armed forces did not wait for Blackett
and his associates to study them, as shown by the example of Captain D. V.
Peyton-Ward. A former submarine officer and naval liaison officer at CC,
Peyton-Ward became a specialist in antisubmarine warfare independently
and before Blackett, and he continued to sit on all committees of importance
along with Blackett. Peyton-Ward and Blackett would often succeed one
another in submitting reports for committee meetings, and it was Peyton-
Ward who made the greatest contribution to the standardization of combat
reports and helped operators realize that the information they contained
was crucial. The production of the two groups showed differences—Blackett
generally produced more "mathematical" treatment, more global quantitative approaches—but a more detailed study would be necessary to judge their
respective impact on the development of wartime operations.

Finally, *Blackett claimed that OR should produce formal tools and methodological
safeguards*—it is probably in these areas that the group made its most original

contributions. Blackett used a pedagogical approach. He showed how one needed to be careful about deductions made from statistical data and how to establish reliable criteria. He was very cautious with formal tools, however. Like he explained in his 1943 memorandum, he preferred variational methods as opposed to theoretical methods constructed from first principles (such as in game theory). As soon as OR was imported to America, however, more sophisticated mathematical tools and modeling techniques appeared.[16] Two types of hypothesis can be advanced to account for this difference. One can first mention the fact that, in the United States, the mobilization of scientists mainly took place in academic settings—where local criteria of excellence tended to prevail. The second reason is the fact, central to this chapter, that the United States was not directly threatened. In preparing for a long-term war, the United States had the time to elaborate more global and sophisticated schemas, such as sequential analysis.

In conclusion, Blackett insisted on always (1) conducting studies in collaboration with the people carrying out operations, (2) conducting experiments in real-life situations to test hypotheses obtained from quantitative studies, and (3) committing oneself to concrete proposals and establishing ways of measuring their effectiveness. While this strict procedure was neither unknown nor exceptional to industrial managers, the constraints it established for military action in 1940–1943 made it a decisive novelty.

On Lindemann, Churchill, and the Prime Minister Statistical Branch

One man is at the origin of the SB: Frederick A. Lindemann.[17] It was created when Churchill became first lord of the Admiralty and it had direct access to him.[18] It brought together a dozen people, for the most part statisticians and economists from Cambridge's great rival, the University of Oxford. Contrary to what the group's name may lead to believe, its goal was not to collect statistical data, even if it consistently sought to do this. De facto, the group aimed at (1) producing analyses for Churchill, (2) showing the necessary change in the operations of the state apparatus, and (3) "liberating" technomilitary innovation from the so-called rigidity of the Ministry of Supply.

The memoranda of the SB

To provide a concrete overview of the SB operations, I will first look at the memoranda regularly delivered to Churchill by Lindemann. Two preliminary remarks should be made: these memoranda were the SB's primary activity and Churchill would often return them with comments and directives.

Between September 20 and December 20, 1939, the SB submitted 16 memoranda. This number rose to 13 in January 1940, 15 in February and March, 25 in April, and 34 in June, until it finally stabilized at 35–40 per month. Seven of the sixteen memos submitted in 1939 dealt with materials available for war (only three concerned the Navy), and five dealt with inventions.

From January to May, one-fourth of the memos pertained to war materials and one-fifth to new inventions. More than a quarter, however, were then dedicated to industrial production and to import/export issues. After Churchill became prime minister, the number of memos pertaining to production, workforce, the supply of raw materials, or imports/exports, approached 40 percent.

Two remarks can be made from the above point. First, the subjects treated by the SB evolved over time: once Churchill became prime minister, questions of large-scale wartime economic equilibria came to the forefront. It is however essential to note that these issues were already addressed by the SB before May 1940, and that it did not limit itself to issues concerning the navy. Lindemann's SB clearly functioned as a sort of general think tank for Churchill.

Second remark: military equipment (production, quality control, availability, and new tactical ideas) always represented a large quarter of Lindemann's memos. In 1939, equipment allocation to the navy and to the RAF was about equal. From January to May 1940, Lindemann was far more concerned by securing equipment for the navy. From June on, he became more concerned with supplies for the air war. Of course, none of these shifts in priority are surprising.

Globally, the SB is best characterized as a sort of gadfly harassing services with its demands, in the name of the drastic choices that had to be made. Trying to detect bottlenecks and difficulties that may appear in the course of events, Lindemann systematically pointed out wastefulness, poor management, and lack of anticipation. The best means used by Lindemann to detect such flaws was to expose internal contradictions and blind spots in the demands sent by the services.

I would like to give a few examples for a better explanation. In a memo dated October 18, 1939, Lindemann begins with the fact that 4,500 aircraft were delivered according to RAF statistics. He then notes that the number of newly created RAF squadrons was significantly lower than it should have been with 4,500 aircrafts. He therefore questioned this discrepancy—his goal being to demand an explanation from the Air Ministry. In another instance of his memos sent on October 10, 1940, Lindemann made comments on a report written by an antisubmarine warfare committee. He drew a diagram of British losses based on the report and noted eight week-long "waves" that were completely unaccounted for. He insisted on the potential interest of this fact, proposed a hypothesis to be tested (this phenomenon was perhaps due to rotations of German crews), and demanded that it be studied.

Lindemann, who generally did not demonstrate great subtlety in his social relations, frequently elicited violent reactions—in contrast to Blackett—and the generated tensions hindered his second objective, which was to create reliable and consensual statistics. Unlike Blackett, he frequently asked services to provide data that they knew would be used against them. Nevertheless, Churchill's support enabled fairly good results in data collection.

Here we are prompted to ask why Churchill mainly relied on a small group of dozen of mostly young economists and academics to do that job?

Two different reasons may be invoked to explain this choice. The first reason is at the heart of my argument, and lies in the fact that Churchill and Lindemann thought that the existing state machinery was ill-adapted to the situation, that it had no sense of urgency and was unable to quickly produce the right data—thus a more determined group was necessary. The second reason is that most indicators in use at the time were no longer relevant, as a war economy basically had to be considered mainly in terms of *physical* supply and authoritative allocation of goods. Production sites had to be displaced, the flow of goods had to be reoriented, blockades had to be anticipated, as was the lack of labor force or raw materials. The problem was that existing statistics were not designed for these types of issues. Certainly, a part of the machinery did understand and did react, but not quickly enough for Lindemann and Churchill. This sluggishness is the reason why they both relentlessly intervened, with Lindemann writing condemning memos, and Churchill demanding explanations for inconsistencies in demands.

Churchill's charts and albums

Writing memos was the first of SB's two major functions. The other important task consisted in preparing graphical and cartographic documents for Churchill. Visual objects were provided to Churchill for his own personal use and for him to use during cabinet meetings. Well-preserved in the archives at Nuffield College, Oxford, these objects are often aesthetically impressive; some of them even graced the walls of meeting rooms when Churchill became prime minister. In some ways, they foreshadowed the computer screens that would later display information in command rooms, and some could still be found in the cabinet war rooms at the Imperial War Museum in London.[19]

Here are a few examples. A beautifully covered marine-blue notebook, covering the period between September 3, 1939, and August 31, 1940, contains a continuous series of accordion-bound histograms that show, for every day, the number of "British merchant ships sunk by U-boat, Mine, Surface & Aircraft". Another set of graphics provides a view of the moments during the day (hour by hour) when German submarines attacked convoys. The entire graph is set on a splendidly painted background where lighter colors are used for daytime and darker colors for nighttime. Lunar phases and weather conditions are indicated. A third example is a series of maps that show where ships had been sunk, the nature of the convoys in which they had traveled, the nature of enemy attacks, the time of such attacks, et cetera.

Many such series of graphs and maps were produced. On December 9, 1943, Lindemann suggested to Churchill (who had developed the habit of looking at them over the weekend) that their number be reduced from 150 to 70—something Churchill refused. At the time, charts and albums were produced for the army (personnel, production, stocks, or mobilization), the Air Ministry (20 albums), the fleet (with information on losses, shipbuilding, repairing, ports, imports, and stocks), the state of the economy, its organization and preparedness for war, and so on.

These graphs and maps were working tools for assessing what was contained in the avalanche of numbers generated by the war. Graphs and maps are more than mere "representations" of numbers—we are aware of this. They are carriers of meaning, they bring to light hitherto invisible correlations—in short, they generate specific knowledge. Nonetheless, from the very beginning, these visual documents also clearly functioned as rhetorical tools used by Lindemann and Churchill in various settings. They were tools used to impose certain views of the situation, to show and locate problems; they also were tools to create an image of Churchill not being satisfied with mere impressions. On April 4, 1940, for example, the first lord of the Admiralty sent copies of albums to Buckingham Palace, to the prime minister, and to his colleagues. This triggered an outpouring of recognition coming from all sides. The king asked to meet Lindemann; the Foreign Office declared that the documents opened up a fascinating field of study; the Treasury said it always had to search masses of documents to obtain the smallest piece of information and that it greatly admired the simplicity and coherence of the documents; the Air Ministry said that the presentation was admirable, but the Ministry of Supply, having been targeted in many documents, declared that it had asked its own statistical department to verify whether there were any mistakes or misleading diagrams.

The following year, in March of 1941, the documents were transmitted to President Roosevelt, who found them admirable and decisive for the battle of the Atlantic. He asked Harriman if Churchill would authorize him to receive copies of these documents, until the president had his own small statistical department, similar to the British one. A significant fact is that the chronology is the same for the promotion and diffusion of OR from Britain to American forces and leadership.

We see thus the force that these documents conferred. Churchill possessed information that others did not have, and it was difficult to refute them at once—notably during meetings. The form of these documents was powerful since they provided targeted information. Finally, using these documents helped Churchill position himself as someone who could understand numbers and their importance in government.

Interest in scientific and technical research, and in invention

Lindemann's activities were not limited to collecting data. He had a strong personal interest in technology and innovation and had no confidence in the services in charge of designing new weapons. In addition to writing memos on these issues, Lindemann thus helped creating an experimental research center for innovation—the MD1 at Whitchurch, in the fall of 1940—something Blackett never envisaged. Lindemann pleaded with Churchill not to have the center attached to the Ministry of Supply but did not succeed. This did not prevent him from being a second-in-command of sorts, intervening at all times with the prime minister approval and in close osmosis with the director. He spent hours at the center, sometimes with an enthusiastic

Churchill. After fighting a war of attrition, he was finally awarded, in 1942, direct control over the MD1.

Lindemann's idiosyncratic side in such affairs has two faces—not only a fascination with "gadgets" in the classical tradition of an inventor but also an obsession with devices that could eventually solve global problems, an obsession that he often succeeded in communicating to Churchill. Between January and May of 1940, for example, Lindemann wrote many memos on air barrages—consisting of dropping hundreds of mines equipped with parachutes, forming a sort of barrage against airplanes—and the practicability of the proposal had to be considered by the services. Another example may be found in the memos dedicated to UP (Unrotating self-propelling Projectiles), which accounted for half of the memos on invention sent to Churchill between June and August 1940. I do not want to say that these obsessions were always unfounded—antitank weaponry largely emerged at MD1. I simply wish to better understand Lindemann's persona, the fact that Churchill was not impervious to these proposals (it was quite the opposite!), and how Lindemann fully exploited his position next to Churchill to ensure that his view prevailed over that of the services and their scientific advisors.

A few summary elements

Overall, it seems like the SB's objectives followed three directions: (1) confronting statistical data on key aspects of the situation and inferring policies to be immediately implemented, (2) updating graphical and cartographical albums for the prime minister, and (3) maintaining a close relationship with invention. A priori, the three aspects did not necessarily or logically have much to do with each other. It is a matter of fact, however, that they defined the hybrid entity that was the SB, an entity that historically existed around Lindemann's unconventional persona.

It is then possible to reconsider what led to the creation of the SB. Retracing its origins first requires to have Churchill's political desire in mind. Diagrams and memos from the first nine months aimed to show that a politics of appeasement was not the solution, that it was important to prepare for more decisive action. The fact that Churchill became first lord of the Admiralty was also decisive: in this position, he had to face global problems such as organizing the blockade, guaranteeing maritime commerce security, ensuring the country's constant supply of materials. That led him to address issues necessary to keep the economy on a sound footing, a task that was under the navy's responsibility. Lindemann's own interests must also be considered. As a good physicist, he was concerned by orders of magnitude and the details pertaining to the materiality of war. If Churchill always publicly declared himself incapable of dealing with numbers, he undoubtedly recognized their importance and appreciated Lindemann's memos. Finally, I should mention the reference Churchill had in his mind since World War I—Lloyd George's "Garden Suburb"—which he sought to replicate with Lindemann at its center.[20]

Conclusions

OR and the SB have many things in common, and I would like to conclude on them. Before proceeding, however, it is important to note that these common practices do not include all that what was done under the OR and the SB labels. Innovation always remained essential for Lindemann, as did producing visual tools. For its part, OR conducted field work and managerial work. A significant part of the work was nevertheless common to both groups. It consisted in establishing norms, measuring efficiency ratios, analyzing and testing hypotheses, with the overarching goal being to help improve action as quickly as possible. The main object of analysis was of "the same nature" for both groups: systems undergoing rapid transformations that had to be optimized in a context of extreme urgency.

The legacy of OR is monumental: from 1941 to 1942, it was recognized by all general staffs; after the war, it was central to most management techniques. In the context of RanD, it evolved into systems analysis, and it then drew heavily on modelization and simulation. Whether or not OR was at the origin of all that was a matter of intense debate after the war, however—which is not surprising considering the similarities with many existing managerial techniques.

At first glance, the legacy of the SB is not as straightforward. This is primarily due to the fact that Churchill's "Garden Suburb" was ousted from power by the secretariat in 1942–1943 (even if Lindemann entered the cabinet). It is also due to the fact that, after 1941–1942, most of the statistics collection suited for total war went in the hands of the newly created Central Statistical Office (CSO). The SB was dissolved in 1945, and it is never cited as a model. But it is fair to view the CSO (Lindemann was more than instrumental in its creation) and many practices of Churchill and the cabinet after 1941, as a direct legacy from the SB. This could also be said of the visual objects promoted by the SB.[21]

Finally, the United Kingdom found itself in a situation in 1940 and 1941 that was no doubt essential in determining the working methods of OR and the SB. Their key words were detailed analysis of situations, pragmatism, and immediate action. Brought to general use and transferred in the context of the American war effort, these techniques would be used to justify strategic bombing and eventually nuclear war and preemptive strikes.[22]

I would also like to end this chapter with a brief comment on John Agar's important book on the *Government machine*.[23] In this book, Agar presents the state apparatus as the primary locus in the constitution of an "infosphere" and describes the variety of reformist movements that played a role in that direction over more than a century and a half of British history. In the context of British historiography, this is an important point. In the footsteps of David Edgerton's work (but before Edgerton's *Britain's War Machine*, published only in 2011), Agar insists on the modernity of the British state apparatus, on its dedication to scientific approaches, and on the weakness of the argumentation on the "two cultures" à la C. P. Snow. For that reason, the SB and OR are, for Agar, two perfect instances in a global movement.

The point is relevant, but I feel compelled to make two remarks. First, I would like to insist on the decisive role played by commercial companies in the development of information processing and the management of and through numbers. The case of the United States between the end of the nineteenth century and World War II is perhaps the most obvious: it was within the commercial and industrial context—and far less in the state apparatus—that a fully developed "infosphere" materialized. In fact, too much focus on the Cold War military complex perhaps led us to forget some of the fundamental forces at work in the long-term transformation of "capitalist" societies.[24]

I would also note that the idea of the expansion of an "infosphere" presents an ambiguity since the notion of "information" has no unifying role before 1942–1943. It may be anachronistic to conceive all human activity, from biology to data-management, in terms of information or 'code' before World War II. Doing cryptography is not "processing information" in the sense in which radar engineers "processed information"; the meaning is again different when one speaks of control rooms, British Operational Research, American Operations Research, or the construction of statistical data by the SB or the CSO. Bringing everything together under a single heading tends to underestimate the diversity of practices and hide the diversity of objectives.[25]

This brings me to my final point. Agar's idea of subsuming various activities under the general expression of infosphere derives from the (perfectly justified) long-term perspective that he adopts. By choosing to follow successive reformist actors, differences are leveled and the highly diverse discontinuities are played down. In this chapter, I adopted a symmetrical position, and I was inclined to insist, within the specific situation of Great Britain, on the emergence of new practices under the pressure of time. Thus, given my specific problem, it was less the forward march of the infosphere "in general" that I wanted to consider than the emergence of new tools and obsessions geared on the situation of urgency Britain found itself in at the time. And all I could hope for is to have been convincing in showing that it was worth making this point.

Notes

1. This work is based on intensive research in the archives of the Navy, Fighter Command, Coastal Command, Anti-Aircraft Command, and the Cabinet at the Public Record Office in Kew; in the Blackett archives in the Royal Society of London; and in the archives of Lord Cherwell at Nuffield College Library, Oxford.
2. Rau (1999, 2001). Quotes from Rau (2001), 225, 226.
3. Weaver (1970). The best analysis on automatic firing control is Galison (1994). See also Owens (1989).
4. For a critical history of the making of the atomic bomb, see Ndiaye (2001).
5. Edgerton (2011).
6. The whole section is based on Kew's archives on the Home Chain, Fighter Command, Anti-Aircraft Command, and Coastal Command.
7. On radar and the Home Chain, see Air Ministry (1963), Brown (1999), Buderi (1999), Latham and Stobbs (1996), Lovell (1991), Rau (1999, 2001), and Waddington

(1973). On the Battle of Britain and the Blitz, see Bungay (2001), Collier (1957), and Deighton and Hastings (1999).

8. For the version of a pilot of Fighter Command under radar guidance, see Rawnsley and Wright (1998).
9. Pile (1949). On Blackett, see Blackett (1962), Lovell (1991), and Nye (2004).
10. Notably, Bragg, Blackett, C. G. Darwin, Tizard, E. S. Pearson, and Cockroft.
11. About Coastal Command, see Goulter (1995). On the Battle of the Atlantic, see Macintyre (1961), Padfield (1995), Roskill (1998), Sirett (1994), and Terraine (1999).
12. These texts are reproduced in Blackett (1962).
13. In short, if Blackett's definition of OR is taken as reference, the claim of HC actors that OR was invented around the HC—the existence of the name being the best proof—is misleading.
14. On the notion of "gadget" as used by Cold War American physicists, see Forman (1989).
15. Soubiran (2002) details these successive steps in the case of the French Navy in the 1920s and 1930s.
16. For details on American Cold War practices, see Dahan and Pestre (2004).
17. This section relies on Lindemann's archives in Nuffield College, Oxford. On Lindemann, see Birkenhead (The Earl of) (1961), Harrod (1959), MacDougall (1951), and Wilson (1995). For more contextual elements, see Chester (1951) and Cairncross and Watts (1989).
18. On Churchill, see for example Bedarida (1999).
19. For the Cold War period, see Edwards (1996).
20. On Churchill and Lloyd George's Garden Suburb, see Hamilton (2001).
21. I do not have time to comment here on the creation of the CSO, but Lindemann was the main force in shaping it. For a faulty introduction (in historical terms), see Ward and Doggett (1991).
22. On strategic bombing, see Facon (1996), Kennet (1982), and United States Strategic Bombing Survey (1945, 1946). For other views, see Lindqvist (2002), Bourke (2001), and Sebald (2003). On nuclear war and preemptive strikes, see Dahan and Pestre (2004).
23. Agar (2003).
24. Gardey (2008) shows that perfectly. See also Chandler (1980).
25. Triclot (2008) proposes the best analysis I know of about the many usages of the notion of information during the 1940s and 1950s. See also Kay (2000).

References

Agar, Jon. 2003. *The Government Machine. A Revolutionary History of the Computer*. Cambridge, MA: MIT Press.

Air Ministry. 1963. *The Origin and Development of Operational Research in the Royal Air Force*. London: Her Majesty's Stationery Office, Air Publication 3368.

Bedarida, François. 1999. *Churchill*. Paris: Fayard.

The Earl of Birkenhead. 1961. *The Prof in Two Words, the Official Life of Professor F.A. Lindemann, Viscount Cherwell*. London: Collins.

Blackett, P. M. S. 1962. *Studies of War, Nuclear and Conventional*. Edinburgh: Oliver & Boyd.

Bourke, Joanna. 2001. *The Second World War, A People's History*. Oxford: Oxford University Press.

Brown, Louis. 1999. *A Radar History of World War II, Technical and Military Imperatives*. Bristol: Institute of Physics Publishing.

Buderi, Robert. 1999. *The Invention That Changed the World, The Story of Radar from War to Peace.*.LONDON: Abacus, first edition: 1996.

Bungay, Stephen. 2001. *The Most Dangerous Enemy, A History of the Battle of Britain.* London: Aurum.

Cairncross, Alec, and Watts, Nita. 1989. *The Economic Section, 1939–1961, A Study in Economic Advising.* London: Routledge.

Chandler, Alfred D. Jr. 1980. *The Visible Hand: The Managerial Revolution in American Business.* Cambridge, MA: Harvard University Press.

Chester, D. N. 1951. "The Central machinery for Economic Policy," in Norman D. Chester, *Lessons from the War Economy.* Cambridge: Cambridge University Press, 5–33.

Collier, Basil. 1957. *The Defence of the United Kingdom.* London: Her Majesty's Stationery Office.

Dahan, Amy, and Pestre, Dominique (eds). 2004. *Les sciences pour la guerre.* Paris: EHESS.

Deighton, Len, and Hastings, Max. 1999. *Battle of Britain.* WARE, United Kingdom: Wordsworth, first edition: 1980.

Edgerton, David. 2011. *Britain's War Machine. Weapons, Resources and Experts in the Second World War.* London: Allen Lane.

Edwards, Paul N. 1996. *The Closed World. Computers and the Politics of Discourse in Cold War America.* Cambridge, MA <AU: Please approve.>: MIT Press.

Facon, Patrick. 1996. *Le bombardement stratégique.* Paris: Éditions du Rocher.

Forman, Paul. 1989. "Social Niche and Self-Image of the American Physicist," in Michelangelo de Maria, Mario Grilli, and Fabio Sebastiani (eds). *The Restructuring of Physical Sciences in Europe and the United States, 1945–1960.* Singapore: Word Scientific, 96–104.

Galison, Peter. 1994. "The Ontology of the Enemy: Norbert Wiener and the Cybernetic Vision." *Critical Inquiry* 21: 228–266.

Gardey, Delphine. 2008. *Ecrire, calculer, classer. Comment une révolution de papier a transformé les sociétés contemporaines (1800–1940).* Paris: La Découverte.

Goulter, Christina J. M. 1995. *A Forgotten Offensive: Royal Air Force Coastal Command's Anti-shipping Campaign, 1940–1945.* Abington, Oxon: Frank Cass & Co.

Hamilton, C. I. 2001. "The Decline of Churchill's 'Garden Suburb' and Rise of His Private Office: the Prime Minister's Department, 1940–1945." *Twentieth Century British History* 12(2): 133–162.

Harrod, R. F. 1959. *The Prof, A personal Memoir of Lord Cher*well. London: McMillan.

Kay, Lily E. 2000. *Who Wrote the Book of Life?: A History of the Genetic Code.* Stanford: Stanford University Press.

Kennet, Lee. 1982. *A History of Strategic Bombing.* New York: Scribner.

Latham, Colin, and Stobbs, Anne. 1996. *Radar, A Wartime Miracle.* Sutton: Stroud : Sutton.

Lindqvist, Sven. 2002. *A History of Bombing.* London: Granta Books.

Lovell, Sir Bernard. 1976. *P.M.S. Blackett, A Biographical Memoir.* London: The Royal Society.

Lovell, Sir Bernard. 1991. *Echoes of War, The Story of H2S Radar.* Bristol: Adam Hilger.

MacDougall, G. D. A. 1951. "'The Prime Minister's Statistical Section," in Norman D. Chester, *Lessons from the War Economy.* Cambridge: Cambridge University Press, 58–68.

Macintyre, Donald. 1961. *The Battle of the Atlantic.* London: Batsford.

Ndiaye, Pap. 2001. *Du nylon et des bombes: Du Pont de Nemours, le marché et l'État américain : 1900–1970.* Paris: Belin.

Nye, Mary Jo. 2004. *Blackett: Physics, War, and Politics in the Twentieth Century.* Cambridge: Harvard University Press.

Owens, Larry. 1989. "Mathematicians at war: Warren Weaver and the Applied Mathematics Panel," in David N. Rowe and John McCleary (eds). *The History of Modern Mathematics*, vol. II : *Institutions and Applications*. Boston: Academic Press, 287–305.

Padfield, Peter. 1995. *War beneath the Sea, Submarine Conflict, 1939–1945*. London: Pimlico.

Pile, General Sir Frederick Pile. 1949. *Ack-Ack: Britain's Defence against Air Attack during the Second World War*. London: Harrap.

Rau, Erik P. 1999. "Combat Scientists, the Emergence of Operational Research in the United States during Worlds War II." PhD Dissertation, University of Pennsylvania.

Rau, Erik P. 2001. "Technological Systems, Expertise, and Policy Making: The British Origins of Operational Research," in M. Thad Allen and Gabrielle Hecht, *Technologies of Power, Essays in Honor of Thomas Parke Hughes and Agatha Chipley Hughes*. Cambridge, MA: MIT, 215–252.

Rawnsley, C. F., and Wright, Robert 1998. *Night Fighter*. Manchester: Crecy, first edition: 1957.

Roskill, Stephen. 1998. *The Navy at War, 1839–1945*. Ware, United Kingdom: Wordsworth, first edition: 1960.

Sebald, W. G. 2003. *On the Natural History of Destruction*. London: Hamish Hamilton, first edition: 1999.

Sirett, David. 1994. *The Defeat of the German U-Boats, The Battle of the Atlantic*. Columbia, S.C.: University of South Carolina.

Soubiran, Sébastien. 2002. *De l'utilisation des scientifiques dans les systèmes d'innovation des marines française et britannique entre les deux guerres mondiales. Deux exemples : la conduite de tir des navires et la télémécanique*, PhD thesis defended on October 25, 2002 at EHESS, Paris.

Terraine, John. 1999. *Business in Great Waters*. Ware, United Kingdom: Wordsworth, first edition: 1989.

Triclot, Matthieu. 2008. *Le moment cybernétique*. Paris: Champ Vallon.

United States Strategic Bombing Survey. 1945. Summary Report (European War) September 30, 1945.

United States Strategic Bombing Survey. 1946. Summary Report (Pacific War) July 1, 1946.

Waddington, C. H. 1973. *OR in World War II*. London: Elek Science.

Ward, Reg, and Doggett, Ted. 1991. *Keeping Score, the first fifty Years of the Central Statistical Office*. LONDON: HMSO.

Weaver, Warren. 1970. *Scenes of Change. A Lifetime in American Science*. New York: Charles Scribner's Sons.

Wilson, Thomas. 1995. *Churchill and the Prof*. London: Cassell.

Part II
Laws

5
Witnessing Astronomy: Kepler on the Uses and Misuses of Testimony

Mario Biagioli

The role of eyewitnessing in science and natural philosophy has been a prominent research question in science studies and history of science in the last two decades. Philosophy too has begun to study its epistemic dimensions.[1] Looking at modern scenarios, scholars have focused mainly on the increasingly extensive role of scientists and scientific evidence in legal proceedings. Historians of early modern science have instead focused primarily on the borrowings of legal witnessing practices and standards of evidence into natural philosophy—borrowings aimed at buttressing the new concepts of experience and experiment being developed by mathematicians and experimental philosophers.[2] In this essay, I analyze the peculiar role of eyewitnessing in Kepler's observational astronomy to revisit and substantially revise some of the received views of the relation between law and early modern science.

We already know that Boyle, Pascal, and Newton had distinctly different uses for witnesses and circumstantial evidence in experimental and observational reports.[3] But if we comb through the texts that Kepler produced in response to Galileo's discoveries of 1609–1610 and through the letters he exchanged with the Florentine astronomers, we find yet another original perspective on the role of witnessing in astronomy—one that is elaborated through some references to procedures and standards of evidence of Roman-canon and inquisitorial law.

Kepler's uses of witnessing

In his 1609 *Phaenomenon singulare*, Kepler described what he took to be the transit of Mercury across the solar disk. (This was a phenomenon he was soon to reinterpret as something quite different—a large sunspot). Kepler calculated

103

that Mercury would enter conjunction with the Sun on May 29, 1607 and planned to observe both before and after that date. At first the weather did not comply with Kepler's wishes but, on May 28 (as he was talking to an unidentified Jesuit about the expected transit), the cloud scattered and out came the sun. Kepler rushed to the attic of his home in Prague where cracks between the roof tiles could function as pinholes for solar observation. Once there, he projected the solar disk on a piece of paper and observed "a small spot the size of a small fly on the lower left side" of the solar disk.[4] After moving the piece of paper around and trying out different pinholes to test whether the spot might be produced by either the paper or spiderwebs dangling from the ceiling, Kepler became convinced that he was not dealing with an artifact.

He immediately started to line up eyewitnesses. The first was Martin Bachazek the—rector of the University of Prague and Kepler's landlord—who wrote on Kepler's own report: "I, M. Martin Bachazek, was present to this observation and vow that this is what happened."[5] Kepler then left the house, went by the court (where he instructed a valet to report the news to the emperor), dropped in on the Jesuit to inform him of the discovery, and finally landed in the shop of Joost Burgi—the court clockmaker. Burgi was not in, but the sun (and the spot) were not going to stay up forever. Having no time to waste, Kepler rounded up two of Burgi's assistants and servants, closed all the doors in the shop, and darkened all the windows, except for a pinhole aperture (about 1/10 of an inch) from which they were able to observe (at about 14 feet from the aperture) the same spot in the same location on the solar disk. Like Bachazek a few hours earlier, one of Burgi's assistant was asked to autograph Kepler's report, which he did (in German): "Heinrich Stolle, junior clockmaker-journeyman, my hand."[6]

In the book, Kepler uses the terms "spectator" and "testis" to identify both Bachazek and Stolle, perhaps to specify that they were testifying to something they had personally seen rather than to something they had just heard and deemed credible (as was the case with so-called hearsay witnesses—an older form of witnessing that was still accepted in the medieval period).[7] While Bachazek's socioacademic status contributed to the credibility of his testimony, Kepler's inclusion of Stolle, a workman, suggests that his search for witnesses was nearly class-blind. That practice fit well with Roman-canon law as practiced in the Hapsburg Empire and the German lands, which stated that "adequate witnesses are those who are without evil repute and who otherwise are unchallengeable for any legal ground."[8] Religious differences also did not seem to matter as Kepler (a Protestant) seemed quite eager to enlist the testimony of a Jesuit.[9]

The typographic features of Kepler's text and the positioning of Bachazek and Stolle's testimonials in it are also important. Kepler does not limit himself to include their names within his printed observational narrative to let the reader know that he has people who can back up his claims. Instead, he asks Bachazek and Stolle to autograph the reports he had just written up—reports he then prints verbatim in the *Phaenomenon singulare* in a distinct format. After bracketing each line of the reports with quotation marks to make them stand out from the rest of his own text, Kepler adds their date (Monday May 28, 1607—the same day on which the observations were conducted), and then appends his witnesses signatures using different fonts (regular for Bachazek's signature and gothic for Stolle's), as if to reproduce as much as possible the "aura" of the original signatures on the handwritten document.[10]

In a legalistic fashion, Kepler then writes in the margin (next to the section signed by Stolle) that, while the printed text appears in Latin, the original was written in German (most likely because of Stolle's limited linguistic range) and then translated into Latin by Kepler himself. Interestingly, Bachazek's and Stolle's signatures include their professional titles—the first a master, the second a clockmaker-journeyman—and Stolle's signature is prefaced by Kepler's description of his identity: "The witness is the assistant of Joost Burgi, the maker of automata, who was a spectator."[11] Because Stolle's modest professional title would have had little to add to the credibility of his testimony, the information about the witnesses' position was probably included not for epistemic reasons but for legal identification.[12] Roman-canon law required that testimonies submitted by the plaintiff be "properly written up and transmitted to the judge, along with the witnesses names and *addresses*" for follow-ups.[13]

Although we can assume that Kepler would have taken Rudolph II over Burgi's assistant as a witness, the observation of the (alleged) transit of Mercury was not a staged experiment but a time-specific and not fully predictable event. Because of the narrow window of opportunity, Kepler seemed just happy to find someone—anyone—who could witness it. As a literary genre, Kepler's narrative is closer to a police report of a crime scene than to the description of an instrument-produced experiment performed at the Royal Society at a preadvertised time, in front of preselected witnesses.

Kepler's legalistic concerns reemerge at the end of his report. While stating that he sought Burgi's testimony (when he got back to the shop) as well as that of the Jesuit (earlier in the afternoon), he reported that the priest was unable to corroborate the discovery because of the constraints imposed by his prayer schedule and his lack of a suitable pinhole, while Burgi's observations were cut prematurely short by cloud cover.[14] Interestingly, Kepler bracketed Burgi and the Jesuit out of the observational report not by saying that they had tried and *failed to witness* the truth of Kepler's claims, but rather because that they had *failed to be witnesses*. Instead of saying that he had two negative testimonies and two positive ones, Kepler wrote that he had only two witnesses (Bachazek and Stolle) because the other two (Burgi and the Jesuit) just did not qualify as witnesses (though we know that they did try to observe).[15] That done, Kepler proudly pronounced: "The testimonials of our witnesses [Bachazek and Stolle] are unanimous." Perhaps Kepler's selective counting might reflect the fact that two fully positive eyewitness reports provided a *probatio plena*—Roman Law's standard of criminal proof.

Kepler's use of witnesses was further refined in the *Narratio*—a short book reporting the observations of the surface of the Moon and of the satellites of Jupiter with a telescope between August 30 and September 9, 1610.[16] References to legal practices are found throughout the book. Kepler opens by acknowledging that some had criticized his *Dissertatio cum nuncio sidereo* (published earlier in May) for uncritically upholding the truth of the observations put forward in Galileo's *Sidereus nuncius* (published in March).[17] Unable to access a suitable telescope to replicate some of Galileo's discoveries, Kepler had indeed endorsed the *Nuncius* prior to being able to replicate its claims.[18] The *Narratio* was written to fill such a gap, providing the testimonials he did not include in the *Dissertatio*. Together with the letters exchanged in those months between Kepler and Galileo, these three books provide a wealth of information about the vastly divergent roles the two astronomers attributed to witnessing.

The *Narratio* presents a series of observations that Kepler and his witnesses conducted following a specific protocol to avoid influencing each other's findings. Witnesses' reports are most credible when independent, that is, when most likely to be unbiased and untampered with. Conversely, witnesses who observed together and discussed what they were seeing might have influenced each other's reports. Attempts to avoid the spreading of biases (observational or otherwise) are mentioned throughout Kepler's book. Kepler wants to show that he and his fellow observers did not influence each other, but also that he and Galileo had not staged his publication by communicating and comparing observations with him beforehand:

> Prague is my witness that these observations have not been sent to Galileo. Actually it is for this reason that I have not written him recently despite the fact that I owe him a letter. And those to whom I have communicated these [observations] in generic terms have not been able to copy anything from my papers kept at my house. Similarly, [Galileo] has not been able to send me his observations because only a few days have passed. You can therefore rest assured that there has been no communication.[19]

When it comes to observing, Kepler reports the provenance, ownership, and optical limitations of the telescope he used; some of the challenges he encountered while observing; the slight modifications he introduced in the apparatus; and the names of his various co-observers and witnesses (Benjamin Ursinus, Thomas Seggett, Frans Tengnagel, and Tobias Schultetus). As in the previous *Phaenomenon singulare*, the *Narratio* does not relate the witnesses' credibility directly to their social status. That was not the result of an egalitarian impulse but of a kind of "actuarial calculus." Kepler does not treat trustworthiness as inherently connected to a positive cause (social status and values of honesty) but to a negative factor such as risk (how much a person would lose were she/he to speak falsely). More precisely, Kepler assesses such a potential loss over time rather than in relation to a witness' status at the moment in which a testimony may be judged to be false. Ursinus is the youngest and least prestigious person among the witnesses, but that does not mean that he has less to lose than a more senior scholar like Thomas Seggett, "an Englishman already well known for his books and correspondence with famous men, who therefore cares dearly about the reputation of his name."[20] According to Kepler, because Ursinus "is passionate about astronomy, loves that discipline and has decided to practice it as a specialist, it would not even cross his mind to ruin, right at the beginning, the credibility necessary to a future astronomer with a false testimony."[21] He would lose not only the modest name he had in the present but also the much bigger name he might have developed in the future—the "integral" of his reputation over the length of his professional life.

Kepler then describes the bias-control protocol that was followed throughout the observations:

> Each of us had to draw, in silence, with chalk on the wall anything he had observed without making it visible to the others. After that, we would look together and simultaneously at each other's picture to check our agreement.[22]

Kepler then maps both the consensus and the disagreement on the various observations, often specifying which observations were produced after being "tipped off" by other team members. For instance, "at the fifth hour, I lost sight of the eastern satellite, which was nevertheless spotted by Sir Tengnagel, secret counselor of Archduke Leopold (who had been instructed). He did not, however, see the western one."[23] Later on, "Seggett saw all three of them, and drew them up in the same configuration [as Kepler's and Ursinus']. Sir Schultetus, Imperial tax collector for Silesia, saw (after been instructed) the most luminous among the western ones."[24]

Kepler's protocol resembles Roman-canon law practices. I say "resemble" because it is important to acknowledge the differences between the scenarios dealt with by natural philosophers making claims about new and hard-to-observe objects and criminal cases where judges did not have to establish the fact of a crime but rather of the author of that crime. Placing a person at the scene of a crime (in the past) is quite different from placing a satellite in orbit around Jupiter (now) or from confirming an experimental finding that (unlike a crime) may be replicated. Perhaps cases involving reproducible evidence (like, say, cases of forgeries, coin clipping, etc.) involved evidentiary challenges much closer to those faced by natural philosophers.

Contrary to common law countries like England where trials took place in an open court, trials in Roman-canon law countries were based on evidence produced by interrogating witnesses in private and then forwarding the transcripts to a closed court. This was not just to maintain the power of the judiciary but to prevent what early modern jurist saw as unlawful storytelling. Defendants were often not told what crime they were accused of prior to being interrogated so that they would not be able to prepare self-exculpating narratives. (The documents of Galileo's trial show that the inquisitors followed such practice).[25] Also, when more than one defendant was imprisoned awaiting trial,

> they should be kept apart from one another to the extent that the gaol cells are available, in order that they may not plot false testimony with one another or discuss how they can explain away their deed.[26]

Denying defendants information about the crime was also seen as a way to prevent them from confessing things they had not done (or, in Kepler's case, to report things they had not observed). Jurists thought that, were defendants to know the circumstances of the crime, they might cobble them up into a confession just to get themselves out of the hands of the torturers (or, in George Bush's parlance, "professionals").[27] While the Royal Society's practice of collective witnessing has been shown to fit well the common law model of open trials in front of a jury, Kepler's *Narratio* seems informed by Roman law scenarios: the observers "interrogate themselves" privately and independently, and then show their independently obtained written evidence to the reader-judge (or to themselves as a collective judging body).

Kepler and the lawyers

If the observational and witnessing protocols described in the *Narratio* were more sophisticated than those in the *Phaenomenon*, it was probably because of the pressure exerted on Kepler by Galileo's uncooperative behavior. On August 9, 1610, just three weeks before conducting the observations eventually published in the *Narratio*, Kepler wrote Galileo pressing him to send testimonials to Prague to help him to silence the remaining critics of the *Nuncius*. Kepler was concerned not only with Galileo's honor but with his own. Having enthusiastically endorsed the *Nuncius'* discoveries in his *Dissertatio*, he was then left to hang when Galileo refused to send him third-party testimonials or a telescope with which to produce his own:

> Although I continue to have no doubts, it nevertheless pains me to remain so long without testimonials by others to convince the remaining skeptics. I am asking you, Galileo, to produce other testimonials as soon as you can. From the letters you have sent to various people, I have learned that you do not lack witnesses. But I cannot cite anyone except you to defend the credibility of my letter [the *Dissertatio*]. The authority of the observation rests solely on you.[28]

In the absence of Galileo's collaboration, Kepler had already lined up all testimonials he could find (including ancient ones) for the irregularities of the lunar surface, and the many fixed stars in the Milky Way, but could find none for the satellites of Jupiter.[29] As he requests testimonials from Galileo, Kepler tries to explain to him why they are necessary to begin with by drawing a difference between philosophical and factual arguments. He tells Galileo that the debate over the discoveries reported in the *Nuncius* "is really not a philosophical problem but a juridical question of fact." The main question on the readers' minds is not whether Galileo is a good philosopher (that is, whether he has correctly identified the causes of the phenomena he presents) but simply whether he has "consciously lied to the world" by making false factual claims.[30]

Prefaced by a reaffirmation of Kepler's support of the *Nuncius*, these remarks do not necessarily convey distrust but rather a nonjudgmental description of the predicament faced by anyone who happens to make statements about facts. Because of the nature of their discipline, early modern astronomers often relied heavily on the observations conducted by colleagues in other places and other times—more so than the practitioners of most other disciplines, including experimental philosophy. Still, Kepler is not lecturing Galileo about some delicate trust-based sociability of the astronomers' community and the need to sustain it through value-confirming behaviors such as the disclosure of the instrument's specification, observational practices, and testimonials. His letter does not intimate that Galileo's refusal to provide testimonials may threaten the stability of the astronomers' "form of life," but simply reminds him that, because of the empirical (rather than philosophical) nature of the claims he made in the *Nuncius*, his readers are expecting him to play by the rules of the legal (rather than philosophical) game.[31] Kepler seems to take for granted that Galileo has testimonials available and tells him that he ought to make them public.

Written a few weeks later, the *Narratio* suggests that Kepler had some dislike for the very rules of the game he is exhorting Galileo to follow—a dislike that resonates with some of the recent critiques of the feasibility of the jury system to judge complicated scientific matters. In the *Narratio* Kepler reports that some critics have dismissed his just-published *Dissertatio* as a rhetorical text: "According to them, [my arguments] are cheap and aimed at pleasing the masses, like those used in a tribunal to respond to questions about fact."[32] (The critics, most likely, were responding to seeing the book endorse Galileo's discoveries without replicating or providing testimonials about them).[33] That put Galileo and Kepler on the same boat. If some accused Galileo of lying about facts, others took Kepler to spread a cognate kind of lie—the kind lawyers tell in court when they cannot produce facts.

Kepler is no antiempiricist. He observes whenever he can, collects observations from wherever and whomever he can get them, and even writes a book—the *Ad Vitellionem paralipomena*—on optics and vision with the goal of improving the reliability of astronomical observations. His derisive association of rhetoric and judgments of fact, therefore, is not a critique of empiricism in general but rather a description of what *other people*—the common readers of Galileo's *Nuncius* and of Kepler's own *Dissertatio*—take to be the appropriate protocols to assess facts. Lacking a philosophical background, these people may assume that the discourse of lawyers and courts is the only way one can talk about empirical evidence.

What seems to bother Kepler is not the strictness or laxity of legal standards about fact but the way discussions about facts are framed in (and by) legal settings. Courts, it seems, are the place where facts are put forward, but they are also the place where their absence is routinely covered up by the lawyers' rhetorical arguments. Facts are indeed opposed to rhetoric, but this is an opposition that is *played out within the same legal discursive game*. Whereas rhetorical spins on evidence (or its absence) are corrupt, statements of fact are limited to effects—not causes. Both options are not terribly appealing to someone who, like Kepler, fashions himself as a philosopher (or as a theologian-turned-philosopher).

Shapin and Schaffer argue that Boyle drew from legal practices to build a methodology of experimental philosophy around the "matter of fact," but Kepler seems to see the law as part of the problem rather than of the solution. (His subsequent long and stressful engagement with the courts to defend his mother from accusations of witchcraft probably did little to make him appreciate the legal institutions' handling of testimony and empirical evidence).[34] Kepler's skepticism does not reflect a worry—shared by other seventeenth-century natural philosophers—that statements about nature have a tendency to turn litigious because of the dogmatism of the philosophical or theological frameworks in which they may be made to operate. Lawyers and courts can make facts litigious no matter what they might be about or what previous connotations they might carry. Kepler's solution is not to go for maximum facticity—matters of fact bleached of any interest or ideology—but rather to adopt a two-tier epistemology that, by separating factual statements from philosophical ones, accepts the sad fact of the lawyers' existence.

If one's claims are primarily about observations (as in Galileo's *Nuncius* or Kepler's *Phaenomenon*) then one has to play by the lawyers' rules and provide testimonials. Although Kepler does not seem to enjoy having to write the *Narratio* to corroborate Galileo's discoveries, he feels compelled to do so to vindicate what he wrote in the *Dissertatio.* Philosophers may not need (or even like) testimonials, but they cannot forget that, infected by the "idols of the tribunal," the common readers do need them.

For instance, in the *Narratio* Kepler states that the "more secret" reasons for his trust in Galileo's observations predated his having "proof of the fact." Even in the absence of empirical corroboration, Kepler states that such reasons were strong enough to "completely satisfy my mind."[35] Having initially withheld those "more secret" reasons, he has decided to make them public now that he can provide empirical testimonials as well. We should not, however, take Kepler to behave like the textbook scientist who puts forward her/his claims only when she/he can empirically support them in front of colleagues.

The delayed publication of Kepler's "more secret" reasons does not result from the delayed availability of corroborative evidence but rather from the features of the audience he had to address in that specific book—an audience that was *not* primarily made up of colleagues. Kepler did not wish to address the "common readers" but was forced to do so because of Galileo's decision to pitch the *Nuncius* to them rather than to professional astronomers. Kepler's earlier decision not to take his reasons "in front of the judges" or to "the masses anxious with doubt" reflected a fear that, unable to understand his "more secret" reasons, they would have made fun of him.[36] Kepler seems less concerned with conveying knowledge to the masses than with avoiding being harassed by them.

This sounds like philosophical elitism of the Pythagorean type (a stance certainly not alien to Kepler), but it carries more mundane implications. Echoing the letter to Galileo from a few weeks earlier, Kepler is now suggesting that while philosophical readers would be able to understand Kepler's "secret reasons," common readers could be convinced (or perhaps just pacified) only by testimonials. Kepler does not provide testimonials to *prove* his "secret reasons" but rather to *shield* philosophical knowledge from the derision of the masses—to keep the readers happy and off the philosopher's back. In this sense testimonials function as the epistemological analog to what we now call "one-liners" or "sound bytes."

If the expectations of philosophically lowbrow readers may have been annoying to Kepler, they also came with some silver lining. The same legal conventions that make people expect testimonies from philosophers when they make statements of fact also places quite a low threshold on credibility: "Such is the way of the law: one is presumed sincere until the contrary is proven."[37] Although any additional circumstantial evidence (like, but not limited to, social status) may add to a claimant's credibility, the principle remains that in disputes over facts (as distinct from disputes over points of law) the burden is not on the claimants but on their critics.[38] This has tremendous consequences for discoverers as it means that, as Kepler often states, Galileo's opponents should not attack him and his claims without introducing empirical evidence to support their challenges. His claim that discoverers should be (legally) entitled to the benefit of the doubt is also traceable to lines such as "Why should not I believe in such a profound mathematician" or "Why should I deny my trust." More than rhetorical questions, such constructions indicate that Galileo ought to be granted credibility to begin with and that Kepler would have to find reasons for taking that credibility away from him.[39]

Kepler's application of the "innocent until proven guilty" legal standard to factual claims about nature may also explain his openness to using lower-class witnesses. High social status does help credibility, but that does not mean that claims put forward by a lower class person are not credible. Technically, even a beggar's claims would have to be refuted to be dismissed—a position quite different from Boyle's who was eager to dismiss as untrustworthy testimony from laborers.[40] To Kepler it is all a matter of balance or, rather, of judgment. Everybody starts with some positive credibility that can be then increased or reduced by circumstantial evidence such as the character of the person, the risks that person would be taking by lying, the nature of the claim, the opposing or supporting testimonies, the modalities of observation, the way the claim is reported, and so on. Kepler does not treat testimonials as proofs, but only as evidence—entries in a long list of additions and subtractions through which credibility is assessed—a practice not unlike the evidentiary arithmetic of Roman-canon law.

Idols of the tribunal

I want to return to Kepler's complaint that the *Dissertatio*—a book that provided many arguments but little empirical evidence to support Galileo's claims—was criticized for putting forward arguments that were "cheap and aimed at pleasing the masses, like those used in a tribunal to respond to questions about fact."[41] Kepler's remark, it seems to me, is that his critics assumed that if a person supports somebody else's claims about facts without introducing testimonies that person must be operating at the other end of legal discourse—that of rhetoric. Conditioned by the "idols of the tribunal," such readers are unable to see that if the *Dissertatio* endorsed Galileo's claims without replicating them, it is because Kepler was supporting those discoveries with arguments that were neither factual nor rhetorical. These were philosophical arguments about the causes of Galileo's phenomena rather than about the phenomena themselves.[42]

Kepler's definition of philosophical claim includes the physical causes of natural phenomena but is broader (and less clear) than that.[43] What remains clear, however, is that Kepler attributes certain a priori features to philosophical arguments. Although they may be refuted by empirical evidence, those arguments do not develop from evidence in an inductive fashion. According to the *Dissertatio*, "it is truly not without reason that we much esteem those who [...] precede the senses with reason."[44] One does not need an hourglass to figure out that summer nights are shorter in England than in Rome because that can be easily derived from geographical and astronomical considerations without any further empirical input.[45] At a much higher level of complexity, a sophisticated astronomer can appreciate the truth of Copernican cosmology even in the absence of conclusive empirical corroborations (which, in fact, became available only much later). Another example is Kepler's own "discovery" of the relationship between planetary orbits and Platonic solids in the 1596 *Mysterium cosmographicum*. Empirical data about planetary orbits is of course crucial here, but what Kepler takes to be the explanation for their distribution stems from an a priori construct: the number and geometrical features of the Platonic solids.

This last example introduces a key feature that Kepler attributes to natural philosophical arguments—a feature that can be used to assess the credibility of factual reports even in the absence of direct or reported empirical evidence. By uncovering some of the causes of observed phenomena, philosophical arguments also point in the direction of yet undiscovered phenomena, relations, or even mechanical inventions. When discovery happens, it derives credibility from having been "predicted." What Kepler means by prediction is much broader than a law's ability to predict a certain event (such as shorter summer nights in England compared to Rome, or an apple departing from a tree branch with a certain acceleration). Philosophical arguments are generative of entire families of new arguments and discoveries. For instance, Kepler suggests that his discovery of the correlation between planetary orbits and Platonic solids is not altogether surprising because it is little more than a "confirmation" of Plato's and Proclus's original "prediction" about the role of the perfect solids in the structure of the cosmos. He goes so far as to suggest that Columbus's discovery of the new world is credible (and perhaps not deserving the extraordinary recognition it had received) because, in the end, his voyage corroborated philosophically reasonable speculations about the existence of other continents on earth dating back to Plato.[46] The same logic applies to Galileo's telescope. Kepler has not seen it but believes that it produces the observations described in the *Nuncius* because its optical principles were already laid out in Kepler's 1604 *Ad Vitellionem paralipomena*.

The telescope, Kepler suggests, is the "effect" of the "causes" discussed in his book—a book that can be now seen as having predicted that invention.[47]

Kepler's characterization of his critics' habitus suggests that they did not understand the epistemic status of philosophical arguments, that they do not need the support of testimonials to accept them, or that Kepler's saying that Galileo's claims were "most certain" was quite different from what *they* would take to be the endorsement of a statement of fact. Kepler found the philosophical arguments about the new discoveries so convincing to compel him to endorse Galileo, but "no one should think that, in my eagerness to endorse Galileo, I intend to take away from others the liberty to reject his claims."[48] Unable to tell the difference between a philosopher and a lawyer, Kepler's critics took him to act as Galileo's attorney, trying to force assent with lawyer-style rhetorical arguments packed with invocations of truth when, in fact, he was simply expressing his philosophical appreciation of the discoveries.[49]

Such misreadings, however, were facilitated by the specific contents and literary genre of the *Nuncius*. It is well known that Galileo's book became a cause celebre by blurring the disciplinary lines between mathematics and natural philosophy through the presentation of astronomical evidence with extraordinary implications for natural philosophy and cosmology. Furthermore, such claims were made with a new and poorly understood instrument —an issue that forced a redefinition of the very meaning of "eyewitnessing." Kepler, however, suggests that the *Nuncius* caused even bigger disruptions, such as the scrambling of distinctions between philosophical discourse and legal arguments about facts.

It would have never crossed the mind of the readers of *De revolutionibus* to ask Copernicus to prove his arguments according to the standards of the court of law. Readers of technical astronomical texts belonged to an elite operating according to its own rules of discourse and evidence—rules that, as shown by the outcome of Galileo's trial of 1632–1633, were difficult to translate into to those of the law. But common readers who would have never picked up a traditional astronomy text bought the *Nuncius* because, in addition to the extraordinary nature of its claims, it was presented as an astronomical news-sheet, with very few technical arguments.[50] Furthermore, the book made philosophical arguments almost without stating them, that is, by presenting stunning new facts while keeping discussions of their philosophical implications to a minimum. It did not only blur disciplinary boundaries between mathematics and philosophy but also mixed "high" and "low" audiences without actually warning the readers that what they had bought was a philosophical bombshell in sheep's clothing. That supported the "common" readers' tendency to see it as a book that was purely about facts—though one that failed to provide testimonials for those facts. (This was, I think, the meaning of Kepler's remark that, from the readers' point of view, the issue was "really not a philosophical problem but a juridical question of fact").[51] The (unacknowledged) scrambling of the boundaries between disciplinary genres and audiences complicated the *Nuncius'* reception as well as that of its defense—Kepler's *Dissertatio*.

Marking truth, marking lies

As he discusses the "secret reasons" for endorsing the *Nuncius*, Kepler makes an intriguing statement: he finds Galileo sincere because his book contains things "that are both credible and incredible."[52] Claims that are too good to be true are likely to be untrue; which means that, to (appear to) be true, a claim needs to simultaneously confirm and subvert the reader's expectations.

In the *Narratio* (but also in the earlier "Defence of Tycho") Kepler remarks that liars need to have excellent memory.[53] Memory is a crucial skill for those who make things up, as they need to ensure that each step of their story is construed to fit the previous one. Liars also have a tendency to find an answer to any question that may be posed to them. By contrast, it is a sign of sincerity to say "I do not know," as well as to report phenomena that are difficult to explain: "Why, I ask, would one willfully complicated matters by inventing such things one would the despair to explain?"[54] Galileo, Kepler argues, reported the surprising variation of the brightness of Jupiter's satellites while failing to properly explain it. It is precisely the fact that Galileo is struggling to explain what he has reported (and that Kepler himself could not do better) that convinces Kepler that this is a real phenomenon. It is real because it is difficult, but not as difficult as to be incredible.[55]

Similarly, Galileo's claim that the satellites' periods around Jupiter are remarkably fast (especially compared to Jupiter's 12-year period) is a surprising statement that has the ring of truth—even more so after reading of Galileo's skepticism about being able to determine their exact periods. Had he been a liar, Galileo could have instead "organized those apparitions imagining them on the basis of precise orbits and periods, as if drawing them from an ephemerides."[56] And if he really wanted to make up new planets, Kepler continues, why not make their number infinitely large and place them around an infinite number of fixed stars so as

> to corroborate Cardinal Cusanus, Bruno, and others, and to say things made credible by their authority? And if he did not like the fixed Stars, why should have he invented them around Jupiter while neglecting Saturn, Mars, and Venus? Why would have he imagined four rather than only one (as only the Moon goes around the Earth) or six (as there are six planets around the Sun?)[57]

To be credible, new claims need to defy the most mechanical of expectations, that is, they need to be a bit incredible. But all this is lost on those whose thinking is conditioned by the "idols of the tribunal." With a mix of perplexity and sarcasm, Kepler reports that some people took the many questions he asks around Galileo's claims—questions introduced to argue that Galileo's claims are true because he could have made different ones more easily—to be a sign of skepticism rather than appreciation. By mistaking questions *around* those claims to be *about* those claims, the critics seemed to conclude that Kepler was treating Galileo as a hostile witness.[58]

Kepler's notion of the "mark of truth" applies to arguments that humans develop about nature but hinges, I believe, on ontological rather than epistemological considerations. Deriving from God's infinite power, the workings of nature always exceed our knowledge and expectations. Philosophical narratives that acknowledge gaps in the philosopher's understanding of nature confirm such ontology and derive a ring of truth from it. Unable to fully comprehend nature, the philosopher can only display the gaps and deferrals she/he incurs with while inexorably failing to keep up with it. While nature shows itself to be natural by displaying its *infinite* creations, the philosopher shows her/himself to be truthful by displaying her/his *finite* ability to grasp such infinite complexity and variety. One kind of mark produces the other as its complement.

Admitting to gaps in one's argument, then, is not so much a sign of personal sincerity—the demonstration of socially sanctioned marks of modesty—as a trace of the ontological gap between what nature does and what humans can understand about it. Unlike good philosophers who know and make visible their limitations, liars invent seamless narratives. But even when most intricate and skillful, the liars' fabrications display the smoothness of artifacts—a smoothness that gives them away as mere simulacra of knowledge or creativity.[59] Gaps or statements like "I do not know" in philosophical arguments are the equivalent to the accidental chisel scratch or brush stroke that sets apart a handmade artwork from machine-made identical multiples. Exceptions that confirm the rule, they are signs of authenticity because they mark excess or unnecessary difference (as opposed to the fake smoothness of the liar that signals only her/his lack of real knowledge or creativity).

This may explain why Kepler is not embarrassed to present partially diverging observational reports in the *Narratio*. Such practice, I argue, is quite different from apparently similar admissions of error found in other texts by Kepler or experimental philosophers. For instance, Kepler's chronicling, in his 1609 *Astronomia nova*,[60] of his many missteps on the way to determining the elliptical shape of Mars's orbit, or the reports of failed experiments found in Boyle's *New Experiments* were meant to demonstrate one's sincerity: "I am admitting to you that I expected X, but got Y instead."[61] Because Y is openly presented as a wrong result, such a tactics might help you win the sincerity contest, but not the one about truth. Such reporting of struggles and false starts needs, in fact, to be followed by the delivery of what is deemed to be the right result. The *Narratio*, instead, puts seemingly analogous discrepancies to a completely different use. We have seen that in that book Kepler describes how different people were often unable to observe the same satellites of Jupiter at the same times. Still, he presents such observations as testimonials to the truth of Galileo's claims. That's no slip of the pen. Right at the beginning of the book (well before he describes the observations), Kepler states that

> if, dear reader, you notice some discrepancy or if, as I believe, you will realize that sometimes I have seen fewer satellites than Galileo, *this should not produce any confusion concerning the fact itself.* These, in fact, are my first experiment with such observations; the sky has been often cloudy; the

presence of the Moon has bothered us; the instrument was not very good nor very easy to use; the telescope mount was fixed; it was very hard to find Jupiter[62]

Giving information about the limits of one's instrument has been discussed as a tactic used by experimenters to avoid "giving the lie" to other practitioners whose results did not match their own.[63] But here Kepler uses the very same kind of evidence to say that, despite the discrepancies caused by constraints in the apparatus and in the conditions of observation, the observation stands corroborated. He invokes observational contingencies not to maintain polite intercourse in the face of disagreements about facts but rather to say that such differences do not amount to actual disagreements.

Kepler's radically different stance in the *Astronomia nova* and the *Narratio* may have to do with the specific differences between the problems discussed in the two books. The error-packed struggle chronicled in the *Astronomia* was a mathematical one. Kepler was trying to detect the orbit of Mars based on a specific set of Tycho's observations—a process he described as having clearly binary outcomes: match or no match. He describes the many mismatches until he reports what he takes to be the one that fits. The corroboration of the satellites of Jupiter is a different problem altogether. As he told Galileo a few weeks earlier, it was not a philosophical but a juridical matter. It did not concern the determination of the true orbit of the satellites but the corroboration of their *existence;* that required producing observations (rather than finding the one geometrical figure that made sense of them). Not only do these two different puzzles require different approaches to their solution, but they also fall into what Kepler takes to be two different epistemic registers. The orbit of Mars is more of a philosophical problem (and he cites no witnesses in support of his discovery), while the existence of the satellites of Jupiter is a straightforward empirical or, as he says, a juridical issue (and he cites three witnesses besides himself). This, however, does not quite answer why Kepler thought that observational discrepancies could add (rather than subtract) from the strength of the collective testimony. To get there, we may have to go back to his remarks about liars.

Liars make up improbably seamless stories. Along those lines, Kepler seems to treat full consensus in observations conducted by different people as suspicious rather than reassuring—as if total consensus about a matter of fact is just too improbable to be true.[64] It could suggest that, Mafia-style, someone got to the witnesses. It could also suggest that Kepler and Galileo had checked their observations (or even coordinated their cooking) before Kepler's publication to make sure that they matched. (That's a possibility that Kepler dismisses by citing that "everybody in Prague" knew that there had been no communication between the two).[65] In sum, Kepler behaves as if differences in the observational log do not imply that the phenomenon is unstable or artifactual but that *other differences are at play*—some of them in nature (changing lighting conditions due to the Moon's position), some in the witnesses' perceptual abilities, and some in the apparatus. Those differences tell the reader that the witnesses have not been tampered with.

Kepler's endorsement of the "innocent till proven guilty" rule is also key here. According to that legal stance, the divergent observations of Seggett, Ursinus, and Schultetus do not refute each other. If Seggett reports one specific satellite but Schultetus does not, that does not mean that Seggett's observation is wrong. It simply means that Seggett's report is credible but not confirmed by other testimonials. When multiple observations of the same object confirm each other, the claim's credibility is reinforced. But when they don't, the claim's epistemic status remains positive, though lower than that of a claim that has been corroborated. In sum, Kepler would have been in trouble if all of the four observers had come up with either completely nonoverlapping drawings or with completely overlapping drawings. The first scenario might have indicated failure, whereas the latter would have looked too good to be true. But as Kepler put it, claims need to be both credible and incredible to be true. Some overlap and some nonoverlap provided just the right mix—a proper "reality effect."

Between prediction and prophecy

There is, I believe, a connection between Kepler's notion of the "mark of truth" and his attribution of additional credibility to claims that were somewhat predicted by philosophical arguments. The emphasis here is on *somewhat*. Kepler does not attach credibility to just all factual claims predicted by philosophical arguments but only to those that have been *predicted imperfectly*. Similarly, he attributes truth to philosophical arguments that have been generative enough to produce *imperfect* predictions. God, I think, is just around the corner in Kepler's argument. Imperfection goes with generativity, but not with the infinite power and generativity of God. If humans were God, they could come up with perfect predictions because they could create what they were predicting. However, not being God, they can only produce partial predictions based on some good hunch about physical causes. A too accurate prediction (by a human) would either predict nothing new or predict too much to be true. A perfect prediction is as mechanical as a copy—like a die striking yet another identical coin—or as dubious as something that has been made up to fit.

Kepler argues, for instance, that his discovery of the relationship between planetary orbits and the Platonic solids in the *Mysterium cosmographicum* has simultaneously confirmed and refuted the ancients' claims about "how the five [Platonic] solids were expressed in the cosmos." Kepler credits the Platonists for attributing a key role to the perfect solids in the structure of the cosmos but disproves the specific role they attributed to them. Galileo's discoveries do the same with regard to claims about the fixed stars having their own satellites—claims that Kepler traces back to Edmund Bruce and Giordano Bruno. Bruno and Bruce, Kepler tells us, were right in arguing that there were more satellites in the world, but Galileo has shown that such additional satellites orbit a planet, not a fixed star: "You correct such a doctrine," while also show-ing that "they generally told the truth."[66] This last example reemerges in the *Narratio*, with a crucially different twist. There, Kepler goes back to Bruno's speculations about satellites orbiting fixed stars, but this time to say that Galileo's claims about the satellites of Jupiter were credible precisely because they did *not literally* confirm Bruno: "Had the author decided to make up new planets, why, I ask, did he not imagine them infinite [in number] around infinite fixed stars, so as to corroborate Cardinal Cusanus, Bruno, and others, and to say things made credible by their authority?"[67]

Observations that match all too well the philosophers' predictions are either redundant or artifactual (in the same way that *exact* consensus over one specific observation may be a mark of fraud). Whether redundant or arti-factual, such observations produce no (new) knowledge and contribute no (new) credit to themselves and to the philosophical arguments that predicted them. But although a discovery that matches only the "spirit" (but not the "letter") of a philosophical prediction cannot count as a proof of the philo-sophical argument underlying such a partial prediction, it still demonstrates something epistemically relevant about that philosophical argument. It dem-onstrates its cognitive productivity, its ability to produce hypotheses aligned with at least some of the causes through which nature has generated the newly discovered phenomenon.[68] The notion of prediction that Kepler uses in these texts is therefore quite closer to prophecy than to law-like forecast.[69] It also bears some resemblance to another form of prediction that occupied Kepler for most of his life: astrology.

Notes

1. Jennifer Lackey and Ernest Sosa, eds., *The Epistemology of Testimony* (Oxford: Oxford University Press, 2006); C. A. J. Coady, *Testimony: A Philosophical Study* (Oxford: Clarendon Press, 1992).
2. Peter Huber, *Galileo's Revenge: Junk Science in the Courtroom* (New York: Basic Books, 1991); Kenneth Foster, ed., *Phantom Risk: Scientific Inference and the Law* (Cambridge, Mass.: MIT Press, 1993); Sheila Jasanoff, *Science at the Bar: Law, Science, and Technology in America* (Cambridge, Mass.: Harvard University Press, 1995); Ken Alder, "To Tell the Truth: The Polygraph Exam and the Marketing of American Expertise," *Historical Reflections* 24 (1998): 487–525; Kenneth Foster and Peter Huber, *Judging Science: Scientific Knowledge and the Federal Courts* (Cambridge, Mass.: MIT Press, 1999); Ian Burney, *Bodies of Evidence* (Baltimore: Johns Hopkins University Press, 200); Simon Cole, *Suspect Identities: A History of Fingerprinting and Criminal Identification* (Cambridge, Mass.: Harvard University Press, 2001); Michael Lynch, "'Science Above All Else:' The Inversion of Credibility between Forensic DNA Profiling and Fingerprinting Evidence," in Gary Edmon et al., eds., *Expertise in Law and Regulation* (Aldershot: Ashgate, 2004), 121–135; Tal Golan, *Laws of Men and Laws of Nature: The History of Scientific Expert Testimony in England and America* (Cambridge, Mass.: Harvard University Press, 2004); David Faigman, *Laboratory of Justice* (New York: Owl Books, 2005); Bruce Sales and David Shuman, *Experts in Court* (Washington, DC: APA, 2005); Ian Burney, *Poison, Detection and the Victorian Imagination* (Manchester: Manchester University Press, 2006); Michael Lynch

et al., eds., *Truth Machine: The Contentious History of DNA Fingerprinting* (Chicago: University of Chicago Press, 2009).

3. Barbara Shapiro, *Probability and Certainty in Seventeenth Century England* (Princeton: Princeton University Press, 1985); Steven Shapin and Simon Schaffer, *Leviathan and the Air Pump* (Princeton: Princeton University Press, 1985); Julian Martin, *Francis Bacon, the State, and the Reform of Natural Philosophy* (Cambridge: Cambridge University Press, 1992); Barbara Shapiro, *Beyond Reasonable Doubt and Probable Cause* (Berkeley: University of California Press, 1993); Albert van Helden, "Telescopes and Authority from Galileo to Cassini," *Osiris* 9 (1994): 9–29; Rose-Mary Sargent, *The Diffident Naturalist: Robert Boyle and the Philosophy of Experiment* (Chicago: University of Chicago Press, 1995), esp. 42–61; Steven Shapin, *A Social History of Truth* (Chicago: University of Chicago Press, 1995); Matthew Jones, "Writing and *Sentiment*: Blaise Pascal, the Vacuum, and the *Pensees*," *Studies in History and Philosophy of Science* 32 (2001): 139–181; Barbara Shapiro, *A Culture of Fact* (Ithaca: Cornell University Press, 1999). The literatures that have studied early modern "science-in-law" rather than "law-in-science" are almost exclusively focused on the medical profession: Silvia de Renzi, "Witness of the Body: Medico-Legal Cases in Seventeenth-Century Rome," *Studies in History and Philosophy of Science* 33 (2002): 219–242; Silvia de Renzi, "Medical Expertise, Bodies, and the Law in Early Modern Courts," *Isis* 98 (2007): 315–322; Alessandro Pastore, *Il medico in tribunale: La perizia medica nella procedura penale d'antico regime* (Bellinzona: Edizioni Casagrande, 1998); and Gianna Pomata, *Contracting a Cure: Patients, Healers, and the Law in Early Modern Bologna* (Baltimore: The Johns Hopkins University Press, 1998).

4. Johannes Kepler, *Phaenomenon singulare, seu mercurius in sole* (Leipzig: Schurer, 1609), reprinted in Johannes Kepler, *Gesammelte Werke*, ed. Max Caspar and Franz Hammer, 20 vols. (Munich: Beck'sche Verlagsbuchhandlung, 1937) (hereafter *KGW*) vol. IV, 79–98, at p. 92.

5. "Ego M. Martinus Bachazek, qui interfui huic observationi, fateor rem ita se habere." Kepler, *Phaenomenon*, 93.

6. "Heinrich Stolle klein Uhrmacher-Gesell / mein handt." Kepler, *Phaenomenon*, 93.

7. "Auriti testes" are mentioned by Scheiner in his 1612 *Accuratior disquisitio*, his second book on the observation of sunspots. Galileo Galilei, *Le opere di Galileo Galilei*, ed. Antonio Favaro, 20 vols. (Florence: Barbera, 1890–1909) (hereafter *GO*) vol. V, 62. Kepler's double designation of his witnesses as both "spectator" and "testis" was to reappear in his *Narratio*. Langbein claims that hearsay witnesses had no value in criminal trials: "Witnesses shall testify from their own true knowledge, declaring the detailed grounds of their knowledge. When they would testify to hearsay, however, that shall be treated as inadequate." John Langbein, *Prosecuting Crime in the Renaissance* (Cambridge, Mass.: Harvard University Press, 1974), 284.

8. *Constitutio Criminalis Carolina* (1532) as translated in Langbein, *Prosecuting Crime*, 284.

9. The fact that the Jesuit is left nameless throughout the book is somewhat puzzling. One interpretation is that Kepler did not mention the Jesuit by name, because he (the Jesuit) did not want to be named in print as an active collaborator of a Protestant.

10. Kepler, *Phaenomenon*, 92–94.

11. "Testis est *JVSTI BYRGII Minister Automatopoeus, qui spectator fuit.*" Kepler, *Phaenomenon*, 93.

12. Both Kepler's book and handwritten report refer to the images of the solar disk projected through the pinholes as "schemata"—a Latin term that refers to images

as well as sketches—so it's not clear whether the witnesses witnessed the projection of the sun on a piece of paper or a drawing of such a projection.

13. *Constitutio Criminalis Carolina* (1532) as translated in Langbein, *Prosecuting Crime*, 285: "We want the complainant to have his articles that he wishes to prove by witness properly written up and transmitted to the judge, along with the witnesses' names and addresses, in order that thereafter witness-testimony be taken in necessary and appropriate manner by several of the judgment-givers or by other proper delegates."

14. Kepler, *Phaenomenon*, 93–94.

15. Still, both had tried to witness, an attempt that Kepler plays down because, if played up, would have made them look like negative witnesses, that is, people who did not see what Kepler was seeing.

16. Johannes Kepler, *Narratio de observatis a se quatuor Iovis satellibus erronibus* (Frankfurt: Palthenius, 1611), reproduced in *KGW*, vol. IV, 315–325. Though the book carries a publication date of 1611, it was already circulating in the fall of 1610.

17. Johannes Kepler, "Ad lectorem admonitio," in *Dissertatio cun nuncio sidereo* (Prague: Sedesanus, 1610), in *KGW*, vol. IV, 286–287.

18. On why Kepler endorsed Galileo's discovery without being able to replicate them see Mario Biagioli, *Galileo's Instruments of Credit* (Chicago: University of Chicago Press, 2007), 27–44.

19. Kepler, *Narratio*, 318.

20. Kepler, *Narratio*, 320.

21. Kepler, *Narratio*, 319.

22. Kepler, *Narratio*, 319–320, 322 and Albert van Helden "Telescopes and Authority from Galileo to Cassini," *Osiris* 9 (1993): 12.

23. Kepler, *Narratio*, 322.

24. Kepler, *Narratio*, 322.

25. See the minutes of the depositions of Ferdinando Ximenes and Giannozzo Attavanti by the Congregation of the Holy Office in Maurice Finocchiaro, *The Galileo Affair: A Documentary History* (Berkeley: University of California Press, 1989), 141, 143. Both Ximenes and Attavanti are asked whether they know the cause of the summons, and both respond negatively. When Galileo is asked the same question on April 12, 1633, he responds that "I imagine that the reason why I have been ordered to present myself to the Holy Office in Rome is to account for my recently printed book." Interestingly, the inquisitor does not confirm this but asks back "that he explain the character of the book on account of which *he thinks* he was ordered to come to Rome." Ibid., 256–257. Quite likely, the inquisitor's decision not to confirm the cause of the summons is to deny Galileo any information that may guide his response.

26. *Constitutio Criminalis Carolina* (1532) as translated in Langbein, *Prosecuting Crime*, 270.

27. "In the previous articles it is plainly set forth how someone who confesses under torture or threat of torture to an unsolved crime shall be questioned about all the circumstances of the said crime and how on that basis subsequent investigation shall take place, In order thus to get to the truth...that would, however, probably be frustrated when the said circumstances of the crime were previously told to the prisoner upon arrest or examination and he thereupon examined. For that reason we want judges to take precautions against such happening; instead we want nothing to be put to the accused before or during examination other than according to the manner plainly written out in the articles just concluded," *Constitutio Criminalis Carolina* (1532) as translated in Langbein, *Prosecuting Crime*, 282.

28. Kepler to Galileo, August 9, 1610, in *GO*, vol. X, 416: "*In te uno recumbit tota obser-vationis authoritas.*"

29. Kepler was growing so desperate so as to consider using Martinus Horky's vitupera-tive attack on Galileo's discovery of the satellites of Jupiter, simply because (obvi-ously without its author's knowledge) it reported observations for April 24 and 25 that matched (and thus confirmed) those of Galileo for those same days (Kepler to Galileo, August 9, 1610, *GO*, vol. X, 416).

30. Kepler to Galileo, August 9, 1610, *GO*, vol. X, 415: "*Et vero non problema philosophi-cum, sed questio iuridica facti est, an studio Galilaeus orbem deluserint.*"

31. One issue that links the acceptance of matters of fact and the acceptance of phil-osophical arguments is credibility. In the case of matters of fact, Kepler aligns himself with Roman law. A person who makes a statement of fact should be deemed truthful unless the opponent can disprove it. The burden of proof is on the accuser. Translating this principle to Galileo's case, Kepler criticized those who assailed Galileo's credibility without doing any work to prove him wrong.

32. Kepler, *Narratio*, 317.

33. Kepler had foreseen this problem (*Dissertatio*, 290).

34. Max Caspar, *Kepler* (London: Adelard, 1959), 240–258.

35. Kepler, *Narratio*, 317. See similar claims in the *Dissertatio*, 304–305.

36. The use of terms like "secret reasons" (and their casting in opposition to what the "masses" can understand) signals Kepler's sympathy for exclusive, Pythagorean-style notions of knowledge—a trait shared by Copernicus himself.

37. "*Quia haec via iuris est, ut quilibet praesumatur bonus, dum contrarium non probetur,*" Kepler to Galileo, August 9, 1610, *GO*, vol. X, 415. Comparable claims are in *Dissertatio*, 290.

38. "*Quanto magis si circumstantiae fidem fecerint?*" Kepler to Galileo, August 9, 1610, *GO*, vol. X, 415.

39. Kepler, *Dissertatio*, 290.

40. Shapin and Schaffer, *Leviathan and the Air Pump*, discusses Boyle's dismissal of the testimony "of ignorant divers, whom prejudicate opinions may much sway, and whose very sensations, as those of other vulgar men, may be influenced by the predispositions, and so many other circumstances, that they may easily give occa-sion to mistakes" (218).

41. Kepler, *Narratio*, 317.

42. "It was a pleasure, Galileo, to discuss with you in these terms, that is, philosophi-cally, about the new doubts that you have triggered with your observations." Kepler, *Dissertatio*, 310.

43. It also includes what, in the "Defence of Tycho," he calls "astronomical hypoth-eses." These are substantially less hypothetical than the term might suggest, and are, in turn, opposed to mathematical hypotheses. Nicholas Jardine, *The Birth of History and Philosophy of Science: Kepler's Defence of Tycho Against Ursus with Essays on Its Provenance and Significance* (Cambridge: Cambridge University Press, 1984), 211–257; Rhonda Martens, *Kepler's Philosophy and the New Astronomy* (Princeton: Princeton University Press, 2000), 57–68.

44. Kepler, *Dissertatio*, 304–305.

45. Ibid.

46. Kepler, *Dissertatio*, 293, 304–305. This additional credibility has to be "paid back" by the discoverer with the credit s/he owes to those who had predicted it. Kepler owes Plato as much as Columbus owes older geographers. Galileo too owes credit to the Copernicans who have preceded him: "You, Galileo, should not deprive our predecessors of the glory they deserve for this, for having told you that things had to be the way you say you have just discovered with your eyes." *Dissertatio*, in

KGW, vol. IV, 306. Kepler seems to suggest that the credit should in fact be shared between the "discoverer" and the "predictor"—between Plato and Kepler, Ptolemy and Columbus, Kepler and Galileo.

47. Kepler, *Dissertatio*, 292–293. There Kepler gives credit to Porta for having been the first to propose the optical scheme of the Dutch or Galilan telescope and to himself for having been the first to give a qualitative description of how that combination of lenses could produce enlargement.

48. Kepler, *Dissertatio*, "Ad lectorem admonitio," 287.

49. Kepler, *Narratio*, 317.

50. On Galileo's textual and pictorial narrative strategies, see Biagioli, *Galileo's Instruments of Credit*, 77–134.

51. Kepler to Galileo, August 9, 1610, *GO*, vol. X, 415.

52. Kepler, *Narratio*, 318: "*An non ingenua est confessio rerum observatarum, qua credibilium, qua incredibilium.*"

53. Kepler, *Narratio*, 318; Jardine, *The Birth of History and Philosophy of Science*, 140.

54. Kepler, *Narratio*, 317.

55. In the postscript to the *Dissertatio*, Kepler jokes that the catalogue of the last Frankfurt book fair lists a new book by Thomas Gephyrandrus on his success at squaring the circle. Although both Gephyrandrus's claims and Galileo's are unheard of, Kepler finds the latter credible and the former absurd. As he puts it: "This was not an issue of trusting my eyes, and reason shut down my ears as soon as soon as the news arrived."

56. Kepler, *Narratio*, 318.

57. Ibid., 317.

58. Credibility and trust, therefore, seem to be negative constructs that Kepler relates to improbability and partiality rather than full consensus or verisimilitude. Sincere claims need to be somewhat incredible. Full consensus is suspect as it suggests an automatic matching of expectations. A person's credibility is related not so much to good values but to how much that person would have to lose by lying; information helps credibility insofar as it contain some difference (a lot of circumstantial detail might mean that someone is trying too hard to go mimetic). Credibility seems to be a function of the improbability of the claim and of the improbability of the reporter's cheating.

59. In his previous *Defence of Tycho*, Kepler compared the seamlessness of a liar's narrative to the accidental dovetailing of two false hypotheses that happen to yield true consequences but only once and by accident, not reliably and on different occasions. When a liar's narrative seems to fit the facts, it is by accident. Its smoothness is precisely a mark of it being false.

60. James Voelkel, "Publish or Perish: Legal Contingencies and the Publication of Kepler's Astronomia Nova," *Science in Context* 12 (1999): 33–59.

61. Shapin and Schaffer, *Leviathan and the Air Pump*, 64–69.

62. Kepler, *Narratio*, 318.

63. Steven Shapin, *A Social History of Truth* (Chicago: University of Chicago Press, 1995), 107–119.

64. Acknowledging problems in the observational context is not a way to avoid giving someone the lie but rather to recognize the truth of the fact. This may perhaps be connected to Kepler's openness to using witnesses from different social and religious backgrounds (as shown in the *Phenomenon singulare*). If a phenomenon was witnessed and drew consensus only by one specific socioreligious group, then that may have made it look like a setup. It's not that he trusts everybody, but rather that he would not trust homogeneity of consensus.

65. Kepler, *Narratio*, 318.

66. Kepler, *Dissertatio*, 305.
67. Kepler, *Narratio*, 317.
68. Symmetrically, a philosophical prediction cannot count as a priori proof of the existence of an object whose discovery it did not predict with precision, but it can nevertheless lend credibility to that discovery by providing arguments about its plausible existence as well as by showing that, by the very fact of its differing from the prediction, it does not bear the mark of a forgery. Kepler's logic differentiates between partiality and imperfection.
69. My point is based on the logic of Kepler's arguments rather than on the terms he uses, but it is perhaps telling that he does use the term "prophet" to refer to his friend Johannes Pistorius's prediction that somebody would have come along with a device that could have improved on the accuracy of Tycho Brahe's observations: *"At nunc demum video verun in parte vatem fuisse Pistorium,"* *Dissertatio*, 295. Also in the *Dissertatio*, after referring to Plato's narrative of Atlantis as a "fable," Kepler presents Seneca's poetic prediction of the discovery of a new world as *"versiculos fatidicos."* Ibid., 304.

6

Improvement for Profit: Calculating Machines and the Prehistory of Intellectual Property

Matthew L. Jones

In early 1675 Gottfried W. Leibniz drew up terms for the French Académie des Sciences concerning his "reasonable" compensation from the French crown upon delivery of well-functioning calculating machines. He would first receive a privilege "such as I can reasonably request," one limited neither by number of years nor any "other reserve"—a bold demand. By virtue of this privilege, and by his contract with the minister Jean-Baptiste Colbert and the Académie, no one would get "complete or partial machines except from me or my designees," at a price set by him. Setting the price of the machine, he wrote, involved two major considerations: first, his past and future expenses, and second, the "reasonable advantage that he could expect [*esperer*] from an invention as considerable and difficult" as the calculating machine.

While quantifying expenses proved relatively straightforward, quantifying "reasonable advantage" was trickier. Inventing the machine, Leibniz explained, "occupied and will continue to occupy me almost entirely for some time"; it will thereby "prevent me from profiting from other opportunities." The crown needed to pay Leibniz's opportunity cost. Using the language of early modern contract law and a dash of the emerging probability theory, Leibniz set out the just compensation given the risks taken and to be taken. Lest the reader forget that reasonable advantage was a legal concept, Leibniz wrote, justice demands recognizing "the risk [*hazard*] inventors expose themselves to, in advancing costs at their own expense, and in putting their reputation in jeopardy." Reasonable advantage also had to incorporate novelty: "For embellishments and curiosities for cabinets, novelty and rarity are paid for, as is seen everyday with the examples of pictures, prints, drawings and medals, [all of] which are but dead beauties lacking action and effect."[1] Perfecting and constructing a calculating machine would monopolize an innovative artisan and his atelier, just as it would monopolize its inventor's time and energy. Leibniz seemed set to stride the profitable world of contracting for the state alongside the honorable world of the institution of the Académie des Sciences. On top of the costs associated with someone of his skills and abilities abstaining from his other opportunities, Leibniz sought to monetize the risk to his reputation—to quantify risk in the nonfinancial sphere of international and local honor or glory. Leibniz's concerns about the danger to his reputation were well founded, for the calculating-machine project quickly tarnished it, in France and England alike.

In making his case, Leibniz defended the novelty and distinctiveness of his calculators compared to the famous machines of Blaise Pascal, built in the 1640s, and those of Samuel Morland, built in the 1660s. This essay studies the calculating machines of Pascal and Leibniz—and more briefly of Charles Babbage—within early modern systems for protecting and encouraging manufactures and, indirectly, invention. The calculating machines were products of an early modern protocapitalism and natural philosophy joined to the subcontracting world that comprised much governance in early modern absolutism.[2] Pascal and Leibniz sought to make the most advanced natural philosophical and artisanal knowledge of the day pay off in practical applications for state and market alike. They were philosophical entrepreneurs who sought to be subcontractors and princely sanctioned monopoly vendors of machines and processes. In contrast to many elite practitioners in the sciences before and after their time, Pascal and Leibniz cast the quest for monetary gain as complementary to natural philosophical and technical achievement, and capable of spurring it. Leibniz explained the possibility of unifying personal gain and charity in a letter to his patron, Johann Friedrich of Hanover: "He is who is happy enough to establish his fortune by advancing the public utility, can unite charity with prudence" (A 1,2: 154).

Early modern calculating machines were initially designed to aid calculation in early modern governance and astronomy. The production of these "philosophical" machines was parasitic on artisanal skill and knowledge; the legal protections afforded philosophical machines were likewise parasitic on legal devices tasked to support artisanal, not intellective, activities—on legal devices produced in no small part to perform industrial espionage by rewarding the movement of artisans, their techniques, and their organization of work into new jurisdictions. Even in 1834, Charles Babbage coupled his lofty talk of natural rights to his invention of the difference engine to a philosopher's version of the traditional artisanal threat of transferring technology to foreign lands. Before inventors got patents, artisans did. Before speculative designs for machines became intellectual property, the actual procedures of production of machines were protected through royal and princely monopolies as well as numerous other forms of preferment.

Creating and legally protecting the modern romantic author meant effacing the craft dimensions of writing in favor of privileging the inspired mind and protecting its written products.[3] In the case of machines, creating and legally protecting the philosophic inventor meant effacing the craft dimensions and the managerial practices traditionally protected by privileges in favor of protecting the ideational designs of philosopher-inventors. "Intellectual" property in the form of patents came by the wayside, a side effect of extending the temporary monopolies protecting manufactures to ever more philosophical instruments and their makers. Denying the value of artisanal insight and labor to the conception of a machine—to its essence—was tantamount to denying the artisans' contribution to that conception and thus their ownership in it; it was to confect a legal and philosophical divide between mere manufacture and creative invention belied by actual processes of innovative making.[4] Countering this denial, however, often entailed an implicit concession that a machine and the process of producing it *could* be understood as having a mentalistic essence independent of that entire process. In defending their "authorship" of machines, artisans and other inventors contributed to an understanding of them that ultimately excluded their entire range of labors and promoted an understanding of property exclusive of their range of competencies. Codifying the productions of philosophical inventors within the system of privileges helped make invention more intellectualized; so too did defending against such codification. The proliferation of new written and visual techniques within the legal and bureaucratic sphere reified this codification.[5]

Calculating machines were not important commercial commodities—like pins, stockings, china, or watches—until the late nineteenth century. The substantial existing documentation about efforts to monopolize the machines permits us to see the process of gaining privileges and patents in unusual detail generally lost from the historical record of more central commodities. The philosophical preoccupations of their makers illuminate some key early moments of the genealogical and contingent history through which it became possible to envision machines in mentalistic terms and to create legal regimes of property protecting such an intellectualist understanding. The clash of interests and jurisdictions, of regimes of glory and of money, within absolutism offered a matrix for the contingent production of mentalistic conceptions of machines and legal techniques for protecting them.[6] The history of calculating machines and likeminded projects lets us glimpse absolutist governance and its limits, in action and inaction, and its dependence on skilled people coupled to its pretensions to near omnipotence and independence.

Pascal: The classic misleading example of seventeenth-century "IP"

In drawing up his terms for the Académie des Sciences and Colbert in 1675, Leibniz referred to language in a printed pamphlet of Blaise Pascal requesting royal protection for his calculating machine. In a request addressed to Chancellor Séguier, Pascal called for a privilege, "far from ordinary" that would "suffocate, before their birth, all these illegitimate abortions that could be engendered otherwise than by the legitimate and necessary alliance of theory and art" (JMII: 340). The privilege, granted in 1649, gave him a monopoly on the production of calculating machines in all the realms controlled by the king of France for an unlimited length of time.[7] More unusually, the privilege offered support for continued development of the machine independent of any demand that the machine be perfected in short order and brought into regular manufacture. Unlike the vast number of privileges of the time, Pascal's seems to have offered him something like a patent protecting an idea. More precisely, the privilege covered all possible machines with any mechanism and material that perform arithmetic with automatic carry.

The awarded privilege remarks that Pascal had made "more than fifty models" with various mechanisms, sorts of motions, and materials.[8] In all of the "different manners the principal invention and essential movement consists in that each wheel or rod [*verge*] of a numerical order [*ordre*] when it makes a movement of ten arithmetical digits, makes the next one move one digit only" (JMII: 713). The heart of the privilege, the invention underlying all the distinct mechanisms, is Pascal's isolation of the key problems of carrying tens (the sufficient force problem and the keeping-it-digital problems).[9] More precisely, the privilege appears to cover the *goal* of automatically performing carries, *and not any particular mechanism for doing so.*

Commentators have seen in Pascal's privilege early glimmers of a necessarily unfolding patent system offering mentalistic or intellectual property to those willing to specify their inventions to the public: "Here we have the vigorous beginnings of specification-writing. The object of the machine is fully suggested; next comes an outline of the variants of the basic model; and most importantly, the recitals end with a clear, generic definition of the invention. This definition appears as a forerunner to modern claims. It was drawn to point out the gist of the invention."[10] In fact, the privilege Pascal requested and received was profoundly *atypical*. To see more precisely how Pascal's privilege was "far from ordinary" requires understanding the difference between modern patent regimes and the system of protecting inventions with privileges. We need to explain the conditions that made the unusual, "modern" qualities of Pascal's privilege *possible*, to contribute to a nonteleological account of the contingent development of the resources necessary for the creation of modern patent systems.

Early modern polities did not have patent systems of the sort known since the nineteenth century. Their closest equivalent—the system of privilege—was not about protecting intellectual property in the modern sense.[11] Privileges covered the introduction of trade or art with government protection; they *could* involve some major technological innovation but did not have to. In contrast, patents provide temporary monopoly rights over principles of a technological invention, as embodied in a specification and/or model disclosing those principles; they *can* lead to an introduction of trade or art but do not have to. In Italy, France, and England, the system of privileges emerged in the late middle ages and Renaissance out of "measures for recognizing and rewarding craftsmen's skills," and in particular for encouraging the transmission of technical "know-how" from one territory to another.

Venice first institutionalized such a model in 1474; France and England copied many of the features of such a system.[12] These privileges involved no
innovative legal doctrines about "intellectual property"; they were the stuff
of royal and princely governance involving "gifts" of economic and political
concessions of all sorts to all kinds of people and corporations. Essential to
the toolkit of late medieval and early modern governance, such privileges
were then applied to craftspeople, their knowledge, their organizational procedures, and their skills.

Early modern privileges for inventions generally served to aid technology
transfer into a territory or to serve as gifts for favored courtiers and bureau-
crats. Modern patents cover some key component mechanism of an invention,
always embodied in some way (not quite an idea, though the law, particularly
in the United States, has been moving ever closer to making ideas and natural laws patentable). Early modern privileges protected not ideas or abstract
designs, but processes of manufacture, and provided for regulation of labor,
religious exceptions, payments, and naturalizations.[13]

Novelty requirements for privileges concerned novelty within a given territory, not absolute, global innovation.[14] Written description and public disclosure of the "essence" of an invention or process were generally not required,
whereas a demonstrable, and quick, "reduction to practice" of an invention
or process was crucial. Failing to produce a working device or process in a
short period of time—typically as part of an entire process of production—
invalidated or nullified most privileges.[15]

In his 1645 pamphlet requesting royal protection, Pascal maintained that
his machine had been reduced to practice and that it was both robust and
accurate (JMII: 340). The royal privilege subsequently granted to Pascal
denied his claims; the machine had not been reduced to practice in a meaningful way:

> And since the aforementioned instrument is now at an excessive
> price... [and is] therefore useless to the public,... and so that it might come
> into regular use, all of which he intends to do through the invention of
> a simpler mechanism,... he works continually in search of such a mecha-
> nism, and in training little by little workers still too little habituated to it,
> which things depend on a time that cannot be limited. (JMII: 713)

The privilege suggests that Pascal could *make* machines, as one-off luxury
items, priced for collectors, but he could not yet *manufacture* them in a standardized way as commodities at a price to make them of more general use. As
a rule, privileges did not protect speculative ideas of projects to be worked
out and then realized at some future time; they protected the manufacture of
particular objects or processes already reduced, or soon to be reduced, to practice. Royal privileges often recognized the exceptions and emoluments necessary to obtain and regular a diversely skilled workforce; they often called for
the continuing perfection of new manufacturing processes; but they were
rarely issued, if at all, for processes not yet even devised. Pascal had neither

the secret of a simpler version of the machine nor an organized work-process necessary to produce them without difficultly.

Pascal's awarded privilege explicitly states that the king's provision of incentives for Pascal to bring the machine to a practical form of perfection is a gift aimed at "exciting him to communicate more and more the fruits of [his capacities] to our subjects." The gift serves further to encourage Pascal to continue to innovate and to share the benefits to be accrued by Louis and his subjects from his mathematical and natural philosophical skills. The privilege is a private economic gift that will serve to support Pascal in his role as a philosopher-engineer working for glory and profit alike; he in time will publicize his discoveries and innovations for the public good. A logic of theoretical discovery and publicity—a logic central to modern IP—intrudes into the logic of the privilege.

The royal gift to Pascal went even further. The logic of the privilege involved no "intellectual property," only the protection of the manufactures.[16] Pascal's privilege endorsed a fundamental and perhaps incommensurable injection of a concept of invention and the inventor's mind into the logic of the privilege. Protecting manufacturers against counterfeiting was a central function for French privileges for inventions; such patents often only covered a limited jurisdiction within France.[17] Rather than simply stealing Pascal's property, counterfeiters preclude the possibility of developing the machine into something more than an expensive curiosity. Even as it chides him for his failure to bring the machines into regular manufacture, the granted privilege accepts Pascal's account of the nature of invention and the successful manufacture of an invention as essentially philosophical and mental, something requiring "a total comprehension [*entière intelligence*] of the artifice of its movement." Making the machine practical required, the privilege argued, an unusual grant, based on an acceptance of Pascal's polemical account of invention and artisanal skill (JMII: 713).

According to Pascal, the success of artisans in making copies of extraordinary machines leads them to believe that they possess genuine creative ability and the theoretical knowledge necessary to guide their skills. His own inabilities illuminated the inabilities of the artisans:

> It is not in my power, . . . to execute myself my own design without the aid of a worker . . . it is equally absolutely impossible to all simple artisans, no matter how skilled in their art, to put a new device into perfection, when that new piece has complicated movements, . . . without the aid of someone who gives them the measure and proportion of all the pieces . . . using the rules of theory. (JMII: 338–339)

Although Pascal cannot physically manufacture something based on a fully specified design, he can specify a design of a complex machine using theory and knowledge of the properties of materials. The misfit of theoretical designs with the material world informs the perfecting of technical designs, but only the philosophical engineer, not the artisan, appears in Pascal's account to be able to recognize, compensate, and overcome such misfit by reference to theory. Insofar as artisanal skill reveals limitations (or even useful properties of different materials), Pascal (or another engineer-philosopher like him) recognizes those limitations or useful properties and adapts his design accordingly. The work of physically producing the machine is collaborative; the work of conceiving it is not. Pascal's privilege affirms the gulf between mere manufacture and creative ingenuity.[18]

In his creation narrative, Pascal relates how hard he worked in moving, always with the guidance of theory, from his first "imagination" of the machine to his various designs. Artisans have no vision of the whole, but believe they can create: artisans "work through groping trial and error, that is, without certain measures and proportions regulated by art." They "produce nothing corresponding to what they had sought, or, what's more, they make a little monster appear, that lacks its principal limbs, the others being deformed, lacking any proportion" (JMII: 338). In the Aristotelian and Horatian category Pascal invokes, a monster is precisely a material thing lacking a unifying form. Savants like Pascal can regulate themselves with theory in making their trials; artisans need to subordinate themselves to a savant to do so. Even as they experiment with new designs, savants remain tethered by theory and art that maintain the *unity* of the design; artisans acting alone merely modify pieces willy-nilly without regard to the whole. Savants ensure that a unified ideational essence undergirds and makes possible a mechanical unification; they ensure that the matter could possibly, in the right conditions of production, embody some unifying form. This account bifurcated form and matter, inventor and implementer, inspiration and implementation in ways foreign to actual early modern manufacture and the legal systems organized around it. The account tears apart the amalgam of form and matter of early modern making and makes it possible to imagine an ideational conception of machine independent of any particular instantiation of it, and an inventor with just such an ideational understanding—an *intelligence entière*.

Pascal depicted a normative hierarchy of invention and production that did not exist as part of an attempt to secure something like that hierarchy in practice.[19] According to Pascal, savants can manage themselves and others to direct the production of new designs of unified machines, as well as the improvement of current designs; artisans cannot regulate themselves to produce unified machines autonomously.[20] Artisanal skills are produced through repetitive practice of actions under the direction of savants possessing theory and are nothing but the making habitual of that practice.[21] However much artisans attempt to conceive, they can only misconceive. When they try to innovate, they upset the epistemic, technical, and social order, and, in so doing, produce only monsters or abortions that are nonfunctional disunities. Their efforts at creation tarnish the reputation of real innovators and preclude reduction to practice.

In his important survey of the development of intellectual property in early modern Europe, Carlo Marco Belfanti argues that the privilege was "a tried and tested instrument" taken from the "institutional 'kit'" available to early modern polities. Privileges were "solely intended to reach a concrete economic policy objective," not to provide "an explicit safeguard for intellectual property." And yet, as Pamela Long has stressed, privileges suggested that craft knowledge—know-how—was a form of intangible property; accordingly privileges were remodeled over time to protect artisanal inventors.[22] In accepting Pascal's account of labor, knowledge, and skill, the privilege for the calculating machine draws upon the implicit concession of intangible property in their skills to artisans and abstracts it from the realm of actual production and tacit how-to, to protect a more mentalistic account of inventive activity. Not quite expressing the concept of a truly intangible idea that is owned, the privilege grants him ownership in all possible machines—all possible expressions—incarnating his essential breakthrough. Pascal's privilege, highly unusual for 1649, allows a glimpse into the process by which the monopoly protection of actual processes of production and the how-to knowledge involved in that protection could be transformed, under certain conditions, into the protection of abstract (but not necessarily functional) designs produced by an intellective author.

How did a logic of an intellectual inventor and noncreative artisans, of a machine conceptualized mentalistically, of support for a project far from reduction to practice, come to figure in a legal document produced within a privilege system in which ideas of machines in inventor's minds had no place? Pascal's privilege declares itself to be a royal gift to a favored client, or rather, a gift to the son of a favored client.[23] Whereas modern patents are rights of citizens, something the government is obliged to issue and protect for those meeting the appropriate criteria, early modern privileges were legally gifts freely presented to preferred subjects or groups of subjects.[24] Although all privileges were legally undeserved and unearned gifts, more generous privileges with long durations or peculiar clauses, such as Pascal's, tended to go to favored clients such as Pascal and his family.[25] The unusual qualities of Pascal's privilege became possible because of his connections at the highest levels of the royal government. However atypical Pascal's privilege, his route to it followed a path well rutted by early modern clients.

Specifying and contesting absolutism: A hypothesis

Enforcing privileges often required the coercive powers of the state, especially when those privileges infringed on the traditional activities and rights of others. As royal gifts, grants of privilege and patents often provoked protest and ire—sometimes leading to lèse-majesté and violence, sometimes held to legitimize rebellion against the abuse of royal authority.[26] In February 1664, "Simon Urlin, at a meeting of wire drawers summoned by him to oppose Mr. Garill's patent, said in passion that the last King lost his head by granting such" patents.[27] Such views of the causes of the English revolution testified to dangers of the untrammeled use of royal prerogative. The wiredrawers were ultimately successful in blocking the patent. In France, local judicial bodies such as the *Parlement* of Paris had to approve privileges; they nearly always modified their terms. On occasion, they rejected them outright. These judicial bodies often also restricted the granting of privileges to court favorites who had no real innovation or locally new process.[28] In 1621, for example, the master baker Denis Mequignon received a royal privilege lasting ten years for a new sort of mill; in registering and approving the grant the *Parlement* reduced the duration to "five years only" as well as limiting the price to 50 sols within its jurisdiction.[29] The *Parlement* of Paris approved the proposal for the Paris bus system with two provisos: first, that there be only a single price, and not one prorated by distance traveled, and second, that soldiers and liveried servants be excluded.[30] In its customary way, the *Parlement* checked the power of the crown to grant privileges by stressing that extant privileges and liberties were not to be infringed. Such changes and reductions served as a daily reminder of the real limits of royal authority. The crown was likewise prone not to upset existing rights. When Christiaan Huygens sought a privilege for his pendulum clock in 1658, Pascal's patron Chancellor Séguier refused three times because he "did not want all the master clockmakers of Paris crying after him." No matter how grounded "in reason" Huygens' appeal, "these difficulties and obstacles" precluded the chancellor from exercising his grace and freely granting the privilege.[31] At the time, Huygens wasn't worth the trouble.

The continual contest over privileges was part and parcel of the quotidian jostling to retain and to gain control over aspects of governance more generally by crown, representative bodies, and various evaluative bodies. In France and England alike, the crown's claimed prerogative to issue patents, privileges, and monopolies was constantly challenged. Early modern princes with pretensions to absolute power liked to present themselves as offering "gifts" freely, unconstrained by obligation, just as their propagandists presented them as able to rule, in principle at least, unconstrained by legal traditions. Early modern monarchs were caught in worlds of traditional obligations and legal constraints, which they could modify only with some difficulty and sometimes only through violent coercion. Royal privileges extended and ratified royal authority; resistance to them checked that authority in the name of traditional prerogatives, local sovereignties, and, more rarely, rights.

Scholars writing teleological histories of patenting have found it easy to find evidence for a Whiggish narrative; they have readily found—or, rather, cherry picked—examples of global novelty, specification, immateriality and intellective invention, and appeals to right. Revisionist historiography downplays these Whiggish examples. We need to go further, to explain the sizeable production of such examples. The dispersed empirical stuff of the teleological account is evidence neither of a hidden inherent "substance" of a modern patent "regime" in the process of unfolding nor of mere accidents within early modern privilege "regime" to be disregarded. Rather, the dispersed evidence of teleological accounts was systematically generated by the clash of interests constitutive of actual governance in early modern sovereign states and from the weakness of crowns invested in portraying their states as strongly centralized and unified around a single power. The modern patent regimes of the rights-oriented states of the late eighteenth century drew upon resources produced within the clash of legal regimes within sovereign states and reinterpreted then within a doctrine of rights.

At the center of the intellectualization of modern patents systems is the specification, a written document, that in principle should enable those with "ordinary skill in the art" to replicate a disclosed invention. Such specifications played no part in the early modern privilege system. The requirements for writing specifications did much to put the "intellectual" into intellectual property.[32] Before the late eighteenth century, written specifications were not required by law or by bureaucracies as a standard procedure to receive a privilege or patent. New forms of specification emerged out of attempts to combine the different sets of logics at play in practically lived absolutism. Two are apparent in the history of calculating machines: first the interplay of economies of glory and of lucre, and second, the interplay of crown prerogative and the defense of privileges and rights already granted.

First, the attempts at synthesizing the international "glory" (symbolic credit) of philosophers with financial credit required a rejection of local, territorial novelty in favor of temporally defined, global novelty. Isolating this global novelty could take a variety of forms: among the most obvious was the written articulation of some mentalistic form held to be essential to the invention. More important than the ability to capture such an essence accurately, through new forms of technical description or drawing, is the belief in such an essence independent of any particular material instantiation. Pascal's privilege isolates an essential ideational core to be protected just as it denies that that the machine actually has been reduced to practice. As we will see, Leibniz had to demonstrate his novelty through written or ostensive specification.

Second, ad hoc forms of specification became more central as a means for protecting the liberties and rights of others to whom privileges were already granted (or understood to have them by custom).[33] Such specification happened largely under duress, as the sovereign's prerogative was checked by extant privileges and their associated rights and liberties, either by representative and judicial bodies or by groups agitating publics or complaining to authorities. Such contestation was as present in the more absolutist France or German principalities as in more representative Britain.[34]

The rights-based patent regime drew upon atavistic products and practices of the tensions within absolutism and centralizing states. The "intellectualization" of patents and privilege stemmed from many sources, which were only later crystallized into formal bureaucratic and legal features.

Leibniz: Protocols of glory, protocols of financial credit

On the back of an undated autograph document entitled "Things to be fixed in the [arithmetical] machines," concerned mostly with carrying mechanisms, Leibniz set out a list of things necessary in order for "the machine to be put into use"—to be brought into practice. His list included what we might call his business model. Like savvy cosmopolitan artisans of his day who possessed valuable techniques, he set out a plan to get privileges from "many republics" across Europe before securing them from the Holy Roman Emperor and the king of France. He outlined the various groups that would buy such a machine: universities and academies, merchants, as well as collectors of curiosities. Leibniz ends his to-do list by noting that "the King"—clearly Louis XIV—"can make it become fashionable."[35] Leibniz foresaw not only symbolic and machine reasoning but also the marketing strategies of a certain modish Cupertino computer concern.

As so often in early modern governance, getting a privilege involved the arduous, personal cultivation of powerful patrons. Leibniz had to cultivate personal connections to powerful people around kings and princes who alone could grant privileges. Leibniz appears to have focused on France, rather than the smaller states, even before his trip to Paris. A correspondent suggested Leibniz contact a key intermediary to Colbert: "M. de Carcavy is a person whom you ought well to cultivate, for he is all powerful around Mgr. Colbert in everything concerning letters."[36] Leibniz sent the mathematician and librarian Pierre de Carcavy news of his calculating machine and several printed works concerning natural philosophy and other instruments he claimed to have invented.

In December of 1671 Leibniz received a letter at once admonishing and encouraging from Carcavy. He advised Leibniz not to send so many unclear and half-baked schemes and proposals to the Académie. Colbert "is satisfied only with what is real and solid," so Carcavy would present something to him only once Leibniz "had begun to send something effective to present to" the minister. Despite his reservations, Carcavy remained interested in Leibniz's plan for a new calculating machine and explained what he knew about Pascal's. Carcavy explained the protocols for evaluating and possibly rewarding Leibniz. If "you wish to send me something worthy of being seen," Leibniz could be "assured about three things" involving the protocols of considering and rewarding invention and new techniques. The suggestions offer a glimpse of the unwritten informal protocols around invention in Colbert and Louis XIV's France.[37] First, Carcavy promised that Leibniz need not fear his invention would be stolen: "no one here will usurp what another has done"; he "pledged to conserve all the glory to whom it is owed." Carcavy carefully avoided granting that Leibniz had in fact something new; he simply maintained that credit would be fairly apportioned and that Leibniz would receive his due if in fact he deserved any. Second, Leibniz could set the conditions for the use and dissemination of any machines or descriptions of them he imparted to Carcavy: "I will absolutely use whatever you send me only as you proscribe." Third, appropriate financial credit would be granted to those worthy: "I will procure from it, for you and for those deserving, the reasonable advantage necessary."[38] Though reason and justice need not constrain the crown, Carcavy assured Leibniz they would. Finally, Carcavy intoned about the dangers of the "amour-propre" of "authors" who overestimate the novelty of their inventions. However skeptical Carcavy may have been about claims to novelty, he explicitly treated inventions as something produced by *authors*, not something as part of an entire manufacture, that is, something produced by skilled artisans.

The clarity of these terms shows Carcavy's gatekeeping function on Colbert's behalf.[39] Previously charged with bringing Huygens to the Académie, he worked to protect Colbert from mere projectors with empty schemes and also to assure inventors, artisans, and savants that their projects, glory, and economic interests would be protected. He needed to insulate Colbert from scams while recognizing and encouraging useful inventions and techniques. Credulity was dangerous; so was too much suspicion. Before any privileges, money, or other gifts were awarded, Carcavy offered a set of guarantees to inventors with potentially important projects. In outlining his protocols for protecting the invention and the distribution of economic and symbolic credit (cash and "glory"), Carcavy was careful to assure Leibniz that he would receive his due, just as others would receive theirs—but nothing more than what they deserved. According to its own self-representation and the logic of the privilege, the crown had no requirement to follow abstract rules of justice in rewarding inventors, but Carcavy assured them that the crown would do so. The protocols Carcavy set forth probably were improvised adaptations of standard procedures for finding and luring innovative craftspeople to France as a matter of economic policy. Adapting these procedures for philosophers and engineers working in realms of glory—international reputation—loosened the locality and materiality central to the privilege system.

Leibniz trusted Carcavy to keep his secrets and to apportion glory and money justly.[40] As Leibniz himself often stressed in his economic and political writings, Colbert had made the granting of privileges to foreign artisans and their workshops a centerpiece of his policy.[41] A privilege for royal manufacture was not so much the protection of intellectual property as a grant of a range of rights associated with a manufacture. These policies applied above all to artisans with technologies thought useful to the crown but also, in modified ways, to foreign savants such as the Protestant star of the Académie, Christiaan Huygens. Like Huygens, Leibniz was to be rewarded by lightly adapting practices designed to entice and keep skilled artisans in France for philosopher-engineer-mathematicians. Carcavy managed much of such recruitment for Colbert.

Since Leibniz's earlier proposed machines to the Académie were seen as vague projections of possible machines, Carcavy pushed him to disclose more about his devices or to send them, so that Carcavy and the Académie could judge whether they could be useful and whether they were, in fact, innovative. Determining how to apportion credit required an account of novelty. For Carcavy, this meant devising some means for comparing Pascal and Leibniz's machines. Carcavy explained that Pascal "had not provided a particular description of his numerical machine," but he offered to have a detailed specification of the device written up—"a more ample description." Alternatively, Leibniz could send an exemplar of his machine: "If you want to send me yours with the manner of working it, I will tell you what is the same and what different."[42] Carcavy's suspicions were well founded. He rightly surmised that Leibniz's machine existed more as an aspiration than as a real device, or even a concrete design for one.

Global novelty was no necessity for most early modern privileges. Carcavy nevertheless explained that Leibniz needed to be providing something globally new and useful to the crown. Carcavy was working in two different economies of credit: that of philosophy reputation ("glory") and that of money ("reasonable advantage"). Had Leibniz been concerned exclusively with monopoly protection and the creation of a manufactory, Carcavy probably would have only required local novelty; but because Leibniz was looking for the international glory that would follow a strong approbation from the Académie des Sciences, he was subject to a stricter requirement of global novelty.

Proving global novelty in a written description promoted a mentalistic conception of inventions as possessing some essence. Among Leibniz's manuscripts from the Parisian period is an autograph assessing his machine, written in the third person. It is likely a fragment of a report Leibniz prepared on someone else's behalf—likely Carcavy or Colbert's—to justify granting him a payment, a pension, a privilege, or some other preferment. In comparing his machine with those of Pascal and Morland, he carefully assigned credit while suggesting the faults of the competing machines.

The beauty and ingenuity of Pascal's machine cannot be dismissed, Leibniz explained. Like almost all serious critics of Pascal's machine, Leibniz noted that it could only be used right to left, so that subtracting could not be done directly. Leibniz likely saw Morland's machine at Whitehall during his visit to London and possessed a copy of Morland's book concerning the machine.[43] Although Morland's machine failed to perform carries, Leibniz noted, it could be used in either direction, unlike Pascal's machine. Morland's machine was therefore distinct enough that there could be no question of the independence of his invention: "We can conclude based on these differences that Mr. Morland is the inventor of his machine without owing to M. Pascal the idea and still less the execution."[44] Leibniz carefully assessed credit for the invention of the essence of various types of calculating machines and their constituent elements. Leibniz claimed, furthermore, that Morland had made clear that he would readily accede glory to Leibniz for his kind of calculating machine, if what he had heard about it were true.[45] Morland was famous across Europe, above all for his speaking tube; not only was he a credible witness to Leibniz's innovations but he could also easily afford to grant Leibniz credit. Leibniz sketched out the ideational essence of the different machines and then used this to partition credit. These distinctions in the realm of glory served to prove global novelty and to justify his demands for a privilege. He would claim for years afterward that both the Royal Society and the Académie des Sciences accepted the "infinite difference" between his machine and others' machines, even if both institutions demanded a reduction to practice.[46]

Much as he was never offered a full, pensioned position in the Académie, Leibniz never received a privilege for his machine. Pascal managed to get a privilege for work that still needed to be perfected to be put into practice. Leibniz did not, despite this precedent. Based on his models and drawings, he received preliminary orders from Colbert for machines for the crown, for the Observatory, and for Colbert himself, contingent upon bringing the machine to practice. Colbert also offered support in the form of payment for the artisan Ollivier to bring the machine into practice. In the documents Leibniz wrote up for Colbert and the Académie, he outlined the package of incentives necessary to motivate a skilled clockmaker to abandon his trade, to concentrate exclusively on perfecting, and then building the machines. Skilled artisans who were willing to innovate needed incentives to give up their profitable accustomed ways: "It is just to pay a skilled master not only for the time he has worked—without speaking of the novelty and the risk of the enterprise—but also his industry and his skill [*adresse*] in discerning himself from an ignorant." The payment of artisans must register the higher creative abilities of the superior sort of artisan. His artisan Ollivier "protests that he prefers to make his living easily in the ordinary way, rather than to embark for nothing in an enterprise full of disquiet and risk, and capable of turning off the most patient man in the world."[47] Leibniz insisted on the importance of such superior artisans for the development of technique and of economy: "An artisan who knows nothing of Latin or Euclid, when he is a skilled man [*habile homme*], and knows the reasons for what he does, he truly has the theory of his art, and is capable of funding expedients for all sorts of events [*rencontres*]."[48] In drafting his brief to Colbert, Leibniz first wrote that he needed a "obedient" artisan, before changing that to a "handy" [*commode*] workman.Leibniz's brief evidently worked—at least temporarily. The accounts of the "Batiments du Roi" record that on December 15, 1674, a "Sieur Ollivier, clockmaker" received 300 livres "in consideration for a numerical machine he has made."[49]

A singular aspect of the briefs is that Leibniz translated arguments about motivating artisans into arguments about philosophers such as himself. Like Ollivier, Leibniz needed incentives to focus exclusively on machines. The first version of his briefs set out Ollivier's arguments about his just compensation; subsequent versions transmuted many of these arguments into the claims about Leibniz's just compensation with which this essay began. Leibniz's manuscripts illustrate how arguments about rewarding the industry of inventive artisans were adapted to justify the rewards for philosophical invention—an improvisational reworking of ideas of proper recompense and incentives.

In his narrative, Leibniz explained that his project was nearing fruition, and that he was moving from active invention to simply directing "the execution of this work."[50] In this claim—soon to be falsified—the ideational work of invention was complete; all that remained was materialization. He claimed to have already reduced his proof-of-concept model to practice—a claim that appears often in his correspondence. Early modern inventors often were given only small windows, often six months, to provide working versions of their model instruments. Leibniz attempted to build more time—and money—for reducing his model to practice into his contracts. Even though it was generous by early modern standards, the extra time Colbert and the Académie gave him proved not nearly enough. At no stage was the model as perfected as Leibniz boasted. His patrons and supporters in France quickly grew irritated, as did others such as Henry Oldenburg and Robert Hooke.

In late October 1675, Leibniz received a summons from the privilege broker Dalencé to the house of the Duc de Chevreuse in Saint Germain the next day. "Given that you have taken the trouble to tell me that the *machine* is *all ready*," Leibniz was "to *bring*" the machine "as it was before one began this fourth wheel; I beseech you *not to fail* to come coming to my place tomorrow at one hour exactly after noon and *to bring the machine*" to take to the Duc's house.[51] The emphasized words, in the original, strongly suggest appointments missed and promises not kept. Leibniz failed to show. A few days later, he lamely wrote through an intermediary that an "indisposition" prevented him from making the appointment, and he dared not write the Duc directly.[52] Leibniz may have presented a model of the machine to Colbert at Saint Germain in late 1675.[53] Despite the support of the Duc de Chevreuse and others, Leibniz did not receive a pensioned position in the Académie des Sciences, and could not remain in France. The failure of a timely reduction to practice likely contributed to undermining Leibniz's candidacy for a rare permanent—and pensioned—position in the French Académie.[54]

Given these failures, Leibniz's overconfident language of the just recompense due someone who contributes to the glory of the crown and the common good disappeared. By the time he wrote to Colbert in a tone of some desperation in January 1676, Leibniz no longer drew on a language of just compensation; he wrote as a submissive and unworthy client begging for any recognition at all. "Some time ago I took the liberty of presenting you a *placet*. It is true that I demanded nothing positive, as my pretensions were founded only on the good will that You could have for me after what you have publicly witnessed for the advancement of sciences, in which my works have not been entirely without success." Addressing Colbert, Leibniz continued, "is a sort of recognition [*reconnoissance*]: we owe you the presentation, but you owe nothing in exchange, and the liberty of choice remains entirely yours."[55] With no timely delivery of the machines, Leibniz had moved from demanding his reasonable due to begging for favor, based on the novelty, promise, and interest of his models. In so doing, he moved from the idioms and practices of workaday legal, financial, and commercial world of early modern France, with its everchecked sovereignty, to the idioms and practices of the self-representations of absolutist France, from an independent contractor to a self-effacing courtier. He moved from representing himself as the sort of person upon which the success of the crown depended, and thus deserving of credit, to seeking favor out of the unconstrained and undeserved goodwill of a patron.

Coda: Babbage

In 1834 Charles Babbage wrote to the Duke of Wellington, "My right to dispose, as I will, of such inventions" as the difference engine, his elaborate

calculating machine for automatically producing tables, "cannot be contested; it is more sacred in its nature than any hereditary or acquired property,
for they are the absolute creations of my own mind."[56] Brave talk of natural
rights in intangible property notwithstanding, inventors still had no such
rights by statute or judicial decision in Britain.[57] Britain retained its privilege
system, even if it included a judicial requirement for specification from 1778.
All patents formally remained monarchical gifts with high fees attached.[58]

Babbage's overconfident and modern-sounding claims about his sacred
property rights appear in the middle of a far more traditional warning to his
government. Like artisans seeking privileges and preferment, Babbage threatened to move his manufacture abroad. Babbage could "collect together all
that is most excellent in our own Workshops—those *Methods* and *Processes*
which are equally essential to the Perfection of Machinery, but which are
far less easily transmitted from Country to Country" and that "would be at
once brought into successful practice under the Eyes and by the Hands of
Foreign Workmen." Creating a new corpus of engineers would give "a lasting Impulse to the Manufactures of that Country, and that the secondary
Consequences of the Acquisition of that Calculating Engine might become
far more valuable, than the primary object for which it was sought."[59] Bluster
about rights in ideas aside, Babbage recognized that transferring manufacturing practices and the people embodying them—not the ideas behind the
difference engine—was the real risk he, philosophical-entrepreneur, could
pose to the state.

Babbage no more had natural property rights in his machine than he had
control over its production or ownership of the procedures. Babbage's confident articulation of natural rights in the products of genius masked more
immediate anxieties about ownership around machines and invention.
Babbage was so uncertain of his rights that he asked the government at one
point: "Suppose Mr. Babbage should decline resuming the machine, to whom
do the drawings and parts already made belong?"[60] Though he held no privilege or patent, Babbage pushed his mechanic, Joseph Clement, to agree, "It
would be manifestly a great injustice for the contriver of such a machine
whose sole risk it was made that any other should be made by the same workman with the same tools."[61] Clement allowed Babbage the drawings and built
bits, but refused not to make additional machines.[62] Why should he? By tradition, his work gave him ownership in the things produced—claims of exclusive ownership thanks to philosophical ideas and romantic authorship be
damned.

Babbage knew this. The technical, organization, and material obstacles to
bringing his difference engine into practice led him to recognize the dis-
persion of creative skills necessary to produce machines in actual practice.
In discussing the vagaries of his own invention, he stressed time and again
the importance of a creative engineer such as Clement. The "the first necessity" for the difference engine was "to preserve the life of Mr Clement...it
would be extremely difficult if not impossible to find any other person of
equal talent both as a craftsman and as a mechanician."[63] Like Leibniz before

him, Babbage left major design decisions up to his engineer.[64] Even as he fantasized about a future division of labor giving all initiative in design to intellective inventors and thus eliminating the need for engineers such as Clement, Babbage warned of the difficulties of transforming ideas into working machines and he retained reduction to practice as the standard of success: "When the drawings of a machine have been properly made, and the parts have been well executed, and even when the work it produces possesses all the qualities which were anticipated, still the invention may fail; that is, *it may fail of being brought into general practice.*"[65]

In his *Economy of Machinery and Manufactures* of 1832, Babbage offered the programmatic dream of a mechanical reproduction of parts as fully specified by theory. "Nothing is more remarkable, and yet less unexpected, than the perfect identity of things manufactured by the same tool."[66] Such a system of manufacture was a *goal* of reorganization of labor and technique, not something yet achieved; the difficulties in producing calculating machines served as an emblem for needed reforms of work and a major spur for the development of new machining techniques and organization of labor.[67] The ability to create a regime of standardized manufacture of form helped legitimate a conception of invention of a mentalistic "form" independent of a process of the actual production. Manufacture so reduced grounded a severing of form and matter, a justification of a division of machines into intangible ideas/essences and mere instantiations.[68] Such manufacture, should it come to pass, would eliminate the sagacious and creative artisan. Only then could his actions be seen as merely repetitive, as machine-like, and thus undeserving of ownership.[69] Recent disputes about traditional knowledge in the global south underscore the continuing political potency of denying and recognizing novelty and innovation.[70]

If much of the dispersed stuff that went into the modern patent regime came out of a divided sovereignty, producing and creating an intellectualist patent regime drawing upon those elements, among others, in the name of rights required a greater realization of sovereign power to overcome traditional valuations of labor, skill, and intelligence. Rights to patents required a regime capable of enforcing them *against* traditional prerogatives and one capable of transforming work and the ownership in it *against* artisanal and traditional practices and understandings of property. The transformation of the patent bargain from an individual gift of a sovereign to a subject into a generalized contract between the public and an inventor was predicated not on the liberation *from* sovereign power but on the (ever imperfect) actualization of that sovereign power.

Abbreviations

A   Leibniz, Gottfried Wilhelm von. 1923-. Deutsche Akademie der Wissenschaften et al. (ed). *Sämtliche Schriften und Briefe*. Berlin, Munich, etc.
Cited A series, volume: page.

JM   Pascal, Blaise. 1964-. Jean Mesnard (ed). *Œuvres complètes*. 4 vols. to date. Paris: Desclée de Brouwer.

LH   Leibniz Handschriften, Gottfried-Wilhelm-Leibniz Bibliothek, Hanover, Germany.

Notes

1. Leibniz for Académie des Sciences (early 1675), LH 42, 1, f. 33r-v; edition in von Mackensen 1968, 175–176.
2. For the growing niches for technical expertise in the Enlightened cameralist state, see Heilbron 1993 and Heilbron 2011; see the considerable recent literature on early modern capitalism, state contracting, and new natural knowledge, including Smith 1994, Nummedal 2007, Cook 2007, and the essays in Smith and Findlen 2002.
3. See Woodmansee 1994, esp. 36; for the ubiquity of the romantic author in modern American intellectual property jurisprudence, see Boyle 1996.
4. For the importance of collective invention in the developments of steam engines after Watt, see, for example, Nuvolari 2004; for the collective production of the Canal de Midi, see Mukerji 2009.
5. Pottage and Sherman 2010.
6. This study takes its cue from J. L. Heilbron's striking account of the productive clash between rationalizing bureaucracy and longstanding metrological localism that yielded an upsurge in numeracy around 1800: "The domestication of the metric system," required the people to work in two systems, which in turn "made reckoning bilingual," Heilbron 1993, 275.
7. "Privilège de la machine arithmétique," May 22, 1649, JMII: 714.
8. See Biagioli 2006a, 172n39 and see the study of the privilege in Gauvin 2008.
9. I treat these two technical questions at length in Jones In preparation, Chapter 1; the most extended treatment of the machine is Mourlevat 1988.
10. Prager 1964, 281–282.
11. The revisionist literature includes Biagioli 2006a, 2006b; Bracha 2004; Bracha 2005; and Johns 2009; these works build on Belfanti 2004; MacLeod 1988; Hilaire-Pérez 2006; and Long 1991, 2001; unlike the Italian and English cases, the French case before 1700 rests on anachronistic, if still useful, scholarship, e.g. Isoré 1937; Prager 1964; Frumkin 1947–1948; and Silberstein 1961, 209–252.
12. Belfanti 2006, 335–336; for artisans, see especially the key study Long 1991.
13. See Cole 1939, vol. 2, 136 and Boissonnade 1932, e.g. 328–332.
14. Compare Isoré 1937, 102–107, and the subsequent examples.
15. For the English case, which is very clear, see Bracha 2004, 12–14 and Bracha 2005, 17–20. For the French case, where working clauses were largely implied, see Isoré 1937, 108–109, but see 115–116 and Hilaire-Pérez 1991, 914. Privileges that did not result in a working manufactory seem to have been simply treated as nullified. For Germany, however, see Popplow 1998, 107, following Silberstein 1961.
16. Biagioli 2006b, 1146.
17. For counterfeiters, see Isoré 1937, 114–115 and esp. Gauvin 2008, 228–234.
18. For this distinction in modern patent jurisprudence, see Pottage and Sherman 2010, 30–32. The best account of Pascal's discussion of artisanal labor is Michaux 2001, 211–212.
19. Compare the case of Babbage in Schaffer 1994.
20. Compare Collins and Kusch 1998, 184.
21. JMII: 339.

22. Belfanti 2006, 335–336; for the acceptance of artisanal knowledge as intangible property, see esp. Long 1991; for incisive remarks on "the idea of intangibility as a form of action" of producing something, see Sherman and Bently 1999, 48.
23. Biagioli 2006a, 143.
24. Strongly emphasized in Bracha 2004 and Biagioli 2006b.
25. For the patronage relationship Meurillon 2001, 96–97; Richelieu called upon Etienne Pascal to help investigate a scheme to determine longitude. See JMII: 82–86; Prager 1964, 273–277.
26. For England, see the treatment in Bracha 2004, 17–18.
27. Calendar of State Papers, Domestic, Charles II, Feb 16–28, 1664. See the discussions in MacLeod 1988, 16, 32 and Hulme 1917, 65–67.
28. See Isoré 1937, 101.
29. Receuil Le Nain, Conseil, v. 49, pt. 2, f. 334v, quoted in Isoré 1937, 112; see also p. 115.
30. Arrêt de *Parlement* de Paris, February 7, 1662, JMIV: 1401.
31. Ismaël Boulliau to Huygens, June 21, 1658, Huygens 1888, vol. 2, 185–186; discussed in Gauvin 2008, 238n112.
32. Biagioli 2006b, 1143; Pottage and Sherman 2010; for an earlier incisive articulation of the shift engendered by specification, see Gomme 1946, 26–27.
33. "The language of privilege, of royal sanctioned rights, often served as the unacknowledged partner to *liberté* ." Smith 1995, 238.
34. See the fine example in Hanley 1997.
35. "De Machina ad usum transferenda," LII 45, 2, 19v. Compare the strategy of Huygens and his partners for a new form of carriage, Jansen 1951, 173–174; see also Huygens' strategies around the pendulum clock, discussed in Howard 2008.
36. Louis Ferrand to Leibniz, February, 11, 1672, A 1,1: 183.
37. Compare the discussion of protocols for priority at the Royal Society and elsewhere in Iliffe 1992.
38. Pierre de Carcavy to Leibniz, 16/1, A 2,1*: 307, 308.
39. For Carcavy's job of running "the tandem collections" of the royal library "as a machine of public administration," see Soll 2009, 99–101.
40. Leibniz to Louis Ferrand, 1672, A 1,1: 452.
41. See the fine example in Leibniz 1906, 141; Leibniz to Christian Habbeus, May 5, 1673, A 1,1: 416.
42. Pierre de Carcavy to Leibniz, December 5, 1671, A 2,1: 307.
43. See Oldenburg to Leibniz, January 30, 1672/3, Oldenburg 1965–1986, vol. 9, 431.
44. [1674], LH 42, 2, f. 10v.
45. [1674], LH 42, 2, f. 10v; compare the similar discussion at LH 42, 2, f. 67v. I have found no comments concerning Leibniz's machines among the known Morland manuscripts.
46. Leibniz to [Elizabeth of Bohemia?], [11.1678], A 2,1*: 661.
47. Leibniz for [Académie?], [before December, 15, 1676], LH 42, 1, f. 37v, 38r; von Mackensen 1968, 172, 173.
48. A 6,4: 712.
49. Guiffrey 1881, c. 781.
50. Leibniz to Johann Friedrich, January 21, 1675, A 1,1: 492.
51. Joachim Dalencé to Leibniz, [October 29, 1675], A 3,1: 303 (emphasis in original).
52. Leibniz to Gallois, November 2, 1675, A 3,1: 306.
53. Salomon-Bayet 1978, 156.
54. For Leibniz and the Académie, see Salomon-Bayet 1978.
55. Leibniz to Colbert, January 11, 1676, A 1,1: 457.
56. Babbage to Duke of Wellington, December 23, 1834, BL (British Library) Add. MS 40611, f. 183v.

57. Brewster 1830, 333; see Johns 2009, 250–258.
58. See Bracha 2004, 33–35.
59. Babbage to Duke of Wellington, December 23, 1834, BL Add. MS 40611, f. 183v (emphasis in original).
60. Babbage to Ashley, BL Add. MS 37184, 459–460; in Collier 1990, 68.
61. Babbage to Bryan Donkin and George Rennie, April 11, 1829, BL Add. MS 37184, 254–255, quoted in Collier 1990, 53–54.
62. Donkin to Babbage, April 22, 1829, BL Add. MS 37184, 266–267, quoted in Collier 1990, 54.
63. Babbage, "Report on the Calculating Machine," BL Add. MS 37185, f. 264, quoted in Schaffer 1996, 291; for Babbage's estimation of Clement's work, Ginn 1991, 172–175.
64. For the evidence of Clement's innovations, see Williams 1992, 74.
65. Babbage 1989, §325, p. 185.
66. Ibid., §79, p. 47.
67. Schaffer 1994, 1996.
68. See Pottage and Sherman 2010, Chapter 2, drawing on Berg 2002.
69. For artisanal ownership in the works of skill, see Rule 1987, esp. 104–105.
70. See, among a large literature, Sunder 2007, esp. 109.

References

Babbage, Charles. 1989. *The Economy of Machinery and Manufactures*. Vol. 8, *The Works of Charles Babbage*. London: W. Pickering, first edition: 1832.

Belfanti, Carlo Marco. 2004. "Guilds, Patents and the Circulation of Technical Knowledge: Northern Italy during the Early Modern Age." *Technology and Culture* 45: 569–589.

———. 2006. "Between Mercantilism and Market: Privileges for Invention in Early Modern Europe." *Journal of Institutional Economics* 2: 319–338.

Berg, Maxine. 2002. "From Imitation to Invention: Creating Commodities in Eighteenth-Century Britain." *The Economic History Review* 55 (1): 1–30.

Biagioli, Mario. 2006a. "From Prints to Patents: Living on Instruments in Early Modern Europe." *History of Science* 44: 139–186.

———. 2006b. "Patent Republic: Representing Inventions, Constructing Rights and Authors." *Social Research* 74: 1129–1172.

Boissonnade, P. 1932. *Colbert, le triomphe de l'étatisme; la fondation de la suprématie industrielle de la France, la dictature du travail (1661–1683)*. Paris: M. Rivière.

Boyle, James. 1996. *Shamans, Software, and Spleens: Law and the Construction of the Information Society*. Cambridge, MA: Harvard University Press.

Bracha, Oren. 2004. "The Commodification of Patents 1600–1836: How Patents Became Rights and Why We Should Care." *Loyola of Los Angeles Law Review* 38: 177.

———. 2005. "Owning Ideas: History of Intellectual Property in the United States." SJD Dissertation, Harvard Law School.

[Brewster, David]. 1830. Review of Charles Babbage's *Reflections on the Decline of Science in England, and on Some of Its Causes*. *The Quarterly Review* 43: 305–342.

Cole, Charles Woolsey. 1939. *Colbert and A Century of French Mercantilism*. 2 vols. New York: Columbia University Press.

Collier, Bruce. 1990. *The Little Engines that Could've: The Calculating Machines of Charles Babbage*. New York: Garland, first edition: 1970.

Collins, H. M., and Kusch, Martin. 1998. *The Shape of Actions: What Humans and Machines Can Do*. Cambridge, MA: MIT Press.

Cook, Harold J. 2007. *Matters of Exchange: Commerce, Medicine, and Science in the Dutch Golden Age*. New Haven: Yale University Press.

Frumkin, Maximilian. 1947–1948. "The Early History of Patents for Invention." *Transactions of the Chartered Institute for Patent Agents* 66: 20–69.

Gauvin, Jean-François. 2008. "Habits of Knowledge: Artisans, Theory and Mechanical Devices in Seventeenth-Century France." PhD, Dissertation, History of Science, Harvard University, Cambridge, MA.

Ginn, William Thomas. 1991. "Philosophers and Artisans: The Relationship between Men of Science and Instrument Makers in London, 1820–1860." PhD, Dissertation, Unit for the History of Science, University of Kent at Canterbury.

Gomme, Arthur Allan. 1946. *Patents of Invention: Origin and Growth of the Patent System in Britain, Science in Britain.* London and New York: Published for the British Council by Longmans Green and Co.

Guiffrey, Jules. 1881. *Comptes des Bâtiments du Roi sous le règne de Louis XIV: Tome Premier, Colbert: 1664–1680.* Paris: Imprimerie nationale.

Hanley, Sarah. 1997. "Social Sites of Political Practice in France: Lawsuits, Civil Rights, and the Separation of Powers in Domestic and State Government, 1500–1800." *The American Historical Review* 102 (1): 27–52.

Heilbron, J. L. 1993. *Weighing Imponderables and other Quantitative Science around 1800.* Berkeley: University of California Press.

———. 2011. "Natural Philosophy," in P. Harrison, R. L. Numbers, and M. H. Shank (eds). *Wrestling with Nature: from Omens to Science.* Chicago: University of Chicago Press.

Hilaire-Pérez, Liliane. 1991. "Invention and the State in 18th-Century France." *Technology and Culture* 32 (4): 911–931.

———. 2006. *L'invention au siècle des lumières.* Paris: Albin Michel.

Howard, Nicole. 2008. "Marketing Longitude: Clocks, Kings, Courtiers, and Christiaan Huygens." *Book History* 11: 59–88.

Hulme, F. Wyndham. 1917. "Privy Council Law and Practice of Letters Patent for Invention from the Restoration to 1794." *Law Quarterly Review* 33: 63–75, 180–195.

Huygens, Christiaan. 1888. *Oeuvres complètes.* 22 vols. The Hague: M. Nijhoff.

Iliffe, Rob. 1992. "'In the Warehouse': Privacy, Property and Priority in the Early Royal Society." *History of Science* 30: 29–68.

Isoré, Jacques. 1937. "De l'existence des brevets de l'invention en droit français avant 1791." *Revue historique de droit français et etranger* 4th Series, 16: 94–130.

Jansen, P. 1951. "Une tractation commerciale au XVIIe siècle." *Revue d'histoire des sciences et de leurs applications* 4 (2): 173–176.

Johns, Adrian. 2009. *Piracy: the Intellectual Property Wars from Gutenberg to Gates.* Chicago: University of Chicago Press.

Jones, Matthew L. In preparation. Reckoning with Matter: Calculating Machines, the Political Economy of Innovation, and Thinking about Thinking from Pascal to Babbage.

Leibniz, Gottfried Wilhelm. 1906. E. Gerland (ed). *Leibnizens nachgelassene Schriften physikalischen, mechanischen und technischen Inhalts.* Leipzig: B.G. Teubner.

Long, Pamela O. 1991. "Invention, Authorship, 'Intellectual Property,' and the Origin of Patents: Notes toward a Conceptual History." *Technology and Culture* 32: 846–884.

———. 2001. *Openness, Secrecy, Authorship: Technical Arts and the Culture of Knowledge from Antiquity to the Renaissance.* Baltimore: The Johns Hopkins University Press.

MacLeod, Christine. 1988. *Inventing the Industrial Revolution: The English Patent System, 1660–1800.* Cambridge: Cambridge University Press.

Meurillon, Christian. 2001. "Le Chancelier, les Nu-pieds, et la machine: Pascal père et fils à Rouen," in J. P. Cléro (ed). *Les Pascal à Rouen 1640–1648.* Rouen: Université de Rouen.

Michaux, Bernard. 2001. "Pascal, Descartes et les Artisans," in J. P. Cléro (ed). *Les Pascal à Rouen 1640–1648.* Rouen: Université de Rouen.

Mourlevat, Guy. 1988. *Les machines arithmétiques de Blaise Pascal, Memoires de l'Académie des sciences, belles-lettres et arts de Clermont-Ferrand, vol. 51.* Clermont-Ferrand: La Francaise d'Edition et d'Imprimerie.

Mukerji, Chandra. 2009. *Impossible Engineering: Technology and Territoriality on the Canal du Midi.* Princeton: Princeton University Press.

Nummedal, Tara E. 2007. *Alchemy and Authority in the Holy Roman Empire.* Chicago: University of Chicago Press.

Nuvolari, Alessandro. 2004. "Collective Invention During the British Industrial Revolution: The Case of the Cornish Pumping Engine." *Cambridge Journal of Economics* 28 (3): 347–363.

Oldenburg, Henry. 1965–1986. A. R. Hall and M. B. Hall (eds). *The Correspondence of Henry Oldenburg.* 13 vols. Madison; London: University of Wisconsin Press; Mansell; Taylor; and Frances.

Popplow, Marcus. 1998. "Protection and Promotion: Privileges for Inventions and Books of Machines in the Early Modern Period." *History of Technology* 20: 103–124.

Pottage, Alain, and Sherman, Brad. 2010. *Figures of Invention: A History of Modern Patent Law.* Oxford: Oxford University Press.

Prager, Frank D. 1964. "Examination of Inventions from the Middle Ages to 1836." *Journal of the Patent and Trademark Office Society* 46: 268–291.

Rule, John. 1987. "The Property of Skill in the Period of Manufacture," in P. Joyce (ed). *The Historical Meanings of Work.* Cambridge: Cambridge University Press.

Salomon-Bayet, Claire. 1978. "Les Académies scientifiques: Leibniz et l'Académie Royale des Sciences," in *Leibniz à Paris (1672–1676).* Wiesbaden: Franz Steiner Verlag.

Schaffer, Simon. 1994. "Babbage's Intelligence: Calculating Engines and the Factory System." *Critical Inquiry* 21: 203–227.

———. 1996. "Babbage's Calculating Engines and the Factory System." *Reseaux* 4 (2): 271–298.

Sherman, Brad, and Bently, Lionel. 1999. *The Making of Modern Intellectual Property Law: The British Experience, 1760–1911.* Cambridge and New York: Cambridge University Press.

Silberstein, Marcel. 1961. *Erfindungschutz und merkantilistische Gewerbeprivilegien.* Zürich: Polygraphischer Verlag.

Smith, David Kammerling. 1995. "'Au bien du commerce': Economic Discourse and Visions of Society in France." PhD, Dissertation History, University of Pennsylvania.

Smith, Pamela H. 1994. *The Business of Alchemy: Science and Culture in the Holy Roman Empire.* Princeton: Princeton University Press.

Smith, Pamela H., and Findlen, Paula (eds). 2002. *Merchants & Marvels: Commerce, Science, and Art in Early Modern Europe.* New York: Routledge.

Soll, Jacob. 2009. *The Information Master: Jean-Baptiste Colbert's Secret State Intelligence System.* Ann Arbor: University of Michigan Press.

Sunder, Madhavi. 2007. "The Invention of Traditional Knowledge." *Law and Contemporary Problems* 70: 99–124.

von Mackensen, Ludolf. 1968. "Die Vorgeschichte und Entstehung der ersten digitalen 4-Spezies-Rechenmachine von Gottfried Wilhelm Leibniz nach bisher unerschlossenen Manuskripten und Zeichenung mit einem Quellenanhanhang der Hauptdokumente." Dr. rer. nat., Fakultät für allgemeine Wissenschaften, Technischen Hochschule München, Munich.

Williams, Michael R. 1992. "Joseph Clement: The First Computer Engineer." *Annals of the History of Computing* 14 (3): 69–76.

Woodmansee, Martha. 1994. *The Author, Art, and the Market: Rereading the History of Aesthetics.* New York: Columbia University Press.

7
Genes, Railroads, and Regulations: Intellectual Property and the Public Interest

Daniel J. Kevles

In 1988, in a report on the emerging Human Genome Project, the National Research Council called for keeping open the data the project would generate, declaring that "access to all sequences and material generated by these publicly funded projects should and even must be made freely available."[1] The admonition to openness expressed the scientific community's long-standing communitarian norm, part ethical and part practical, that knowledge of nature is to be publicly shared. But in 1991, J. Craig Venter, a biologist at the National Institutes of Health (NIH), in Bethesda, Maryland, struck a blow for privatization of the genome by proposing the wholesale patenting of human gene fragments called "expressed sequence tags," or ESTs. Genes comprise a sequence of DNA (Deoxyribonucleic acid) base pairs, some of which code for amino acids, most of which do not. Those that do code are said to be "expressed" when they are active. An EST comprises a short sequence of the expressed base pairs, a form of the gene's DNA that is called a "cDNA." Although just 150–400 base pairs long, each serves to identify the gene of which it is a part. Venter claimed that ESTs would have utility as diagnostic probes for genes, but he also seemed bent on using the fragments to gain control of the intellectual property in the entire gene that the EST identified even though the EST revealed nothing about the gene's function. Within a year the number of ESTs covered by the Venter/NIH patent application had multiplied to almost 7,000. A lawyer for the leading biotechnology firm Genentech noted, "If these things are patentable, there's going to be an enormous cDNA arms race."[2]

Much to the relief of most academic scientists and a sizable fraction of the biotechnology industry, the US Patent and Trademark Office (USPTO) rejected the Venter/NIH application, holding that ESTs were not patentable proxies for entire genes.[3] But the episode reveals that, from the beginning, human genomics has been torn between a commitment to serving a public interest, in medicine as well as in science, and an impulse to privatization and profit.

Public-interest advocates have persistently contended that the human genome is the birthright of all human beings and that its parts ought not to be privately owned. Many of them also point out that knowledge of the human genome has been gained as a result of huge public investments and that the public has a right to reasonable use of the results of this research in both science and medicine. Advocates of privatization, in contrast, insist that private investment has been required to transform the basic knowledge of human genes into the biotechnology industry, a major contributor to the nation's economic development and medical well-being that has generated products ranging from diagnostic tests to pharmaceuticals; and that the private investment that made all this possible would not have occurred without the guarantee of private genomic ownership, usually in the form of patents on individual genes.

Both sides of the issue converge on several essential questions: Where is the boundary in human genomics between public interest and private property rights? How has it come to be drawn historically? And where should it be now? Useful guidance in exploring these questions can be obtained from a brief comparison of contemporary human genomics with the early history of the American railroad industry.

The comparison may seem improbable. Railroads are huge and genes are tiny, but the processes by which they came to figure in the American economy are marked by significant similarities. In the latter third of the nineteenth century, the transcontinental railroad system was developed with hefty state and munificent federal patronage in the form of grants of rights of way and tracts of land along them to private railroad companies.[4] Washington provided the existing states with federal lands for railroad subsidies and in the territories it granted vast lands to the railroad companies directly. By 1871, when the last grant was given, the federal government had transferred to the railroads some 130 million acres of land, which, in early twenty-first-century dollars was worth at least $14 billion. In return, the companies built the transcontinental railroads and grew rich by serving the day's national interest, joining the East and the West in a system of rapid transport of people and goods and creating a national economy out of what had been a loosely linked network of local and regional economies.

Railroads were at the heart of the economy. The enterprises that built and operated them were the century's largest. By 1900, they had laid nearly two hundred thousand miles of track, having consumed Croesus-like quantities of capital. They carried most of the nation's freight and employed more than 1 million people. Adumbrating the creation of the biotechnology industry, the investments to create this system had come from private American sources, overseas investors, and state and local governments as well as the federal government.

The railroads also foreshadowed some of the biotech industry's tribulations. They courted instability by overbuilding in many areas. Competition was fierce, and business downturns reduced revenues and devastated stock prices. Mark Twain noted of one company that "this is the very road whose stock always goes down after you buy it, and always goes up again as soon as you sell it." Nonetheless, railroading yielded great fortunes not only from the operations of the roads but also through stock speculations, mergers, and construction-finance schemes. Like the biotech industry, the expansion of the railroads depended on technological innovations, including sturdy steel rather than iron rails, the more efficient "compound" (or two-cylindered) locomotive, and the air brake as well as a new coupler that greatly increased safety for both passengers and crew.

The impact of railroads on the nation's economy was immense, reaching into almost every sector, with the result that Americans came to believe it necessary to subject the railroads to public oversight. Although private corporations, the railroads performed public functions. They played too significant a role in American life and the economy to be permitted absolute control over their private corporate property rights. By 1897, 28 state railway commissions had been created, mainly to investigate and publicize concerns about railroad practices.

Public scrutiny of the railroads intensified after the Civil War both because they were wielding increasing power over shippers and consumers and establishing pricing policies that seemed discriminatory. These policies were not necessarily the product of greed, primarily. They were the result of the railroads acting as profit-making institutions, lowering prices on long-haul routes where they faced competition and compensating for the reduced revenues by raising them on short-haul routes where they were often monopolies. Their policies and practices disadvantaged small farmers and other suppliers of freight. Thus, as the railroads diverged from the service of an equitable public interest, increasing demands were raised for regulation of them. The companies objected, insisting that such regulation would interfere with their private property rights, but the demands were sufficient to result in state and then federal action.

In the 1870s, a number of midwestern states passed the "Granger" laws, so-called because their advocates were farmers who belonged to an organization called the Grange. Regulating railroad property rights, these laws created railway commissions, empowered them to set maximum or "reasonable" rates, and prohibited price discrimination. They constituted a major initiative on the part of public authorities to regulate private corporate behavior. In 1877, the Supreme Court, in *Munn v. Illinois,* upheld the constitutionality of the Granger laws, concluding that private property, when "affected with a public interest...must submit to be controlled by the public for the common good."

In 1887, two years after the court partially reversed itself in *Wabash v. Illinois,* Congress passed the Interstate Commerce Act (ICA), which prohibited discriminatory pricing policies, required published rate schedules, and insisted that all railroad rates be "reasonable and just." Enforcement of the law was entrusted to an Interstate Commerce Commission (ICC) whose members were appointed by the president. Although much of the law was imprecisely worded and there was little agreement about what constituted "reasonable and just" rates, the ICA was a pathbreaking piece of legislation that established the right of the federal government to actively regulate some private enterprise whose operations affected the public interest.

Like the railroads in the late nineteenth century, the field of molecular biology grew and flourished in the late twentieth century in no small part as a result of federal, state, and municipal patronage, notably through the NIH, combined with private investment. Research in the field produced increasing knowledge of human genes, especially after the creation of the Human Genome Project, which was eventually fostered by the National Human Genome Research Institute and the Department of Energy. Particularly important, progress was made in identifying genes responsible for, or at least implicated in, diseases.

As a child of the federal government, human genomics resembled the high-energy particle physics with its giant accelerators and vast laboratories, the most prestigious and expensive area of physics after World War II. Despite the secrecy and security imposed on parts of science during the Cold War, high-energy particle physics was marked by openness in the development of its technologies under the policies of the Atomic Energy Commission (AEC).[5] Unlike participants in human genomics, the accelerator scientists and engineers worked in an environment largely free from patent constraints that greatly speeded accelerator development. Both law and policy tended to vest in the AEC ownership of patentable inventions made in its laboratories or under its contracts and to make freely available the technologies of particle physics to scientists engaged in basic research.[6] A similar freedom characterized the exchange of basic data among high-energy physicists. They went on to achieve a formidable level of integration, now via the Internet, in respect of creating, evaluating, and banking data about the properties of elementary particles.[7]

In the life sciences, circumstances have long contributed to a strong anticommercial orientation. With some exceptions—for example, hybrid corn—most university research, especially in the basic life sciences, yielded little that was commercializable or patentable, and of that, less that commanded significant, if any, market value. Although fruit fly geneticists developed *Drosophila*, the workhorse of classical genetics, into standardized strains at the cost of much time and painstaking effort, no one attempted to profit from them; indeed, fruit fly stocks were freely exchanged among genetics laboratories on an international basis.[8] Similarly, in the middle third of the twentieth century, bacteriophage were also standardized and made widely available among geneticists. In these cases cooperation worked because there was little reason not to cooperate, and many reasons to cooperate, including the prospect of professional rewards. Besides, most living organisms and their parts were held not to be patentable as a matter of law.[9]

Academic culture's resistance to commercialization was particularly strong in the life sciences related to health and medicine. The University of Toronto scientists who were responsible for the isolation of insulin excluded themselves from shares in revenue from the insulin patent, assigning their rights to the University of Toronto for one dollar each. Ditto for Harry Steenbock, at the University of Wisconsin, who ceded his patent on a process for producing vitamin D to the institution, which made millions on it until it was declared invalid. In the mid-1930s, Harvard promulgated the explicit policy that innovations in medical research arising from its laboratories must not be patented or, if they were, should be given freely to the public.[10]

A member of a British group in particle physics, once asked why the field was so cooperative, responded, "Particle physics data have no economic or strategic worth."[11] In this respect, particle physics was an outlier in the physical sciences. Beginning in the late nineteenth century, physics and chemistry had fueled what is known as the second industrial revolution that has continued through our own day. In branches of these fields, commercial competition penetrated academic science far more widely than it had hitherto.

John Heilbron and his collaborator Robert Seidel pointed out a number of years ago that the trend made itself felt during the 1930s even in the field of accelerator physics. A not-for-profit organization, the Research Corporation, obtained rights to the cyclotron from its inventor, the Berkeley physicist Ernest O. Lawrence, on the understanding that his Berkeley laboratory would continue to be a beneficiary of the Corporation's policy of investing proceeds from its patents in university research. The Corporation hoped that these proceeds would include royalties from licenses to commercial firms using cyclotrons to make radioisotopes for biological and medical applications. No radiopharmaceutical industry developed before the war, however, and after the war, owing to inventions made to exploit atomic energy, the cyclotron appeared to have little commercial value. The Research Corporation then wrote all cyclotron laboratories to grant royalty-free use of the machine, formally sanctioning the practice already in place in particle physics that continued in the postwar period.[17]

The interleaving of commercial and academic enterprise grew substantially after World War II, when cutting-edge advances in the physical sciences and engineering—the products of research supported by the federal government, mainly the military—were spun out into and developed by the industrial sector. Prominent in the trend were MIT and Stanford, both powerhouses in the new branches of engineering and physics that the military generously supported. Although a number of the laboratories and projects were classified, they provided ample opportunities for unclassified training and thesis writing for hundreds of doctoral students—and also advanced instruction for staff from military agencies and industrial firms. Professors and students together produced an enormous amount of significant research and a panoply of textbooks that quickly became classics. The Stanford and MIT programs spun off knowledge and trained people who turned the knowledge into a plethora of new companies. By the early 1960s, the MIT Instrumentation Laboratory alone had stimulated the formation of 27 firms, with 900 employees and total sales of $14 million, and Stanford Industrial Park, which bordered the campus on university land that had been designated for the purpose, had 27 tenants, with some 8,600 employees.[13]

The drive to commercialize the results of academic research spread into the life sciences and was given an enormous boost in 1976, when Herbert Boyer, one of the coinventors of the technique of recombinant DNA, joined with a venture capitalist named Robert Swanson to form the biotechnology firm Genentech—short for "genetic engineering technology." The company set out to produce human insulin, a protein in which diabetics are deficient and the demand for which was projected to exceed the supply of substitute insulin, which was obtained largely from cattle and pigs.[14]

In early September 1978, at a press conference crowded with media and held at the City of Hope, a research hospital in Southern California, Genentech announced to the world at large that it had bioengineered human insulin and that, about two weeks earlier, it had entered into an agreement with Eli Lilly & Co. whereby the pharmaceutical concern would manufacture and market the hormone. The breakthrough was heralded in every major newspaper and magazine in the United States except *The New York Times*, which was on strike. Reports of dramatic technical progress multiplied, and the interest of the financial markets in biotechnology grew feverish. When in mid-October 1980, Genentech—assigned the stock symbol "GENE"—went public, its shares were snapped up at a more than twice the offering price of $35, astonishing Wall Street observers, not least because Genentech's earnings for 1979 had totaled a mere two cents a share.[15]

By then, the fledgling biotechnology industry was attracting broad attention among federal policymakers. It seemed likely to increase the United States' international trade surplus in high-technology goods, which since the mid-1970s had been offsetting a sizable trade deficit in other types of manufactures. "Innovation has become the preferred currency of foreign affairs," a patent lawyer advised a committee in the House of Representative. In 1980, the government granted the biotechnology industry a triple boost: NIH, which had been easing restrictions on recombinant research, ended them almost entirely. Congress passed the Bayh-Dole Act, which explicitly encouraged universities to patent and privatize the results of federally sponsored high-technology research. And in June, the United States Supreme Court ruled in *Diamond v. Chakrabarty* that a patent could be issued on a genetically modified living organism, holding, over the legal and moral objections of critics, that whether an invention was living or not was irrelevant to its qualification for intellectual property protection. In 1985, the USPTO expanded patentability to include any kind of plant, and in 1987 it declared that patents were allowable on animals although not on human beings.[16]

In 1986, to promote the commercialization of the practical results arising in federal research laboratories, Congress authorized governmental agencies to license patents on these results to private industry.[17] All the while, public and private investment in biomedical research mushroomed. Inspired by examples such as Genentech, new biotechnology companies sprang up to exploit the accumulating genetic knowledge. Together with major pharmaceutical firms as well as a number of oil and chemical giants, they formed a burgeoning biotechnology industry in the United States that was strongly interleaved with academic and federal biomedical research.[18]

The 1986 law provided the legal and policy foundation for Craig Venter's effort to patent ESTs identified at NIH. If he failed at the wholesale patenting of human genes, he remained eager to capitalize on human genomics. He left NIH in 1992 to head a new private venture, The Institute for Genomic Research (TIGR), that would be devoted to DNA sequencing and that would turn over results useful for commercialization to a new company, Human Genome Sciences. In January 1998, Venter resigned from TIGR to join in the formation of a new company, the Celera Genomics Corporation, that aimed to sequence the entire human genome using a new, recently developed fast-sequencing technology. Celera's original business plan called for its data to be held as proprietary by the company and released at first only to paying subscribers, while patents would be sought on genes of interest.[19] After the human sequence was completed in 2001 jointly by Celera and the National Human Genome Research Institute at NIH, Celera allowed academic scientists to download data only on a restricted basis—for example, requiring that they not be given to anyone else.[20]

Other firms in the United States and Europe have managed to achieve exclusive control over genomic databases. Perhaps the best known is the arrangement of deCode with the Icelandic government: the company was granted exclusive access for commercial purposes to the national medical database via a 1998 agreement with the government that was to last for 12 years. The drug firm LaRoche, which financed deCode, got exclusive rights to develop pharmaceuticals for 12 diseases, in exchange for which it contracted to provide the Icelandic population with any such drugs free of charge.[21]

The principled objections to the privatization of the genome have been largely ineffective against the commercial drive, but the mutual self-interest of most genomic researchers in access to basic scientific information has kept genomic databases largely public. Several models demonstrated how this could be done. Among them was the *Centre d'etudes du polymorphisme humain* (CEPH), established in 1984 in France with genetic material from French and American families that was made freely available to scientists constructing a human genetic map.[22] There was also the *Worm Breeder's Gazette*, a record of the worldwide effort to map and sequence and characterize the C. elegans genes, including their multiple mutations. The worm breeders shared data, methods, instruments, and stocks, including mutants. Within this community John Sulston began construction of a physical map of the worm's genome, and the community at large linked this map to the genetic map it had been developing collectively.[23] The enterprise was characterized by the award of credit within communitarian norms.

The worm model influenced representatives of the multinational human-genome enterprise when they met in Bermuda in 1996 under the sponsorship of the Wellcome Trust, a biomedical philanthropy in Britain, to strategize the project scientifically and draw up rules for the treatment of data. The rules, which were proposed by Sulston, were clearly a response to the growing commercialization of the genome, with its tendency to keep genomic data under wraps until patents could be filed. Adopted unanimously, the rules stated in their ultimate polished form: "All human genomic DNA sequence information, generated by centers funded for large-scale human sequencing, should be freely available in the public domain in order to encourage research and development and to maximize its benefit to society."[24] The publicly funded human-genome effort, which since the early 1990s has operated on an international scale, has undercut privatization somewhat by retaining its commitment to openness in its databases. Since the beginning of the sequencing phase of the Human Genome Project, all the data generated by the participants have been deposited in publicly available databases every 24 hours. By 2003, the human genome sequence, essentially complete, was posted on the Internet with no barriers to use, no subscription fees, and no obstacles.[25] A growing number of journals will not publish genomic articles without proof that the authors have submitted their data electronically to GenBank, in Los Alamos, the central genomic database in the United States. The National Center for Biological Information, which runs GenBank, places no restrictions on reasonable use and distribution of its data.[26]

Large, well-established pharmaceutical firms have recognized the value of publicly available databases. Ten of them were instrumental in the establishment of the SNP (single nucleotide polymorphisms) consortium, in 1999. Far more interested in using genomic data than in generating it, they saw in the consortium a means of reducing costs for the employment of such data and recognized that making it freely available to all would accelerate the growth in the knowledge base and benefit the public good.[27]

But despite the ubiquitous availability of genomic data, openness and profit-making in human genomics have remained in conflict. The key reason is patents.

In 1996, the call of the Bermuda rules for making genomic sequence data part of the public domain implied that DNA sequence data should not be patented. But even if academic and biotech scientists submitted genomic data to the public databases, they were free to file patents on it first.

Indeed, by 1996 several private corporations—notably Human Genome Sciences—had filed patent applications on thousands of ESTs, claiming various useful functions for them such as genomic probes. In 1997, the USPTO announced that it would allow EST patents for such purposes. The shift in policy aroused opposition from both Harold Varmus, the then director of the NIH and from the international Human Genome Organization's International Property Rights Committee (HUGO IPR Committee). The objections were grounded in apprehensions that EST patent holders, by having a claim on even just a small part of a gene, would discourage research addressed to the discovery and characterization of the entire gene.[28] They might also hinder the patenting of the gene, and with the loss of patentability the incentive to invest in the development and commercialization of therapeutics and diagnostics specific to it.

Although some scientists such as Sulston objected to the patenting of genes even if they were fully characterized as to structure and function, neither Varmus, the HUGO IPR Committee, nor for that matter many other biomedical scientists did. On the contrary, many biomedical scientists supported full-gene patenting. The HUGO IPR Committee, for example, hoped that free publication of DNA sequence data "will not unduly prevent the protection of genes as new drug targets" because it was "essential for securing adequate high-risk investment."[29]

What is wrong with patenting fully characterized human genes? Nothing, many say, adding that everything is right with it—not only because it encourages investment and innovation in genomics but also because it falls within the USPTO's definition of what is patent-eligible. Many people assume that a patent on a gene covers the gene in the body, but this is not the case. Genes in the body are products of nature and as such, in accord with a longstanding doctrine of patent law, are not patentable. What is patentable according to the core of the statute (U.S.Code Title 35, Section 101) includes new and useful compositions of matter made by man. A gene can qualify for patentability if its native DNA, including all the base pairs that comprise it, is isolated from the body or if it is produced as a cDNA, which includes only the base pairs that are expressed. In either version, according to the USPTO, it constitutes a new composition of matter and is therefore patent-eligible. The USPTO had been issuing patents on such isolated genes since the 1980s, and it had turned its practice into formal policy in 2001.[30]

But critics countered that patenting human genes is at the least problematic because the practice entails costs to the enterprise of research, biomedical innovation, and the delivery of medical services.[31] In contemporary academic research, the expectation of patentability discourages open discussion of technical detail during the critical R&D phase before patent filing. Then, too, patented genes are potential research tools, and such tools— according to a decision by a federal court in 2002, in the case of *Madey v. Duke University*— are controlled by the patent holder, who may restrict and charge for their use because research even in its most abstract form is part of a university's "business" and as such is not exempt from threats of patent infringement suits.[32] And although the gene in the body may not be owned, the patent holder can exclude all others from using the extracted genomic DNA. Since this is the only form in which it can be studied, analyzed, or made the basis of a diagnostic test or a new therapeutic, the patent holder enjoys a complete lock on the field of biomedicine that depends on the gene.

A human gene patent establishes what has been called "a chain of dependency" in biomedical research that includes efforts to characterize the gene and its functions more fully and to develop diagnostic tests based on it. It thus has a chilling effect on all research that involves the gene.[33] One firm patented a gene encoding the CCR5 lymphocyte receptor without any knowledge of its link to HIV infection. When the latter was established by another laboratory, the patent holder declared that it would enforce its patent against anyone making use of the discovery in the development of any pharmaceutical to combat HIV. Patent law supported the threat because it gives the patent holder rights over all uses of the invention, including those not claimed by the original inventor. In 1999, a survey of 74 clinical labs revealed that a quarter of them had abandoned a clinical test they had developed because of pending patents and almost half had decided not to develop a clinical test because of the patent.[34]

Deeply troubling problems in the delivery of medical diagnostic services have arisen from the control by Myriad Genetics, a biotechnology company based in Salt Lake City, Utah, of the patents on BRCA1 and BRCA2, the two genes known to dispose women to hereditary breast cancer. Myriad's BRCA1 patent covers the sequence not only as a descriptor of the gene but also as the physical substance in and of itself and its mutant forms. The patent also covers the uses of the gene as a probe or a primer and its protein. Myriad's patent claims cover all diagnostic methods that use the gene, including those developed by others.[35]

For various reasons, by the end of the 1990s Myriad held monopoly control through patents and exclusive licenses over the DNA sequence of both BRCA1 and BRCA2.[36] Myriad demands that all commercial testing for the two genes be done in its lab. It will not license the test to anyone, with the result that a woman diagnosed by Myriad cannot obtain a second opinion from an independent laboratory.[37]

Myriad has enforced its patent rights against various universities, a hitherto exceptional practice. In 1999, for example, it notified Arupa Ganguly, of the University of Pennsylvania clinical genetics lab, that she was infringing the Myriad patents, because she had independently developed a test to screen for mutations in the BRCA genes and, to cover her clinical costs, was charging her patients a fee to undergo the test. Myriad advised the university to halt Ganguly's activities or risk suit. To meet criticism from academic researchers, Myriad negotiated an agreement with NIH in 2000 whereby NIH-funded researchers would be charged $1200 per test instead of the usual $2580 so long as the purpose was research. In exchange, Myriad would have access to the resulting research data.[38]

Such practices threaten, among other consequences, to limit research on disease-related genes, to concentrate expertise in only a few institutional centers, to fragment molecular medical services, to elevate the prices consumers pay for diagnostic tests, and to make doctors vulnerable for infringement suits. The denial of access to second and independent diagnostic opinions also flies in the face of sound medical practice.

Resistance to the BRCA patents has been high in Europe and gathering force in the United States. The European Patent Office (EPO) granted Myriad Genetics three BRCA1 patents in 2001, but in 2008 according to Gert Matthijs, head of the Centre for Human Genetics at the University of Leuven, in Belgium, no European clinic was paying royalties for BRCA1-related diagnostics.[39]

Many scientists and clinicians objected to the patents on varying general grounds—that it was unethical to grant patents on a human gene, especially one for disease; that the patent was unwarranted in any case because a gene is a product of nature and because obtaining the sequence was obvious.

The objections led not only to defiance of Myriad's patent rights in the laboratory and the clinic but also to legal challenges to the patent. In Europe in 2004, a technical legal argument won the day. This was that the patent had been improperly granted because Myriad had submitted an incorrect sequence when it first filed for the patent, in 1994. A board in the EPO revoked the patent on BRCA1, holding that a perfect sequence is required to make a full diagnosis. In the face of the mounting opposition to the BRCA1 patents, Myriad transferred ownership of them to the University of Utah in November 2004.[40]

However, on November 19, 2008, the EPO's highest board of appeals countermanded the 2004 decision after the patent owners said that they would reduce the scope of the patent to cover only frame-shift mutations—that is, the deletion or insertion of one or two nucleotides so that the gene generates the wrong series of amino acids. These frame-shift mutations represent only about 60 percent of the mutations associated with breast and ovarian cancer, and the board held that an exact sequence of the gene is not required to detect them.[41]

In principle, the ruling meant that the University of Utah had gained the right to collect royalties on the tests that tens of thousands of women in Europe were undergoing every year. The royalties are potentially very substantial. In the United States, Myriad now charges $3,500 for a full analysis of both BRCA1 and BRCA2 and $460 for a single-mutation test. In Europe, the test for both genes can come to as much as $1,900. However, it seems likely that scientists and clinicians in Europe will continue to defy Myriad's patent rights. Dominique Stoppa-Lyonnet, a clinical geneticist at the Curie Institute in Paris, expressed disappointment at the EPO's ruling, having fought Myriad's patents for seven years. She declared, "We will wait to see what royalties the University of Utah might demand of us, but [the ruling] won't stop us testing the gene in France."[42]

Myriad's rights in the patent for BRCA2 are also under challenge. A broad patent on the gene has been granted in Europe to a consortium that is partly owned by the charity Cancer Research UK, in Britain. One of the inventors behind the patent explains that the charity obtained the patent "to defend the gene against other patent approaches," adding, "We offer free licensing to any reputable laboratory who wants to use it."[43]

In the United States, on May 12, 2009, the American Civil Liberties Union (ACLU) and the Public Patent Foundation, an advocacy group associated with the Benjamin Cardozo Law School, in New York, filed a landmark lawsuit in federal district court challenging the legitimacy of both Myriad's patents and the policy of the USPTO that allowed them. The suit was filed on behalf of a coalition of parties—several women with breast cancer or those at risk for it; various scientists and clinicians, including Ganguly and her collaborators;

and several biomedical organizations, including the Association for Molecular Pathology and the American College of Medical Genetics—claiming that they were or would be injured by Myriad's management of its patents.[44] The suit raised several legal and even constitutional issues, but the key question of whether isolated DNA was patent-eligible centered on whether such extracted DNA was, as the patent statute required, a new composition of matter rather than a product of nature. According to Myriad Genetics, what made BRCA1 and BRCA2 patentable was that they had been isolated from their natural state in the body and were thus no longer natural products. They were akin to a purified chemical molecule and merited a patent as such. Not so, argued ACLU et al. The sequence of cancer-disposing base pairs in the isolated gene was identical to that in the natural gene. It encoded specific genetic information whether it was in the body or removed from it. It thus remained a product of nature and was unpatentable.[45]

Judge Robert Sweet, presiding over the federal district court in which the case had been brought, agreed with the plaintiffs. On March 29, 2010 he struck down the two patents, explaining:

> The resolution of these motions is based upon long recognized principles of molecular biology and genetics: DNA represents the physical embodiment of biological information, distinct in its essential characteristics from any other chemical found in nature. It is concluded that DNA's existence in an "isolated" form alters neither this fundamental quality of DNA as it exists in the body nor the information it encodes. Therefore, the patents at issue directed to "isolated DNA" containing sequences found in nature are unsustainable as a matter of law and are deemed unpatentable subject matter under 35 U.S.C. § 101.[46]

Myriad Genetics appealed Judge Sweet's ruling to the Court of Appeals for the Federal Circuit, which is based in Washington, DC and which possesses sole jurisdiction over all appeals concerning patents arising from decisions in the federal district courts. On July 29, 2011, by a vote of two to one, the court upheld Myriad's challenged patents on BRCA1 and BRCA2—the majority finding that the isolated DNA from each of the two genes was patent-eligible.[47]

On December 7, 2011, the ACLU and the PPF petitioned the Supreme Court for a review of the case.[48] On March 26, 2012, the U.S. Supreme Court vacated the finding of the Court of Appeals in the BRCA DNA case, instructing it to reconsider that ruling in light of a decision the high court had announced a week before in *Mayo Collaborative Services v. Prometheus Laboratories*. In that case, the justices unanimously struck down a patent that covered the relationship between the size of a drug dose and the level of certain metabolites in the blood. Speaking for the Court, Justice Stephen Breyer, held that the relationship was unpatentable because it constituted a law of nature. [49]

The relevance of the decision to the BRCA DNA case seemed evident from Bryer's noting the Court's repeated emphasis "that patent law not inhibit future discovery" or "impede innovation more than it would tend to promote it" by granting monopolies over use of laws of nature, natural phenomena, and natural substances.[50]

The ACLU argued that Myriad's patents did thus inhibit and impede, but the Court of Appeals majority was not persuaded. On August 16, 2012, by two to one, it again upheld Myriad's patents on the DNA isolated from the BRCA1 and BRCA2 genes. On September 25, 2012, the ACLU and the PPF again petitioned the Supreme Court to review the case.[51]

However the case turns out, it has exposed with sharpness and clarity a fundamental difficulty in the extension of patent protection to isolated human genes. Among the justifications of patents is that the processes and the inventions they protect must be published, with the result that other inventors can be enabled in attempting to invent around and improve upon what is protected. In defending human gene patents, the USPTO has affirmed that view, saying that if genes are treated as are "other chemicals, progress is promoted because the original inventor has the possibility to recoup research costs, because others are motivated to invent around the original patent, and because a new chemical is made available as a basis for future research."[52] Myriad and its allies advanced similar arguments in defending the BRCA patents.

In fact, human gene patents establish no such incentive because no one can invent around a gene, including the mutated forms that cause disease. Unlike, say, carburetors, a gene that disposes a person to a disease is unique. Finding another gene that predisposes a woman to breast or ovarian cancer will not help identify whether she is at risk for either the BRCA1- or BRCA2 induced illnesses. Human disease genes thus constitute a kind of material good akin to those in which the public has a stake and for which by reason of circumstances there is no, or no competitive, alternative. Society excludes or allows only very limited private property rights in some such goods,—for example, Yellowstone National Park or the Cape Cod Seashore. It allows private property rights in others—say, railroads or the radio spectrum—but, as in the case of the railroads beginning in the late nineteenth century, it does not permit the property holders to use their rights of ownership absolutely. It regulates the property rights in service of a public interest.

There is ample foundation in the structure of American law for the regulation not only of companies but also of patented innovations that are essential to public interests, including health. Congress may grant the federal government "march-in" authority to license a patent to third parties if the patent holder has not made the invention available within a reasonable time or does not reasonably satisfy needs of health or safety.[53] Congress could extend regulation to human gene patents. Such regulation might take the form of compulsory or voluntary licensing, patent pools, or exemptions for research. Congress might also go further, modifying the patent statute to deny patentability to human gene sequences, which would then make them available to anyone for research into the gene, development of diagnostic tests for it, discovery of its functions and malfunctions, and creation of pharmaceuticals based on it. This is a position advocated by many scientists, patient groups, and medical practitioners, including the American College of Medical Genetics. The strategy would allow for the patenting of the tests and the drugs while leaving the gene freely available for research.[54] Whatever the particulars of the proposals, it is evident that a growing number of analysts and policymakers on both sides of the Atlantic hold that human genes are too essential to health—just as in the nineteenth century the railroads were too crucial to the economy—to allow private control of the intellectual property rights in them to be absolute and unregulated.

Notes

1. Committee on Intellectual Property Rights in Genomic and Protein Research and Innovation, Board on Science, Technology, and Economic Policy, Committee on Science, Technology, and Law: Policy, and Global Affairs, National Research Council, *Reaping the Benefits of Genomic and Proteomic Research: Intellectual Property Rights, Innovation, and Public Health* (Washington, DC: The National Academies Press, 2005), 22. Hereafter, NRC, *Reaping the Benefits*. Shortly thereafter, the National Institutes of Health (NIH), the lead agency in the project, chimed in, holding that the data should be "in the public domain, and redistribution of the data should remain free of royalties." Ibid. Parts of this chapter first appeared in Daniel J. Kevles, "Genes, Disease, and Patents: Cash and Community in Biomedicine," in Caroline Hannaway, ed., *Biomedicine in the Twentieth Century: Practices, Policies, and Politics* (Amsterdam: IOS Press, 2008), 203–216. I am grateful to Peter Westwick for research assistance and for support from the Andrew W. Mellon Foundation.
2. Daniel J. Kevles and Ari Berkowitz, "Patenting Human Genes: The Advent of Ethics in the Political Economy of Patent Law," *Brooklyn Law Review* 67 (Fall 2001): 237.
3. Ibid., 239.
4. This section is based on Pauline Maier et al., *Inventing America: A History of the United States*, 2nd ed. (New York: W.W. Norton, 2006), 518–523.
5. Government Accounting Office, *DOE's Physics Accelerators: Their Costs and Benefits* (GAO: RCED-85–96, April 1, 1985), 45.
6. Atomic Energy Act of 1946, Secs. 4, 6; Atomic Energy Act of 1954, Sec. 152. Executive Order 10096, January 23, 1950, gave the government rights to all inventions made by government employees during working hours or while using government facilities. Case law originating in implementation of the Order is reviewed by John O. Tresansky, "Patent Rights in Federal Employee Relations," Patent and Trademark Society, *Journal* 67 (1985): 451–488.
7. F. D. Gault, "Physics Databases and Their Use," *Computer Physics Communications* 22 (1981): 125–132.
8. Robert Kohler, *Lords of the Fly: Drosophila Genetics and the Experimental Life* (Chicago: University of Chicago Press, 1994), Chapters 2, 3, 5; Ernst Peter Fischer and Carol Lipson, *Thinking About Science: Max Delbrck and the Origins of Molecular Biology* (New York: W.W. Norton, 1988), 153–154.
9. Daniel J. Kevles, "Ananda Chakrabarty Wins a Patent: Biotechnology, Law, and Society, 1972–1980," HSPS: *Historical Studies in the Physical and Biological Sciences* 25, no. 1 (1994): 111–112.
10. Daniel J. Kevles, "Principles, Property Rights, and Profits: Historical Reflections on University/Industry Tensions," *Accountability in Research* 8 (2001): 12–26.
11. Gault, "Physics Databases and Their Use."

12. J. L. Heilbron and Robert W. Seidel, *Lawrence and His Laboratory* (Berkeley: University of California Press, 1989), 192–193, 196–199.

13. See Stuart W. Leslie, *The Cold War and American Science: The Military-Industrial Academic Complex at MIT and Stanford* (New York: Columbia University Press, 1993).

14. Daniel J. Kevles, "The Battle over Biotechnology," in Alan Brinkley and James McPherson, eds., *Days of Destiny* (New York: Agincourt Press, 2001), 460–463.

15. Ibid.

16. Ibid.

17. Ibid.

18. Daniel J. Kevles, "Of Mice and Money: The Story of the World's First Animal Patent," *Daedalus* 131, no. 2 (Spring, 2002): 78, 81–88; Daniel J. Kevles and Glen Bugos, "Plants as Intellectual Property: American Law, Policy, and Practice in World Context," *Osiris*, 2nd Series, VII (1992): 88–104; Kevles and Berkowitz, "Patenting Human Genes," 233–248.

19. Kevin Davies, *Cracking the Genome: Inside the Race to Unlock Human DNA* (New York: The Free Press, 2002), 64–65, 146–149, 208; NRC, *Reaping the Benefits*, 29.

20. Maurice Cassier, "Private Property, Collective Property, and Public Property in the Age of Genomics," *International Social Science Journal* 171 (2002): 87.

21. Ibid., 85. The Iceland courts later disallowed deCode's access to the health records of the country's citizens, finding it an invasion of privacy. The company then turned to building a database for genetic disease using volunteers. The venture has proved scientifically very useful but not commercially successful. In the summer of 2009, deCode was running out of money. Jocelyn Kaiser, "Cash-Starved Decode Is Looking for a Rescuer for Its Biobank," *Science* 325 (August 28, 2009): 1054.

22. Cassier, "Private Property," 84.

23. John Sulston and Georgina Ferry, *The Common Thread : A Story of Science, Politics, Ethics and the Human Genome* (New York: Bantam, 2002), 38–55; NRC, *Reaping the Benefits*, 45–46.

24. NRC, *Reaping the Benefits*, 46; Eliot Marshall, "Genome Researchers Take the Pledge," *Science* 272 (April 26, 1996): 477–478. Sulston proposed the rules as three rough principles, jotting them down on a blackboard. Sulston and Ferry, *The Common Thread*, 145–146.

25. Ibid.

26. Ibid., 27. See the statements of policy on sequencing and other data on the National Human Genome Research Institute web site: www.genome.gov/

27. Cassier, "Private Property," 84, 94.

28. David Dickson, "HUGO and HGS Clash over 'Utility' of Gene Sequences in US Patent Law," *Nature* 374 (April 27, 1995): 751; Eliot Marshall, "Intellectual Property: Companies Rush to Patent DNA," *Science* 275 (February 7, 1997): 780–781; Claire O'Brien, "US Decision Will Not Limit Gene Patents," *Nature* 385 (February 27, 1997): 755; Meredith Wadman, "NIH Is Likely to Challenge Genetic 'Probe' Patents," *Nature* 386 (March 27, 1997): 312; and Alison Abbott "Hugo Warning over Broad Patents on Gene Sequences," *Nature* 387 (May 22, 1997): 326.

29. Abbott, "HUGO Warning," 326.

30. United States Patent and Trademark Office, "Utility Examination Guidelines," *Federal Register* 66, no. 4 (January 5, 2001): 1092–1099.

31. Lori Andrews, "Genes and Patent Policy: Rethinking Intellectual Property Rights," *Nature Reviews Genetics* 3 (October 2002): 803–806.

32. *Madey v. Duke University*, 307 F3d 1351 (Fed. Circuit 2002); NRC, *Reaping the Benefits*, 23.

33. Cassier, "Private Property," 90.

34. Ibid.; NRC, *Reaping the Benefits*, 44.

35. Cassier, "Private Property," 89–90.
36. NRC, *Reaping the Benefits*, 52.
37. Cassier, "Private Property," 88.
38. NRC, *Reaping the Benefits*, 52.
39. Alison Abbot, "Europe to Pay Royalties for Cancer Gene," *Nature* 456 (December 4, 2008): 556.
40. Ibid.
41. Ibid.
42. Ibid.
43. Ibid.
44. American Civil Liberties Union, "ACLU Challenges Patents on Breast Cancer Genes," ACLU et al., "Complaint," May 12, 2009, http://www.aclu.org/freespeech/gen/brca.html, accessed September 5, 2009.
45. "Plaintiffs' Memorandum of Law in Support of Motion for Summary Judgment," August 26, 2009; "Myriad Defendant's Memorandum of Law (1) in Support of their Motion for Summary Judgment and (2) in Opposition to Plaintiffs' Motion for Summary Judgment," December 23, 2009, United States District Court for the Southern District of New York, Civil Action No. 09–4515, Case Docs. 62 and 153. Mark Skolnick, a professor at the University of Utah and a founder of Myriad Genetics, contested the idea that DNA is information, insisting that it is a chemical and must be treated as such for patent purposes. He said, "If you discover a new molecule, whether it's a pharmaceutical or a paint or a dye or a gene, it's a new molecule, you should be protected; ... genetic patents really follow the model that's been set up in organic chemistry." Cassier, "Private Property," 88.
46. Judge Sweet, "Opinion [re: Summary Judgment], March 29, 2010, Case Doc. 255, p. 4, *Association for Molecular Pathology et al [Plaintiffs] v. United States Patent and Trademark Office; Myriad Genetics {Defendants}* . United States District Court for the Southern District of New York. Civil Action No. 09-4515 (RWS), suit filed May 12, 2009.
47. United States Court of Appeals for the Federal Circuit, *The Association for Molecular Pathology et al. v. United States Patent and Trademark Office and Myriad Genetics, Case 2010–1406, Appeal ... Decided July 29, 2011.*
48. "ACLU and PUBPAT Ask Supreme Court to Rule that Patents on Breast Cancer Genes Are Invalid," ACLU press release, Dec. 7, 2011.
49. U.S. S.Ct., Court Orders, Case 11-725, at: http://www.supremecourt.gov/orders/courtorders%5C032612zor.pdf; *Mayo Collaborative Services v. Prometheus Laboratories*, 566 U. S. 1-4, (2012); Adam Liptak, "Justices Back Mayo Clinic Argument on Patent," *The New York Times*, March 20, 2012.
50. U.S. S.Ct., Court Orders, Case 11-725, at: http://www.supremecourt.gov/orders/courtorders%5C032612zor.pdf; *Mayo Collaborative Services v. Prometheus Laboratories*, 566 U. S. 1-4, (2012)..
51. ACLU and the Pulic Patent Foundation, "Supplemental Brief for the Appellees," June 15, 2012, *Association for Molecular Pathology et al [Plaintiffs] v. United States Patent and Trademark Office; Myriad Genetics {Defendants,* United States Court of Appeals for the Federal Circuit, No. 2010-1406. "ACLU and Pubpat Ask Supreme Court to Rule...," ACLU press release, Sept. 25, 2012.
52. Ibid.
53. Andrews, "Genes and Patent Policy," 806. Andrews notes that "under the Clean Air Act, courts can, when necessary, order compulsory licensing of patents on equipment or technology used in air pollution control on reasonable terms to ensure competition." Ibid.
54. Cassier, "Private Property," 84, 95.

8

Epidemiology, Tort, and the Relations between Science and Law in the Twentieth-Century American Courtroom

Tal Golan

This chapter follows the intertwined careers of epidemiology and toxic tort litigation, and examines their effects on the relations between science and law in the late twentieth-century American courtroom. Epidemiology's career in the American courtroom has been short but brilliant. Until the 1970s, epidemiological evidence could hardly be found in the legal system. By the 1980s, it was already announced "the best (if not the sole) available evidence in mass exposure cases,"[1] and by the start of 1990s, judges were dismissing cases for not supporting themselves with solid epidemiological evidence.[2] Epidemiology, I argue below, owed much of this prosperity to the equally meteoric career of mass tort litigation—a late-modern American species of litigation involving crowds of plaintiffs, all claiming to be harmed by the same exposure or mass-marketed product. Ever since the 1980s, dangerous drugs, industrial accidents, design defects, environmental pollutants, radiation exposure, and other species of technological breakdowns, have all become the subject of prolonged mass tort litigation with ever-escalating financial stakes.[3] Questions about risk and causation have been central to a great majority of these cases, and when a direct proof of cause and effect has proven elusive, the courts turned to statistical evidence to resolve these questions.[4]

Tort is a branch of private law that deals with personal injury claims. Early in the twentieth century tort still prided itself on its long tradition of personalized services. Its clients were wilful and rightful citizens whose causal agency could not be subsumed mechanically, without the careful exercise of human judgment on a case-by-case basis.[5] But as the twentieth century progressed, tort law became less private and more public, and by the end of the century the "statistical victim" became tort's biggest client, and epidemiology its favorite science.[6] With the new client came new practices: individual care gave way to economy of scale and direct testimony to statistical evidence. These were uneasy changes for tort law and they presented the legal mind with a host of difficult problems regarding the differences between statistical correlation and legal causation; the circumstances in which we could pass from one to the other; and how and by whom should these be decided.

By the end of the twentieth century, this set of problems had reshaped the relations between law and science. The warning was sound that the courts have been infested with junk science, and a chorus of commentators urged the judiciary to tighten their control over science admitted into the courtroom. In response, the US Supreme Court, which had never before addressed the practices of scientific evidence, found it necessary to visit the topic on three separate occasions during the 1990s, all of them tort cases.[7] Christened as the "Daubert Trilogy," the three Supreme Court opinions announced the arrival of a new era in the relations between law and science. The traditional legal deference to scientific expertise was overruled. Instead, the trial judge, who had long been passive in the play of science in the adversarial courtroom, was newly charged with the responsibility of preventing junk science from entering the courtroom and bamboozling the lay jury.[8] This new role of the judge as a gatekeeper of true science, I suggest below, was corelated with the new role of the statistical expert as the gatekeeper of true causes in mass tort litigation.

The rise of epidemiology

Modern science has offered public decision-makers two distinct modes of calculating risks and constructing causality: toxicology, an experimental reductionist science, built on the strength of the laboratory; and epidemiology, an observational statistical science, built on the power of big numbers. Earlier in the twentieth century the toxicity of things was checked in the laboratory. One strategy, called in vitro studies, examined the effects of chemical agents on various organic materials ranging from DNA and proteins, to cells, bacteria, and even embryos, in attempt to understand the biochemical mechanisms involved. Molecular structural analysis was also called upon to gain clues from structural resemblance to other, better known, chemicals.[9]

It is a long way, however, from molecules to humans, and other researchers have taken a shortcut by performing in vivo studies. This reduced some difficulties but introduced new ones. Unable to experiment directly with humans, the toxicologists run their studies on other mammals. But even though much is common across the mammalian species, much is also different, and scientists were not always sure which is which. In addition, in vivo studies typically involve larger-than-life doses, to shorten the experiment and to augment the effects. To make these studies policy relevant, toxicologists must then extrapolate from the short and intense exposure of the tested mammals to a chronic low-level exposure of humans.[10] The extrapolation is dubious, but it allows to work the numbers into a dose-response curve that allows calculating the risks per any given dose and any given period, and most importantly for the setting of exposure standards, with appropriate safety factors to protect the more susceptible subpopulations.[11]

During the 1970s, as environmental regulation took central stage in Western polity, the capacity of this laboratory science to provide reasons good enough to legitimize administrative action was closely scrutinized. As the young regulatory agencies began to churn out their safety standards, both industry and civil action groups challenged the science behind the standards—industry in attempt to moderate the standards; civil activists, to step them up.[12] The ensuing legal battles revealed to all the fragility of the science involved. What had thrived in the temperate climate of the laboratory did not survive the adversarial heat of the courtroom. The notorious nonlinearity of physiological systems was mobilized to undermine the extrapolations from high to low doses and from short to long exposures, and the poorly understood interspecies and intrahuman variations were called upon to show that the justification of the standards went beyond scientific and technical competence.[13]

Eager to protect the regulatory regime, the legal system responded by adopting the powerful precautionary doctrine, which admitted the fragility of the science involved but justified the right of the authorities to act upon it, based on the ever-pressing need to regulate potential risks before they turn into actual harms.[14] The legitimacy of such a regulatory regime, the courts prescribed, resided in its deployment of the best scientific tools available. These tools, the judges also increasingly suggested, may no longer be found in the laboratory but in the arsenal of epidemiology.[15]

Earlier in the twentieth century, epidemiology served public policy as a form of surveillance technology.[16] Medical attention was focused on infectious diseases—each caused, it was generally held, by a specific microbiological agent. Fighting infectious diseases was a job for the laboratory—to isolate the specific causal organism, study it, and devise the best means to fight back.[17] Epidemiology served in this campaign merely by informing of geographical and social patterns of the disease. But by the middle of the twentieth century the balance had begun to shift. The battle against infectious diseases seemed to have been won in the developed world, and public and medical attention increasingly turned to a new pattern of diseases: noninfectious, chronic, with long latency, and poorly understood etiology; diseases such as blood pressure, cancer, or heart problems—all of which were previously considered inevitable failures of the aging organism—now began to top the medical charts.[18]

Experimental science, with its reductionist logic, made little progress with these so-called diseases of civilization. They seemed to involve multiple causes and effects; their long latency made experimentation difficult, and their mechanisms kept eluding the researchers. Epidemiology, however, proved much more flexible. A postfacto observational science that relates exposure to outcome, it did not have to ponder too much over the illusive biological mechanisms involved. Instead, epidemiologists adapted their computational strategies to a distributed, multivariate model of causation that seemed to better fit the nature of these new diseases, where a cause could have many effects and an effect many causes.[19]

The power of epidemiology to make causal claims in this new weblike universe of irreducible, chronic health problems was first demonstrated during the late 1950s and early 1960s, when a cluster of British and American epidemiological studies first implicated cholesterol and smoking as significant causal factors for heart disease, and in the case of smoking, also for lung cancer.[20] Running ahead of experimental research, these studies made no appeal to concrete biological mechanisms.[21] Instead, they introduced a new lexicon that appealed only to what came to be known as "risk factors"— environmental, social, and other patterns that are statistically correlated with higher incidence of disease; the more robust the correlation the more certain the association. Nevertheless, or precisely because of it, many medical scientists went up in arm. At stake, they cautioned, was no less than the scientific essence of modern medicine, which was very much rooted in the laboratory. Epidemiology, they pointed out, was not an experimental science. It could neither sufficiently control its data nor test the veracity of its conclusions. Thus, while epidemiology remained useful in generating causal hypotheses, only experimental science could reliably validate them.[22]

Criticism of the newfangled epidemiology was by no means limited to die-hard experimentalists. Geneticists faulted epidemiology for focusing attention on environmental effects, while social scientists blamed it for concentration on individual factors abstracted of social context. The most damaging critique came from within—from biostatisticians anxious to protect the integrity of their science and from epidemiologists who were concerned that too much would be claimed for their fledgling science that was just starting to make inroads into medicine. These sophisticated critics were able to point out various methodological difficulties inherent to epidemiological research, from selection biases to confounding variables, all of which further undermined epidemiology's capacity to establish authoritative causal claims.[23]

The proponents of the new risk-factors epidemiology responded by appealing to usefulness rather than truthfulness. They pointed out that although a clear experimental demonstration of a concrete causal relation may indeed constitute a higher form of proof, it was nevertheless hard to come by in this new era of chronic diseases. In the absence of such strong proof, they prescribed a diet of epistemological modesty and methodological flexibility. The distributed nature of the problem was to be matched by an equally distributed scientific effort. The epidemiologist's search for health risks was still to be based on the strength of carefully constructed statistical studies, but they should remain mindful of the limitations of their method and be careful to support it with other types of evidence. In the absence of a concrete demonstrable mechanism, they should nevertheless look for a plausible biological explanation. In the absence of direct experimental control, the epidemiologists should support their causal hypothesis by plausible temporal and dose-response curves, and indeed by any other coherent source of evidence. Neither of these explanatory factors was sufficient or necessary, nor could any of them bring forward indisputable evidence for or against the causal hypothesis researched. Epidemiologists should therefore qualify their confidence with appropriate confidence margins, and single studies should be treated skeptically until their results are verified by other studies, conducted by different persons, in various places, circumstances, and times. The combined weight of these studies, they maintained, was in a growing number of cases the best science could offer public health decision-makers in this new era of latent and irreducible causes and chronic diseases.[24]

Disdained by scientific purists, this pragmatic program of epidemiology was warmly embraced by the expanding regulatory regimes of the late twentieth century. Practical by nature, judges, legislators, administrators, and public health officers were less concerned with the rigorous pursuit of experimental design and more with the pressing businesses of public policy, which often necessitated judgment made with less than perfect information.[25] They found epidemiology with its quantified logic and its focus on the population as the unit of investigation perfectly placed to provide them with potent tools to estimate the prevalence of otherwise irreducible health problems, investigate their probable sources, identify those groups with elevated risks, and target them with preventive measures.[26]

The later part of the twentieth century therefore saw the flourishing of the so-called black-box epidemiology—a technical, policy-driven epidemiology that shunned biological hypotheses and concentrated on computing the risks facing taxpayers from a myriad of modern conditions.[27] The parallel growth of medical registries and computer technology allowed for the deployment of increasingly complex statistical techniques in the search for smaller and smaller risks in larger and larger populations. The epidemiologists traded up their mechanical rulers first for punch cards and then for software programs, and got comfortable with the new tools of multivariate correlation and regression, and exotic tests of statistical significance and confidence intervals. By the end of the twentieth century, the reduction of causes to a distributed network of risk factors had become prevalent and increasingly informed medical research as well as regulatory and legal action. In theory, some continued to insist that this was not a science of causation. In practice, however, it was exactly this—a hunt for causes; if not for science then certainly for administrative and legal action.[28]

The rise of mass toxic torts

Tort's tradition of private, individualized justice cultivated a theory of causality as reductive as that of the science of infectious disease. To exist, a legal cause had to be reduced to a causal agent.[29] This causal agent was a human being, not a microbe, a fact that added a moral dimension and much complexity to the process of proof. Nevertheless, the plaintiff's burden of proof, like that of the medical experimentalist, was to single out the causal agent and demonstrate the chain of events that linked the agent's actions to the plaintiff's injury. If a specific causal agent could not be uniquely determined,;

if the plaintiff could show only that the defendant's action might have caused the harm; or if another indistinguishable potential cause existed, the courts dismissed the claim for the failure to prove specific causation.[30]

This reductionist model of specific causation has worked quite well in traditional tort cases, such as accidents or assaults. The defendant's identity and conduct could be verified by direct evidence such as eyewitness testimonies, and the causes for a black eye or a flooded house were understood well enough to allow the courts to decide liability based on whether those causes were controlled by the defendant. This was not the case, however, in a growing range of environmental, work-safety, and product liability cases that came to be known by the end of the 1970s as "toxic tort" cases.[31] These cases involved injuries of the kind that has frustrated experimental science—chronic, with long latency, and poorly understood etiology; injuries that could not be comfortably reduced to a single cause. In the absence of direct or experimental proof of cause and effect the courts increasingly turned in these cases to epidemiological evidence. That was particularly true for the new and emerging phenomenon of mass toxic tort litigation that clustered together large crowds with various case histories, all claiming to be harmed by the same exposure or by the same standardized, mass-marketed product. Here, lawyers and judges—just like legislators, administrators, and public health officers—found epidemiology's quantified logic and population-based analysis particularly conducive to their needs.[32]

The helpfulness of epidemiology in deciding the slippery question of causation in toxic tort cases was first demonstrated in the late 1970s by two massive mass tort litigations involving asbestos and the first synthetic hormone, diethylstilbestrol (DES).[33] In both litigations epidemiological evidence played a major role in establishing causation by demonstrating a strong correlation between the exposure and a unique 'signature' disease among the exposed. Mesothelioma, a rare form of cancer, was alleged to be uniquely associated with asbestos exposure, and Adenocarcinomas of the vagina and uterus was claimed to be almost unknown among women whose mothers had not taken DES.[34] These exclusive relations allowed the plaintiffs to argue that their exposure to asbestos or DES was responsible for their specific ailment and to win decisive legal victories against the manufacturers.[35] The successes of the asbestos and DES plaintiffs brought a rising tide of toxic tort actions to the courts in the early 1980s. The two largest actions were *Allen v. United States* and *Agent Orange*, and each of them presented fresh challenges to the judicial embrace of epidemiology.

Allen v. United States

For 12 years, between 1951 and 1963, the US government detonated more than 100 atomic bombs at test sites above and below the southern Nevada desert. Three decades later, in the early 1980s, civilians who lived in the neighboring regions entered 1,192 individual lawsuits against the government, accusing it of negligence and carelessness in carrying out the tests and demanding hundreds of millions of dollars in damages for hundreds of radioactive-related deaths and injuries.[36] By the 1980s, scientific research, including studies of surviving victims of World War II atomic warfare, had left little doubt that ionizing radiation can indeed cause cancer.[37] Still, the downwinders, as the plaintiffs came to be known, suffered from all kinds of cancers, many of which could be found also in the general population and could have resulted from causes other than the exposure to radioactive fallout. The downwinders found it therefore extremely difficult, if not impossible, to satisfy the legal demand for a proof of specific causation and to persuade the court that their ailments would not have occurred but for the radioactive fallout from the nuclear testing.

Despite the lack of an adequate proof of specific causation, Bruce Jenkins, the federal district judge who tried the litigation, refused to dismiss the case.[38] In a 489-pages massive opinion, Jenkins assembled ample precedents to show that in cases in which the defendant's conduct was manifestly tortuous but the plaintiff had no means of identifying the specific cause of injury, the courts had taken steps to ease the plaintiff's burden of proof by shifting some of it to the defendant.[39] Jenkins considered *Allen v. United States* to be such a case. He found the government negligent not only for failing to provide off-site civilians with adequate warnings and protection from the radioactive fallout, but also for failing to adequately monitor and record off-site exposures, thereby depriving the plaintiffs of information crucial for the proof of causation. In such circumstances, Jenkins reasoned, causal analysis using "but–for" tests in any form falls short of the mark. Instead, the requirements should reflect both the objective difficulties involved in the proof of causal relation between radiation and nonspecific cancers, and the government's responsibility for encumbering these difficulties. Thus, Jenkins ruled, it was sufficient for the downwinders to present properly-supported epidemiological evidence "from which reasonable men may conclude that it is more probable that the event was caused by the defendant than it was not." Once the plaintiff had done so, Jenkins prescribed, the burden of proof will shift to the government to produce evidence extricating itself from the tangle of causality.[40]

Allowing each of the 1,192 individual plaintiffs to have his or her day in court was a tall order, especially since it was not clear whether an appellate court would not later dismiss the whole litigation and exempt the government from claims for damages caused by policy decisions. Jenkins decided therefore to test the water first by trying a group of 24 cases, selected out of the nearly 1,200 claims on his docket.[41] In deciding these "bellwether" cases, Jenkins relied heavily on the epidemiological studies available. In nine of these cases that involved leukemia and thyroid cancer, the numbers demonstrated a significant increase in the incidence of the cancer within the exposed population, and Jenkins held for the plaintiffs. In 14 cases that involved other cancers and lacked convincing statistical evidence, he ruled against them.[42]

As expected, in 1987, Jenkins's ruling was overturned by the Tenth Circuit Court of Appeals on the basis that the United States was protected by the legal doctrine of sovereign immunity.[43] However, the appellate decision did not discuss Jenkins's innovative decision to rely on epidemiological evidence in establishing factual causation, even when other possible causes could not be excluded, and it remained standing. Still, the success of Jenkins's strategy depended on the availability of an authoritative body of epidemiological research that could compensate for the lack of direct evidence and allow for a causal determination even in the presence of alternative causes. But this could hardly be expected in many mass tort actions, given the scarcity of even the most basic toxicity data.[44] How was the court to decide causation then, in the absence of an authoritative scientific advice? This question stood at the center of the largest and most publicized mass toxic tort litigation of the 1980s—the *Agent Orange* case.

Agent Orange

The *Agent Orange* action was brought by many thousands of Vietnam veterans who believed they had suffered or might suffer a variety of diseases due to their war-time exposure to Agent Orange—an herbicide the US military spread widely on Vietnam's jungles to destroy the advantages they afforded to the enemy. Agent Orange contained minute quantities of dioxins, a family of highly toxic compounds that the veterans believed were responsible for their health problems, which included cancers, heart attacks, a suppressed immune system, hormonal imbalances, diabetes, menstrual problems, increased hair growth, and weight loss.[45]

Much was in common between *Allen v. United States* and *Agent Orange*. As with ionizing radiation, little doubt existed about the severity of dioxins at high doses but far less was clear about their impact at lower doses. Like in *Allen*, the specific levels of individual exposure to Agent Orange were unknown and had to be reconstructed from insufficient military records and from personal memories, many years after the fact. Like the downwinders, the Vietnam veterans suffered from a variety of ailments that could be found in the general population and could not be reduced exclusively to dioxin exposure. And like Jenkins, Jack Weinstein, the federal district judge who managed the *Agent Orange* case, was willing to rely on epidemiological studies alone to establish factual causation. "We are in a different world of proof than that of the archetypical smoking gun," Weinstein noted. "We must make the best estimates of probability that we can, using the help of experts such as statisticians and our own common sense and experience with the real universe."[46]

However, unlike *Allen v. United States*, the best estimates of probability in *Agent Orange* left much in doubt regarding the capacity of Agent Orange to cause the alleged harms. The epidemiological studies undertaken by the federal government and various state agencies failed to demonstrate a statistically significant increase in the rate of relevant ailments among the veterans and their families. The only alleged injury that was demonstrably correlated with exposure to Agent Orange was chloracne, a disturbing but hardly fatal form of acne.[47] In other words, the statisticians failed to find a causal connection between Agent Orange and the veterans' ailments, leaving Judge Weinstein with the following dilemma: Can he find a causal connection where the statisticians failed to find one?

Two kinds of error can be made in the quest for true causes: a false cause can be found (false positive) and a true cause can be overlooked (false negative). Epidemiologists have always been more vigilant about the first kind. To guard against the possibility of claiming associations where they do not exist, they adopted a two-tier defense strategy called the *null hypotheses*. This strategy operates under the presumption that no causal connection exists between the exposure and the disease under study, and demands a strong proof to reverse this presumption. The strength of such proof depends on two things: the measurement of a high enough risk and the assurance that this measurement is not false, the fruit of chance alone.

Epidemiologists use a simple relative index to measure risk in exposure cases. The index is defined by the ratio of the measured incidents of the disease in the exposed (numerator) to the unexposed (denominator) groups tested. A risk ratio of one signifies that the incidence rate is the same among the exposed and the nonexposed and thus indicates a lack of association between the suspected exposure and the alleged disease; a risk ratio greater than one suggests that the exposed are in higher risk of disease than the nonexposed; and a risk ratio greater than two indicates that the exposed more than doubled their chance to contract the disease. From a population perspective, this means that more than half of the exposed owed their disease to the exposure. From the individual's perspective, the epidemiologists suggested, one could interpret it to mean that the exposure was more likely than not responsible for his or her specific disease.[48]

But the epidemiologists are not satisfied with measuring the strength of the risk. They demand an assurance that their measurements do not lead to a false association. Statistical theory provides such an assurance by calculating the probability of false association, and epidemiological dogma demands it to be smaller than 5 percent (i.e., less than 1 in 20) for the association to be considered statistically significant. This "statistical significance" standard is far more demanding than the "preponderance of the evidence" (or "more likely than not") standard used in civil law. It reflects the cautious attitude of scientists who wish to be 95 percent certain that their measurements are not spurious. But such prudence comes with a price. The rates of false positives and negatives are inversely related. Hence, the more you guard against false causes the more you are bound to miss true ones.[49] Epidemiologists have considered the price well worth paying. So has criminal law, which emphasizes the minimization of false conviction, even at the price of overlooking true crime. But civil law does not share this concern. Unlike science or criminal law, it has no preference for either false positives or negatives. It only cares for the preponderance of the evidence.

Both Jennings and Weinstein were aware of this incommensurability between epidemiology and civil law. But they differed in their reactions. Jenkins noted that the statisticians' 95 percent probability requirement for significance was arbitrary and stringent, and cautioned his colleagues not to constrain themselves by simplistic models of causal probability imposed upon the judicial preponderance of the evidence standard. "Like statistical significance," he wrote, "mathematical probability aids in resolving the complex questions of causation raised by this lawsuit, but is not itself the answer to those questions."[50] Judge Weinstein, however, was far less concerned with the strictness of the epidemiology. A scholar of evidence law, and a known critic of the deployment of science in the adversarial courtroom, Weinstein embraced the stringent 95 percent significance threshold as a ready-made admissibility test that could validate the veracity of the statistical evidence used in court. Thus, although he referred to epidemiological studies as "the best (if not the sole) available evidence in mass exposure cases," he nevertheless refused to accept them in evidence, unless they were statistically significant.[51]

In the absence of statistical significance, the veterans' lawyers based their proof of factual causation on animal studies and supported it with occupational studies of industrial accidents involving dioxin that demonstrated the potential of dioxins to cause many of the ailments involved. But Weinstein discounted both types of evidence. The differences in species tested and in the high levels of exposure examined, he maintained, undermined the significance of these studies, and without the support of epidemiology, they did not suffice to prove causation in tort. Still, like Jenkins before him, Weinstein was reluctant to allow the strict views on causation in tort prevent the veterans from recovering. Unable to satisfy the stringent standard of proof required in tort, Weinstein chose to question the applicability of this standard, and by implication, the applicability of the entire traditional tort system to the late-modern phenomenon of mass toxic tort litigation.[52]

To remind you, under the traditional causation doctrine in tort, statistical correlations alone were insufficient, even if indicating that the probability of causation exceeds 50 percent (e.g., a risk ratio greater than 2). Some additional proof was required to shift the legal mind to one side or the other; preferably some direct testimony about the causal relationship between the defendant's conduct and the plaintiff's injury. Weinstein, like Jenkins, noted that the chance for such evidence is very small in mass toxic tort cases, and that the consequence of retaining this requirement might allow defendants whom, "it is virtually certain, have injured thousands of people and caused billions of dollars in damage to be free of liability."[53] Jenkins, in his bellwether cases, modified the causation requirements to allow a verdict in mass tort cases chiefly on statistical evidence. Weinstein seemed to side with Jenkins's modification, but argued that its successful adoption required further procedural adjustments.

Weinstein pointed out that the application of epidemiological evidence in a mass tort action on a case-by-case basis will not only be an administrative nightmare but will also almost always result in either under or overcompensation. If the probability calculated is a hair less than 50 percent, each and all plaintiffs will lose and a clearly tortuous defendant could walk away. And if the probability be a hair over 50 percent, each and all plaintiffs will win, including those not injured by the defendant. Shifting the burden of proof does not solve the problem. A defendant would still have to compensate all or no one, depending on which side of the 50 percent threshold the probability fell. This made no sense to Judge Weinstein. Given the unprecedented scale of mass tort and its financial stakes, he was worried about the potential implications of this problem, which could lead to the financial ruin of an entire industry or the deprivation of a large number of injured people from proper compensation.[54]

Weinstein's solution was as straightforward as it was radical: given the necessarily heavy reliance on statistical evidence in mass exposure cases, the time-honored tort practices of plaintiff-by-plaintiff and winner-takes-it-all will have to go. Mass tort cases should "try all plaintiffs' claims together in a class action thereby arriving at a single, class-wide determination of the total harm to the community of plaintiffs...The defendant would then be liable to each exposed plaintiff for a pro rate share of that plaintiff's injuries." In short, if mass toxic torts are to allow verdicts based on statistical evidence, the courts need to match it with the equally aggregative mechanisms of class action and proportional liability.[55]

Weinstein was aware that his cutting of the Gordian knot of mass toxic tort ran against the legal grain and would probably fail if the *Agent Orange* action would go to trial. He therefore pushed the parties to sign an out-of-court class-action settlement he engineered. He cajoled the industry to put together a modest $180 million fund and ordered its distribution among the 250,000 Vietnam veterans on the degree of disability alone, regardless of cause. True to his analysis, Weinstein later summarily dismissed without a trial the individual claims of those veterans who chose to opt out of the agreement and insisted on their day in court. Under the existing tort doctrines, he ruled, it was unfeasible to causally connect their individual ailments to Agent Orange exposure without solid epidemiological evidence.[56]

Allen v. United States and *Agent Orange* were key chapters in the adaptation of late twentieth century American tort law to the challenges of mass toxic tort litigation. They put on display the inadequacies of the traditional tort doctrine of causation in dealing with mass toxic torts litigation and clarified many of the differences between the questions asked by law and the answers given by science. Jenkins and Weinstein, each was able to fashion a remedial process to compensate for the evidentiary complexities inherent in mass toxic tort litigation. Both solutions acknowledged the central role epidemiological evidence came to play in the resolution of mass toxic tort cases, but neither of them seemed general enough. Jenkins's solution depended on the unlikely availability of an authoritative body of epidemiological research, and his case by case approach was inapplicable enmass. Weinstein's solution, fusing probabilistic causation with class action and proportional liability, was equally inapplicable for the everyday businesses of tort. Nevertheless, his dismissal of animal studies as of "so little probative force and are so potentially misleading as to be inadmissible," and his championing of epidemiology and its strict statistical-significance test proved remarkably influential in the years to come.[57]

Junk science, epidemiology, and legal reform

The rapid growth of mass tort litigation and its unprecedented financial consequences bred much anxiety and contention.[58] Not surprisingly perhaps, the balk of the criticism was directed at the science involved. By the early 1990s, the alarm was sounded that America's courts were being swamped by junk science, produced by unscrupulous experts hired by opportunistic attorneys aiming for the deep pockets of America's corporations.[59] The legal embrace of epidemiology was central to this growing debate over junk science. This time around, the critics were concerned less with the scientific nature of epidemiology and more with the ability of the courts to handle its ruse.[60] Respectable judges found themselves more confused than enlightened by technical terms, such as significant levels, confidence margins, and P-values, and made embarrassing mistakes.[61] Still, judges can be trained and procedures can be improved.[62] The real concern lay with the lay jury and their ability to handle the rich subtleties produced by the exploding market of expert epidemiological advice. The distrust in the jury's capacity to handle complex evidence runs long and deep in American legal culture, and the well-financed junk science campaign gave it new energy and focus.[63] To shield the credulous jury from pseudoscientific expertise and protect corporate America from greedy lawyers, the judges were urged to become more vigilant with the new science they let into their court.[64]

The complimentary debates about the proper role of judges and epidemiologists in mass toxic tort litigation and the standards each of them should follow in their own art have crossed paths in another mass toxic tort litigation that has occupied the courts since the early 1980s. This one involved Bendectin, a drug that was widely prescribed during the 1960s and 1970s for pregnant women to combat nausea. Ultimately, approximately 2,000 suits were filed against the drug manufacturer, Merrell Dow Pharmaceuticals, Inc., asserting that Bendectin caused a wide variety of birth defects, ranging from limb reductions to heart defects to neurological problems.[65] Merrell Dow denied, refused offers for aggregated settlement, and instructed its lawyers to fight every case in court.

Like in other mass toxic tort litigations, the crucial battles of the Bendectin litigation were over the causal relation between Bendectin and the plaintiffs' illnesses. On one side, to prove a causal link, the plaintiffs offered toxicological evidence that included in-vivo and in vitro studies that found links between Bendectin and malformation, and chemical analysis that pointed to structural similarities between Bendectin and other substances known to cause birth defects. On the other side, Merrell Dow's lawyers based their defense strategy on the failure of a growing number of epidemiological studies to demonstrate a statistically significant causal connection. Citing Weinstein's in *Agent Orange*, they discounted the relevancy of the animal studies and chemical analysis, and claimed that in the absence of solid epidemiological support, the scientific evidence was insufficient to show causation in tort.[66]

By the end of the 1980s, the ongoing evidentiary battle between the plaintiffs and Merrell Dow in the Bendectin litigation was tilting toward the later. Finding themselves increasingly dependent on epidemiological evidence, the courts responded to the growing criticism against junk science by dismissing Bendectin cases for lack of statistically significant epidemiological evidence.[67] This culminated the remarkable legal career of epidemiology. At the start of the 1980s the courts still debated whether to allow epidemiologists to weigh in on the issue of causation. A decade later, they were summarily dismissing suits, and even reversing jury verdicts, when they could not support themselves with statistically-significant epidemiological evidence.[68]

Epidemiology was not given a free hand in the courtroom though. To fit it into tort, the courts divided the proof of causation into two: general and specific. General causation referred the potential of a given exposure to cause injury; specific causation, to the actual harm claimed by the plaintiff. The theoretical distinction between these two types of causation was not new, but it began to play an important role in tort litigation only in early 1990s, with the growing concerns with the ability of the lay jury to handle the complexities of the scientific evidence in toxic tort litigation.[69] The proof of general causation was increasingly provided by epidemiology and was checked by the judge before the trial, during the admissibility stage.[70] Only upon the judge's satisfaction that the potential for harm was proven, could the legal action move forward to the trial stage, where the issue of specific causation could be examined by the jury.[71]

Daubert and the legal standards of admissibility

Prior to the twentieth century, there was no special admissibility test for scientific evidence. Like every other type of evidence, scientific evidence was evaluated according to its relevancy, helpfulness, and the qualifications of the witness.[72] Wary of the need to give preference to one kind of science over another, nineteenth-century judges followed a lenient admissibility policy in the case of expert witnesses, and left it for the lawyers to expose quackery during cross-examination, and for the jury to be the judge of the ensuing battle between the lawyers and the experts.[73] No one, of course, trusted the jury to be able to do this job properly. Still, the courts considered it a fair price to pay for a free market of expertise that was considered the best protection from the abuse of political and executive powers.[74]

It was only after World War II that American courts began to consistently apply a distinct standard for the admissibility of scientific evidence, and even then it was only in criminal cases.[75] To that end, the courts resurrected a 1923 opinion of the DC Court of Appeals that rationalized the decision of the lower court to exclude a prominent expert in scientific lie-detection from testifying in a murder case to the veracity of his client's alibi. "While courts will go a long way in admitting expert testimony deduced from a well recognized scientific principle or discovery," the DC Court of Appeals prescribed, "the thing from which the deduction is made must be sufficiently established to have gained general acceptance in the particular field in which it belongs."[76] The lie-detector technology, the appellate court reasoned, did not receive such general acceptance and was therefore properly excluded.

Known as the "general acceptance" standard, or simply as *Frye* (after the defendant's name), the courts increasingly used it during the 1960s and 1970s to decide the admissibility of an array of technologies that was offered by the up-and-coming crime laboratories: voice prints, neutron activation analysis, gunshot residue tests, bite mark comparisons, scanning electron microscopic analysis, truth sera, and others.[77] By the 1980s, *Frye* was well established as the general standard for the admissibility of novel scientific evidence in criminal trials.

Still, *Frye* was not the only user manual in town. In 1975 the *Federal Rules of Evidence* (FRE) were enacted and prescribed no special test to ensure the reliability of scientific evidence, new or old. Instead, the *FRE* cast the widest net possible and provided that "if scientific, technical, or other specialized knowledge will assist the trier of fact to understand the evidence or to determine a fact in issue, a witness qualified as an expert by knowledge, skill, experience, training, or education, may testify thereto in the form of opinion or otherwise."[78] The *FRE* was generally interpreted as the more liberal of the two standards, encouraging a more flexible judicial consideration of scientific evidence. However, since the *FRE* did not state an explicit intent to abandon *Frye*, some federal, and almost all state courts, remained committed to the general acceptance criterion as the prerequisite to the admissibility of scientific evidence, at least in criminal cases.[79]

The tensions between judges and experts, between experimental and statistical science, and between *Frye* and the *FRE*, all came to a head in 1993, in yet another Bendectin federal case in which a minor named Jason Daubert sued Merrell Dow for his birth defects.[80] Daubert's lawyers offered the court the usual toxicological mix of in vitro, in vivo, and structural evidence that pointed to links between Bendectin and the birth defects. In light of the growing judicial emphasis on epidemiological evidence, Daubert's lawyers were careful to support their cause with a well-qualified statistical expert, who pooled together data collected by previous epidemiological studies, reanalyzed it, and was able to detect statistically significant links between the drug and the birth defects. Alas, this so-called meta-analysis was rejected by the trial judge. Prepared especially for the trial, he reasoned, the study was never subjected to peer-review and thus could not be considered under *Frye* as generally accepted. Stripped of epidemiological support, the judge then concluded, the plaintiffs' scientific evidence was insufficient to prove causation and gave a summary judgment for the defendant, Merrell Dow.[81]

Daubert's lawyers appealed all the way to the Supreme Court, arguing that the *FRE* superseded *Frye*, and that according to the *FRE*, it is for a jury, not a judge, to determine the sufficiency of their scientific evidence. To the surprise of many, the Supreme Court, which had never before taken interest in the procedures of scientific evidence, agreed to review the Daubert case to clarify the proper admissibility standard of scientific evidence. With the stakes raised and the focus of the debate shifted from causation in tort to the admissibility of scientific evidence, a new cadre of expertise was called forth by the parties. The scientific experts were replaced by experts on science, and the experimentalists and the statisticians gave way to scientific laureates, historians, sociologists, and philosophers, whose advice on the nature of science and the best way to deploy it in court was presented to the Supreme Court in a large set of friends-of-the-court briefs.

Alas, the advice of the new experts was as contradictory as that of the ones they replaced. Established science stood firm with Merrell Dow and argued that the courts should stick with *Frye* (if not more) and admit scientific evidence only in accordance with laws of nature laid down by scientific authorities and enforced by peer review.[82] In return, Daubert's scientific friends reminded the Supreme Court of the contingencies of scientific knowledge and pleaded with it to adopt the liberal stand of the *FRE* and not reject a scientific opinion only because it lacks consensus.[83] Upon review, the Supreme Court agreed with the petitioners that *Frye* was superseded by the *FRE* but felt compelled to address the widespread concerns with the reliability of the scientific evidence admitted under the *FRE* standard. To that end, the Supreme Court rejected the let-it all-in interpretation of the *FRE* and instead read the *FRE* as authorizing a more active role for the trial judge—to ensure that the scientific evidence admitted into the courtroom is reliable.[84]

To help the judges with their new gate-keeping function, the Supreme Court used of the rich advice it was given by the philosophers to equip the trial judges with a flexible, multifactors recipe they should use in determining the quality of the scientific evidence proffered:[85]

1. Falsifiability and Testability: whether the theory or technique can be falsified and had been tested.
2. Peer Review: whether the theory or technique had been subjected to peer review.
3. Error rate: known, or potential, error rate.
4. Standardization: the existence of control standards.
5. General Acceptance (the *Frye* test): the degree to which the theory or technique has been accepted by the relevant scientific community.

The Daubert decision, which was quickly followed by two more Supreme Court decisions that further expanded the new role of the judiciary as a gate-keeper of good science, generated an unprecedented tide of legal commentaries.[86] Some historians and philosophers of science also took notice. One of them was John Heilbron who wondered whether the advice of the historians, sociologists, and philosophers placed the law in a better or worse situation in addressing the challenges of expert testimony. Should the courts stay with *Frye* and admit only science certified by the scientific community, he asked, or should they follow Daubert and also allow deviant science from any credential expert whom a judge finds plausible? Heilbron was certain neither of the right answer nor that it was for historians and sociologists to provide one. Being a true historian, he nevertheless made sure to demonstrate that the dilemma was anything but new. Recounting the nineteenth-century history of spontaneous combustion as a scientific theory used in courtroom to explain the puzzling burning of rich widows and young women, Heilbron reminded everyone that distinguishing good from bad science can be easy only in retrospect.[87]

Notes

1. Jack Weinstein, "In re: Agent Orange Product Liability Litigation," 597 F. Sup. 749 (1984).
2. Richardson v. Richardson-Merrell, Inc., 857 F.2d 823, 825 (D.C. Cir. 1988); Brock v. Merrell Dow Pharmaceutical, Inc., 884 F.2d 166, 167 (5th Cir. 1989), cert. denied, 110 S. Ct. 1511 (1990); Daubert v. Merrell Dow Pharmaceuticals, Inc., 509 US 579 (1993); General Electric Co. v. Joiner, 522 US 136 (1997).
3 See generally P. Huber, *Galileo's Revenge: Junk Science in the Courtroom* (New York : Basic Books, 1991); M. Angell, *Science on Trial: The Clash of Medical Evidence and the Law in the Breast Implant Case* (New York: Norton, 1996); Joseph Sanders, *Bendectin on Trial: A Study of Mass Tort Litigation* (An Arbor: University of Michigan Press, 1998); Peter Schuck, *Agent Orange on Trial: Mass Toxic Disasters in the Courts* (Cambridge, Mass. : Belknap Press of Harvard University Press, 1986); K. R. Foster and P. Huber, *Judging Science: Scientific Knowledge and the Federal Courts* (Cambridge, Mass.: MIT Press, 1997).
4. Michael Green, "The Impact of Daubert on Statistically-Based Evidence in the Courts," *Proceedings of the American Statistical Association*, Statistics in Epidemiology Section (1998): 35.
5. Morton J. Horwitz, "The doctrine of objective causation," in D. Kairys, ed., *The Politics of Law* (New York: Pantheon Books, 1992), 201–213.
6. Sheila Jasanoff, "Science and the Statistical Victim: Modernizing Knowledge in Breast Implant Litigation," *Social Studies of Science* 32 (2002): 37, 38–40.
7. *Daubert*, 509 US 579; *General Electric*, 522 US 136; Kumho Tire Company, Ltd. v. Carmichael, 526 US 137 (1999).
8. Cf. Symposium, "Scientific Evidence After the Death of Frye," *Cardozo Law Review* 15 (1994); B. Black, J. F. Ayala, and C. Saffran-Brinks, "Science and Law in the Wake of Daubert: A New Search for Scientific Knowledge," *Texas Law Review* 72 (1994),:715–802; Angell, *Science on Trial*, 127.
9. Roger O McClellan, "Human health risk assessment: a historical overview and alternative paths forward," *Inhalation Toxicology* 11 (1999): 477–518; n14; "Criteria for Evidence of Chemical Carcinogenicity. Interdisciplinary Panel on Carcinogenicity," *Science* 225 (1984): 682, 683.
10. The choice of a specific extrapolation method is problematic. Linear extrapolation is prevalent but there are sometimes good reasons to apply nonlinear extrapolations.
11. Office of Sci. and Tech. Policy, Chemical Carcinogens: A Review of the Science and Its Associated Principles, 50 Fed. Reg. 10,372, 10,379 (1985); Sylvia Tesh, *Uncertain Hazards* (Cornell: Cornell University Press: 2000), Chapter 2: "Environmental Health Research," and Chapter 5: "Environmentalist Science."
12. Sheila Jasanoff, *Science at the Bar: Law, Science, and Technology in America* (Cambridge, Mass: Harvard University Press, 1995), 69–92.
13. See Reserve Mining Co. v. Environmental Protection Agency, 514 F.2d 492 (1975) and Ethyl Corporation v. Environmental Protection Agency, 541 F.2d 1 (1976).

14. Jasanoff, *Science at the Bar*;Ref 9.; Id.
15. David Rosenberg, "The Causal Connection in Mass Exposure Cases: A 'Public Law' Vision of the Tort System," *Harvard Law Review* 97 (1984): 856–857, *Agent Orange* 597 F. Sup. 749.
16. Stephen Thacker and Ruth Berkelman, "Public health surveillance in the United States," *Epidemiological Reviews* 10 (1988): 164–190.
17. Koch Robert, "Über den augenblicklichen Stand der bakteriologischen Choleradiagnose," *Zeitschrift für Hygiene und Infectionskrankheiten* 14 (1893): 319–333; A. S. Evans, "Causation and Disease: The Henle-Koch Postulates Revisited," *Yale Journal of Biology and Medicine* 49, no. 2 (1976): 175–195.
18. M. Susser, "Epidemiology in the United States after World War II: The Evolution of Technique," *Epidemiological Reviews* 7 (1985):147–177.
19. Vincent M. Brannigan, Vicki M. Bier, and Christine Berg, "Risk, Statistical Inference, and the Law of Evidence: The Use of Epidemiological Data in Toxic Tort Cases," *Risk Analysis* 12, no. 3 (1992): 343–351.
20. Thomas R. Dawber, MD; Gilcin F. Meadors, MD, MPH; and Felix E. Moore, Jr., National Heart Institute, National Institutes of Health, Public Health Service, Federal Security Agency, Washington, DC, *Epidemiological Approaches to Heart Disease: The Framingham Study*, presented at a Joint Session of the Epidemiology, Health Officers, Medical Care, and Statistics Sections of the American Public Health Association, at the 78th Annual Meeting in St. Louis, Mo., November 3, 1950; R. Doll and A. B. Hill, "Smoking and Carcinoma of the Lung," *British Medical Journal* 2 (1950):740–748; Doll and Hill, "A Study of the Aetiology of Carcinoma of the Lung," *British Medical Journal* 2 (December 13, 1952): 1271–1286; Doll and Hill, "Lung Cancer and Other Causes of Death in Relation to Smoking: A Second Report on the Mortality of British Doctors," *British Medical Journal* 2 (November 1, 1956): 1071–1081; Doll and Hill, "The Mortality of Doctors in Relation to their Smoking Habits: A Preliminary Report," *British Medical Journal* 1 (June 26, 1954): 1451–1455.
21. See for example the language of the landmark 1964 Surgeon General's Advisory Committee report on smoking: "It should be said at once, however, that no member of this committee used the word 'cause' in an absolute sense in the area of this study. Although various disciplines and fields of scientific knowledge were represented among the membership, all members shared a common conception of the multiple etiology of biological processes. No member was so naive as to insist upon mono-etiology in pathological processes or in vital phenomena. All were thoroughly aware of the fact that there are series of events and developments in these fields and that the end results are the net effects of many actions and counteractions." *Smoking and Health: The Report of the Advisory Committee To The Surgeon General of the Public Health Service* (US Department of Health, Education, and Welfare, Washington, DC, 1964), 23.
22. H. M. Marks, *The Progress of Experiment: Science and Therapeutic Reform in the United States, 1900–1990* (Cambridge: Cambridge Univ. Press, 1997); Allen Brandt, "The Cigarette, Risk, and American Culture," *Daedalus* 119 (1990): 155–176.
23. R. A. Fisher, "Alleged Dangers of Cigarette-Smoking," *British Medical Journal* 2, no. 43 (1957): 297–298; Alvan R. Feinstein, *Clinical Judgment* (Baltimore: Williams and Wilkins, 1967); Feinstein, "The Epidemiologic Trohoc, the Ablative Risk Ratio, and 'Retrospective' Research," *Clinical Pharmacology and Therapeutics* 14 (1973): 291–307; Feinstein, "Methodological Problems and Standards in Case-Control Research," *Journal of Chronic Diseases* 32 (1979): 3541; Feinstein, "Scientific Standards in Epidemiologic Studies of the Menace of Daily Life," *Science* 242 (1988): 1257–1263.

24. Bradford Hill, "The Environment and Disease: Association Or Causation?" *Proceedings of the Royal Society of Medicine* 58 (1965):295–300; Leon Gordis, "Challenges to Epidemiology in the Next Decade," *American Journal of Epidemiology* 128 (1988):1–9.

25. Advisory Committee to the Surgeon General of the Public Health Service, *Smoking and health*, PHS Publication No. 1103 (Washington, DC: Department of Health, Education, and Welfare, 1964).

26. Milton Terris, "Epidemiology and the Public Health Movement," *Journal of Public Health Policy* 8, no. 3 (1987): 315–329; Stephen Thacker and Ruth Berkelman, "Public Health Surveillance in the United States," *Epidemiological Reviews* 10 (1988): 164–190; Hertz-Picciotto, "Epidemiology and Quantitative Risk Assessment: a Bridge from Science to Policy," *American Journal of Public Health* 85 (1995):484–491.

27. Neil Pearce, "Traditional Epidemiology, Modern Epidemiology, and Public Health," *American Journal of Public Health* 86, no. 5 (1996): 678–683; M. Susser and E. Susser, "Choosing a Future for Epidemiology: II. From Black Box to Chinese Boxes and Eco-Epidemiology," *Am Journal of Public Health* 86, no. 5 (1996): 674–677.

28. M. Susser, "Epidemiology in the United States after World War II: The Evolution of Technique," *Epidemiological Reviews* 7 (1985):147–177; Susser, "Choosing a Future for Epidemiology: I. Eras and Paradigms," *American Journal of Public Health* 86, no. 5 (1996): 668–673.

29. Joseph H. Beale, "The Proximate Consequences of an Act," *Harvard law Review* 33 (1920): 633–658.

30. R. W. Wright, "Causation in Tort law", *California Law Review*, 73 (1985): 1737–1828. For a classic discussion of this approach, see O. W. Holmes, *The Common Law* (Boston: Little, Brown and Co. 1881), 88–90.

31. The first case to be termed toxic tort was an early *Agent Orange* case from 1979. See Robert F. Blomquist, "American Toxic Tort Law: An Historical Background, 1979–87," *Pace Environmental Law Review* 10 (1992): 85–173, on p. 86.

32. B. Bert Black and David E. Lilienfeld, "Epidemiologic Proof in Toxic Tort Litigation," *Fordham Law Review* 52 (1984):732–785; Schuck, *Agent Orange on Trial*, 26–28.

33. Diethylstilbestrol (DES) was widely prescribed during the 1950s and 1960s to pregnant women to prevent miscarriage.

34. See Rosenfeld, D. L., et al., Reproductive problems in the DES-exposed female, *Obstetrics and Gynecology* 55 (1980) 453–456; Schmidt, G., et al., NIOSH/OSHA Asbestos Work Group Recommendation, DHHS (NIOSH) Publication No. 81-103, 1980.

35. In both litigations there was a problem in identifying the particular manufacturer that caused the injury, forcing the courts to allocate the damages accordingly among the manufacturers, according to their market share.

36. "Trial to Open Today in lawsuit Over Nuclear Fallout," *New York Times,* September 14, 1982. Cf. Philip L. Fradkin, *Fallout: An American Nuclear Tragedy* (Tucson : University of Arizona Press, 1989).

37. For the earlier period, see John Beatty, "Genetics in the Atomic Age: The Atomic Bomb Casualty Commission, 1947–1957," in K. Benson, R. Rainger, and J. Maienschein, eds., *The American Expansion of Biology* (New Brunswick, NJ: Rutgers University Press, 1991), 284–324.

38. Allen v. United States, 588 F. Supp. 247 (D. Utah 1984), 415. This was a juryless action, conducted under the Federal Tort Claims Act, which authorizes suits for damages against the United States. See 28 USC 1346(b), 2401(b), 2671–2680.

39. Jenkins' two main precedents were Summers v. Tice, 33 Cal.2d 80, 199 P.2d 1 (1948) and Basko v. Sterling Drug Co., 416 F.2d (1968).

40. *Allen*, Ref. 38.

41. Ibid., 247.
42. Ibid., on pp. 446–447. One case remained unresolved.
43. Allen v. United States, 816 F.2d 1417 (10th Cir. 1987)
44. Basic toxicity data was lacking for 75% of the 3,000 top volume chemicals in commerce. See National Research Council, *Toxicity Testing: Strategies to Determine Needs and Priorities* (Washington, DC: National Academy Press, 1984), 84. Not much has changed since. See D. Roe et al., *Toxic Ignorance: The Continuing Absence of Basic Health Testing for Top-Selling Chemicals in the United States* (Washington, DC: Environmental Defense Fund, 1997).
45. *Agent Orange*, 597 F. Sup. 749; Id., 818 F.2d 145 (1989); Cf. Schuck, *Agent Orange on Trial*.
46. *Agent Orange*, 597 F. Sup. 749, section B.3: Possible Solution in Class Action.
47. National Academy of Science, *Veterans and Agent Orange: Health Effects of Herbicides Used in Vietnam* (Washington, DC: National Academy Press, 1993).
48. Kenneth J. Rothman and Sander Greenland, eds., *Modern Epidemiology* (Philadelphia: Lippincott, 1998), 79–93; Paula A. Rochon et. al., "Readers Guide to Critical Appraisal of Cohort Studies: Role and Design," *British Medical Journal* 330 (April 2005): 895.
49. On the relations between false positive and negatives, See J. Neyman and E. S. Pearson, "The Testing of Statistical Hypotheses in Relation to Probabilities a Priori," reprinted in J. Neyman and E. S. Pearson, *Joint Statistical Papers* (Cambridge: Cambridge University Press, 1967) (originally published in 1933), 186–202. For modern theory, see Rothman and Greenland, *Modern Epidemiology*.
50. Ref. 40.
51. *Agent Orange*, 597 F. Sup. 749.
52. Ibid.
53. Ibid.
54. Ibid.
55. Ibid.
56. Schuck, *Agent Orange on Trial*, 188–189.
57. Michael Green, "Expert Witnesses and Sufficiency of Evidence in Toxic Substances Litigation: The Legacy of Agent Orange and Bendectin Litigation," *Northwestern Law Review* 86 (1992): 643.
58. Sanders, *Bendectin on Trial*; Schuck, *Agent Orange on Trial*.
59. The leading text was Huber, *Galileo's Revenge*. See also Angell, *Science on Trial*; and Foster and Huber, *Judging Science*.
60. President's Council on Competitiveness, *Agenda for Civil Justice Reform in America: A Report from the President's Council on Competitiveness* (Washington, DC: US GPO, 1991); Carnegie Commission on Science, Technology, and the Government, *Science and Technology in Judicial Decision Making: Creating Opportunities and Meeting Challenges* (New York: Carnegie Commission, 1993).
61. Tom Christoffel and Stephan Teret, "Epidemiology and the Law: Courts and Confidence Intervals," *American Journal of Public Health* 81, no. 12 (1991): 1661–1666; Vincent, Bier, and Berg, "Risk, Statistical Inference, and the Law of Evidence," 343–351.
62. The Federal Judicial Center put out in 1994 a 637-page *Reference Manual on Scientific Evidence* with separate chapters devoted to DNA evidence, epidemiology, toxicology, statistics, and economic damage. A second updated edition came out in 2000.
63. Cf. Tal Golan, *Laws of Men and Laws of Nature: The History of Scientific Expert Testimony in England and America* (Cambridge, Mas.: Harvard University Press, 2004).

64. see refs 58 and 59.
65. Cf. Michael D. Green, *Bendectin and Birth Defects: The Challenges of Mass Toxic Substances Litigation* (Philadelphia: University of Pennsylvania Press 1996) and Sanders, *Bendectin on Trial.*
66. The first case was Mekdeci v. Merrell Nat'l Lab., 711 F.2d 1510 (11th Cir. 1983). The second plaintiffs' verdict was Oxendine v. Merrell Dow Pharmaceuticals, 506 A.2d 1100 (D.C. 1986). A mass trial was conducted in Cincinnati in 1985 on behalf of over 1,000 infants, on the sole question of Bendectin's capacity to cause eight broad classes of birth defects. See In re: Richardson-Merrell, Inc. "Bendectin" Prods. Liab. Litig., 624 F. Supp. 1212 (S.D. Ohio 1985).
67. *Richardson,* 857 F.2d 823, 825; *Brock,* 884 F.2d 166, 167.
68. Ibid.
69. Joseph Sanders, "The Controversial Comment C: Factual Causation in Toxic-Substance and Disease Cases," *Wake Forest Law Review* 44 (2009):1029–1048.
70. See *Daubert,* 509 US 579; Margaret Berger, "What Has a Decade of Daubert Wrought?" *American Journal of Public Health* 95 (2005): 59–65. William O Dillingham, Patrick J Hagan, and Rodrigo Salas, "Blueprint for General Causation Analysis in Toxic Tort Litigation," *FDCC Quarterly* (Oct 1, 2003).
71. Dillingham, Hagan, and Salas, "Blueprint for General Causation Analysis."
72. C. T. McCormick, *McCormick's Handbook of the Law of Evidence,* 2nd ed. (St. Paul: West Publishing Co., 1972), 489.
73. Golan, *Laws of Men and Laws of Nature,* 136–140; D. L. Faigman, E. Porter, M. J. Saks, "Check Your Crystal Ball at the Courthouse Door, Please: Exploring the Past, Understanding the Present, and Worrying about the Future of Scientific Evidence," *Cardozo Law Review* 15 (1994):1799–1835.
74. Golan, *Laws of Men and Laws of Nature.*
75. David E. Bernstein, "Frye, Frye, Again: The Past, Present, and Future of the General Acceptance Test," *Jurimetrics Journal* 41 (2001): 396–400.
76. Frye v. United States, *Federal Reports* 293 (1923): 1013–1014.
77. Paul C. Giannelli, "The Admissibility of Novel Scientific Evidence: Frye v. United States, A Half-Century Later," *Columbia Law Review* 80: 1198–1200; Edward J. Imwinkelried, "A New Era in the Revolution of Scientific Evidence—A Primer on Evaluating the Weight of Scientific Evidence," *William and Mary Law Review* 23 (1982): 261–263.
78. *Federal Rules of Evidence* (New York: Federal Judicial Center, 1975), Rule 702.
79. Golan, *Laws of Men and Laws of Nature,* 261.
80. "Daubert v. Merrell Dow Pharmaceutical, Inc.," *Federal Reports* 951 (1991):1128.
81. Ibid.
82. These included American Association for the Advancement of Science and the National Academy of Sciences, American College of Legal Medicine, American Medical Association, American Medical Association/Specialty Society Medical Liability Project, American Academy of Allergy and Immunology, American Academy of Dermatology, American Academy of Family Physicians, American Academy of Neurology, American Academy of Orthopedic Surgeons, American Academy of Pain Medicine, American Association of Neurological Surgeons, American College of Obstetricians and Gynecologists, American College of Pain Medicine, American College of Radiology, American Society of Anesthesiologists, American Society of Plastic and Reconstructive Surgeons, American Urological Association, and College of American Pathologists. New England Journal of Medicine, Journal of the American Medical Association, and Annals of Internal Medicine added their own brief, and so did the Pharmaceutical Manufacturers Association, and a group of 15 scientists led by Nobel laureates and Professors Nicolaas Boembergen, Dudley Herschbach, and Jerome Karle.

83. The most influential "friend of the court" brief for Daubert was filed by a group of 12 physicians, historians, and sociologists of science, led by Ronald Bayer, Stephen Jay Gould, Gerald Holton, and Everett Mendelsohn. *Daubert v. Merrell Dow*, 113 S.Ct. 2786 (1993), Amici Curiae Brief filed by Physicians, Scientists and Historians of Science on behalf of Daubert. Doc. No. 92–102, A1–A5.

84. "Daubert v. Merrell Dow Pharmaceutical, Inc.," *United States Law Week* 61 (1993): 4805–4811.

85. Ibid. The petitioners won the battle in the Supreme Court but lost the war. The court reconsidered their evidence and excluded it again, this time under the new Daubert criteria. See, *Daubert v. Merrell Dow* (1995). For a detailed analysis of the Daubert decision and its criteria see Foster and Huber, *Judging Science*.

86. In General Electric Co. v. Joiner, 522 US 136 (1997), the Supreme Court held that an abuse-of-discretion standard of review was the proper standard for appellate courts to use in reviewing a trial court's decision of whether expert testimony should be admitted. In *Kumho Tire*, 526 US 137, the Supreme Court held that the judge's new gatekeeping authority applies to all expert testimony, including technical and experience-based expertise.

87. John Heilbron, "The Affair of the Countess Görlitz," *Proceedings of the American Philosophical Society* 138 (1994): 284–316.

Part III
Histories

9

Mercator Maps Time

Anthony Grafton

In the 1570s and 1580s, Jean Bodin ranked with Europe's greatest authorities on the study of history. His *Methodus ad facilem historiarum cognitionem*, first published in 1566, was reprinted in 1572, and it figured as the first and most prominent piece in Johannes Wolf's anthology, the *Artis historicae penus*, published at Basel in 1576 and 1579. Prominent readers from Philip Sidney and Michel de Montaigne to Girolamo Cardano took the time to wade through his prolix book and profit from his wide reading and critical judgments.[1]

Around 1580, a friend of Sidney's, the English humanist Gabriel Harvey, felt unable to decide which modern authorities to trust in the technical domain of chronology. As he remarked, "There are still many difficulties about the correct connection of dates."[2] In 1581 he turned to Bodin and a colleague for advice. The French jurists couched their answer in the form of a short bibliography of trustworthy chronologers: "I was greatly aided," Harvey recorded, "by my conversation with two very expert Frenchmen, Jean Bodin and Peter Baro. They consider Glareanus, Funck, Mercator and Crusius more industrious and precise than any of the ancient chronologers—not to deprive any classical author of his due."[3]

Bodin identified Mercator as a member of a small but vital scholarly movement: one of a group of scholars who had transformed chronology in the last five decades. The Freiburg professor Heinrich Glareanus drew up a series of increasingly detailed chronologies to accompany the Roman historians Livy and Dionysius of Halicarnassus, making every effort to keep abreast of new publications. The Wittenberg-trained theologian and historian Johann Funck was one of the first to build a chronology around the astronomical data preserved by Ptolemy. The cartographer Gerardus Mercator and the Jena chronologer Paulus Crusius also tied their accounts of ancient history to astronomical eras. Both also went further. Taking up a program formulated by Roger Bacon and occasionally pursued by later writers, they used the eclipses mentioned by historians, to provide what they saw as absolute dates for events.[4] Bodin appreciated the way in which the last three had connected history with astronomy. In his *Methodus*, he noted, following Funck, Copernicus, and others, that the Assyrian king Salmanassar known from scripture was also the Babylonian king Nabonassar, for whose accession Ptolemy gave the precise epoch date February 26, 747 BC—a date that served as a cornerstone for three of the four recent chronologers whom he praised to Harvey, and that enabled them, and Bodin, to give the history of the last seven centuries BC a precise order it had previously lacked.[5] Bodin clearly kept up with the field: Crusius's *Liber de aeris seu epochis temporum et imperiorum*, which he cited to Harvey, had appeared as recently as 1578, too late for Bodin to integrate it even into the revised versions of his *Methodus*. He was, in other words, well qualified to attest that Mercator, like his colleagues, had done a skillful job of applying scientific methods to historical problems.

For generations, historians of science and scholarship have repeated, like a mantra, that Joseph Scaliger was the first to attempt this sort of interdisciplinary chronology. As Mark Pattison put it, with characteristic eloquence:

> Hitherto the utmost extent of chronological skill which historians had possessed or dreamed of had been to arrange past facts in a tabular series as an aid to memory. Of the mathematical principles on which the calculation of periods rests, the philologians understood nothing. The astronomers, on their side, had not yet undertaken to apply their data to the records of ancient times. Scaliger was the first of the philologians who made use of the improved astronomy of the sixteenth century to get a scientific basis for historical chronology.[6]

In fact, in his *De emendatione temporum* of 1583, Scaliger built at every point on the work of Mercator and the others whom Bodin singled out, even as he revised them. The wedding of history with science took considerably longer than Pattison thought, and examining Mercator's *Chronologia*, which appeared in 1569, gives us a way to watch part of the process taking place.

Neither Bodin nor Mercator believed that chronology should rest on astronomical and historical evidence alone. For both men, the Bible, properly understood, provided almost all of the solid information about the first three millennia and more of human history. In the chapter on time and chronology in the *Methodus*, Bodin argued, at length, that all of the "Orientals"—the Chaldeans, Persians, Indians, Egyptians, and Hebrews—had believed that God created the world in the fall. After all, he noted, Rabbi Eleazar (correctly Eliezer) had shown that "the secret of the month" of the Creation "was revealed by transposing the letters" of the first word in the book of Genesis, בראשית (*Bereshit*, "in the beginning"), to make בתשרי (*beTishri*, "in the month of Tishrei"). Tishrei, the Jewish month that begins with the festival of Rosh Hashanah, falls in the early autumn: clear confirmation for Bodin's view.[7]

Like many other Christian scholars, in other words, Bodin accepted the view that the Hebrew letters contained special mysteries, and that knowledge of these made it possible to decode the deepest messages—including chronological ones—of the biblical text.

Bodin knew that he was far from the first to build elements like this into the foundations of chronology. Other ancient authorities—notably one Rabbi Joshua—had placed the Creation at the diametrically opposite point of the year, on the first of Nisan, the first month of spring.[8] In the 1572 edition of the *Methodus*, he added a particularly critical note: "Mercator also goes wrong when he holds that the sun was in Leo when the world was born. Because he does such a bad job of laying this foundation, ruin threatens his entire effort to give history certainty by referring to the motion of the stars."[9] When it came to the date of creation and the proper way to elicit from Genesis, Bodin saw Mercator not as the model of a modern chronologer, well equipped with astronomical techniques and data, but as an example of bad method, to be shunned by the serious reader.

Like Bodin, Mercator drew his evidence for the date of creation from the biblical text. He began with an argument based on the literal sense. According to Genesis 8:10, on the twenty-fifth day of the eleventh month, when Noah sent forth his dove for the second time, the bird returned with a flowering olive-leaf in its beak. The testimony of Pliny and others showed that this must have happened in the late spring. Hence the eleventh month must have been Sivan, which falls in the spring, when the sun is in Gemini. Counting backward in the list of Jewish months, Mercator concluded that "Ab, to which Leo corresponds, must have been the first month of the year and the beginning of the world."[10] After all, he argued, before the Flood the Patriarchs would never have dared to change the ordering of the months that God or Adam had ordained: the eleventh month of Noah was that of Adam.

But he added a second argument of a different kind:

> The Hebrew Cabala provides another argument for the first month. It attests that the letter *Aleph* denotes the first and eternal principle, God, and *Beth* the beginning of creation. Hence by the mystery of the letters *Ab* means the beginning of creation carried out by the eternal principle, that is the time when the world was first created.[11]

Not unlike Bodin, in other words, Mercator treated the powers ascribed to the individual Hebrew letters by Jewish exegetical traditions as data of value to the chronologer. It seems quite possible that Bodin took this passage, which used such similar means to arrive at a radically different end, as a personal challenge. And it seems certain that though the two men's conclusions differed so radically, their eclectic methods were in fact quite similar. Mercator's enterprise raises as many questions as it answers about the early modern marriage of science and history—a union that Mercator, as well as Bodin, pursued in ways that reflect period, rather than modern, definitions of both enterprises. What follows is a first effort, not to climb the vast mountain of

Mercator's work, but to reach a base camp and survey some of its foothills and lower slopes.

Bodin was by no means the only informed reader, from his day to ours, who has found it hard to assess Mercator and his work, or to fix his confessional and intellectual identities. A polymath, a typical figure of the early decades of the Latin Republic of Letters, Mercator did original work in fields that ranged from cartography and calligraphy to history. Born a Catholic and educated at Louvain, a great center of both Catholic theology and scientific activity, where he studied with the mathematician and instrument-maker Gemma Frisius, Mercator expressed discontent with narrow Catholic orthodoxy and moved to Duisburg, in the more tolerant Duchy of Cleves. There he served the Protestant William the Rich, Duke of Jülich-Cleves-Berg. Yet Mercator continued, as we will see, to work for Catholic patrons, to maintain relations with Catholic scholars, and to push the borders of what Catholics saw as acceptable argument and opinion. At once a humanist and a creative cartographer, oriented to both the lasting value of tradition and the need for the freshest, most precise information, Mercator produced both a critical edition of Ptolemy's *Geography* and a modern *Atlas* of his own devising. Beloved in his own day even by his greatest competitor, Abraham Ortelius, he left as his most famous gift to posterity his projection, which has given rise to endless controversies.[12]

Mercator's *Chronologia* is particularly hard to pin down. At the outset, it presents itself as a work of up-to-date Catholic scholarship. Its publisher, the firm of Birckmann, was based in Cologne, where the university was a center of orthodox theology. The first text in the volume, a prefatory "judgment" that comes after the title page, seemingly confirms the reassuring impression made by Mercator's choice of publisher. This detailed response to Mercator's work takes the form of a long letter, written in January 1568, from the Roman scholar Onofrio Panvinio to the jurist Jean Matal, who lived in Cologne. Perhaps Matal had been the agent who sent Panvinio what he described as "some gatherings"—very likely partial proofs—of Mercator's work. Both Matal and Panvinio were expert antiquarians, who had learned in the lively Roman scholarly world of the 1540s and 1550s how to decipher and copy inscriptions, and then to use their testimony, along with that of literary works and histories, to reconstruct the living civic and institutional world of ancient Rome.[13] At the start of his judgment, Panvinio thanked Matal for having prodded him, with jokes and persuasion, to finish his own major work, an edition of and commentary on the Roman consular *Fasti* and *Triumphs* that had been found in the Forum in the 1540s and reconstructed by Michelangelo in the Palazzo dei Conservatori. He thus reminded readers that he had long been recognized as an expert on the chronology of the Roman Republic and Empire. In the new climate that took shape during the later years of the Council of Trent, Panvinio had moved more and more into the study of church history, which involved a different range of chronological questions.[14] (He also made money by selling aristocrats spurious genealogies that traced their ancestry back to ancient Rome, a pursuit he did not find it necessary to mention in this context.)[15] If anything, he was better equipped than Bodin to evaluate what Mercator had done. Panvinio did not offer unqualified praise. He noted the complexity of the subject, especially for the period before the birth of Christ, and suggested that if he and Mercator could only meet, he would offer his colleague much advice and help—too much to convey in ordinary letters, even the data-dumping letters of an antiquary. But he also professed that he found Mercator superior, in content, order, scholarly judgment, erudite industry, and "the examination of celestial motions, so far as they are relevant to this," to all of his rivals.[16] By posting this note before his own preface, Mercator implicitly claimed a connection with the weighty learning of the Roman experts, and perhaps with the immense scholarly projects beginning to take shape in Rome, such as Baronio's *Annales ecclesiastici*.

Yet Panvinio's letter makes clear that he was troubled by parts of what he had read. He criticized "those who boast that they use the Cabala of the Hebrews to restore chronology"—a negative reference to the passage at the very start of Mercator's text that would also displease Bodin, for different reasons, not long after.[17] He also confessed that he "retained some doubts about the Egyptians," though he agreed with Mercator that it was necessary to "restore their years to the proper computation using the motion of the sun and moon."[18] Here he referred to an unusual passage in the body of Mercator's chronicle. Christian chronologers had complained since ancient times that Egyptian history was far too long. The boastful priests who had compiled the Egyptian records claimed that their history had lasted several thousand years. The implication was that Egypt had somehow existed before the universal Flood described in Genesis 6–8. When Eusebius composed his great *Chronicle*, he used as one of his sources the Egyptian history composed in Greek around 300 BC by the priest Manetho.[19] He expressed his disbelief in the thousands of years of Egyptian history that Manetho inventoried, but nonetheless he preserved Manetho's list of the members of 31 dynasties.[20]

At the start of Eusebius's chronological tables, which began shortly after the Flood, he had to confront a problem: What to do about the earliest Egyptian dynasties, which seemed to predate the Flood? He offered only a laconic remark: when Abraham was born, "the Egyptians had their 16th power, which they call a dynasty, at which time the Thebans ruled for 190 years."[21] The status of the first 15 dynasties remained unclear: Had they somehow existed before the Flood swept all of humanity away? Only this part of Eusebius's text was translated and adapted in Latin by Jerome, in whose version it dominated much of historical and chronological scholarship for centuries to come.

The most inventive and resourceful of Renaissance chronologers, the Dominican antiquary, theologian, and forger, Annius of Viterbo, abhorred all historical vacuums. So he found a simple way to provide the vital details Eusebius had omitted: he made them up. By doing so, he hoped to ensure that no one could call the Christian history of the world into question on the basis of pagan testimony.[22] In his commentary on Manetho—not the genuine Egyptian Manetho cited by Josephus and Eusebius, but his own unhappy invention—Annius interpreted Eusebius's obscure words as a precise statement about the first dynasties: "As Eusebius writes in his Chronicle, there were 15 early Egyptian dynasties between the flood and the 43rd year of Ninus," when Abraham was born.[23] Annius supported this bold argument by claiming that each dynasty in turn had lasted only a short period of seven years or so. Most sixteenth-century chronologers accepted some form of this argument and contentedly crunched everything they knew about Egyptian history into the portion of their tables that began with the Flood.[24]

Mercator, however, believed in going directly to as wide a range as possible of sources, as is clear from the bibliography he included in the front matter to the *Chronographia*, which provided publication details for a number of the works he cited, an unusual decision for the time.[25] Without making a meal of it, he rejected the Annian texts as obvious forgeries, clearly the work of a single hand.[26] And he confronted the problems of Egyptian history more directly and more frankly than most of his colleagues. Since the later fifteenth century, thanks to Marsilio Ficino, Latin readers had been able to read the passages in the *Timaeus* and the *Critias* in which Egyptian priests, drawing on their superior records, recounted almost 10,000 years of history—a period far too long to be reconciled with any version of the Old Testament chronology. Here as in many other cases, the rich tradition of Neoplatonic scholarship enabled Ficino to eliminate an apparent contradiction between a Platonic statement and Christian doctrine. Eudoxus, perhaps working at Plato's request, had argued that the years computed by the Egyptians had actually been lunar years—that is, months. By doing so, he reduced the length of early Egyptian (and Atlantean) history to the point where it seemed plausible. Ficino, who relied heavily on Proclus's commentaries on Plato, found Eudoxus's arguments there and revived them. For centuries to come, chronologers would follow him and his Greek models, treating Egyptian boasts about thousands of years of history as inflated references to thousands of lunar months, to be divided by twelve and made harmless.[27]

Mercator laid out all of these materials in an unusually long note. He found it implausible that the Egyptian dynasties could actually have been so short as Annius claimed: "It seems awkward to confine them in such a short span of years, since power does not pass lightly and immediately from one lineage or nation to another."[28] But if one imagined that the first 15 dynasties had been as long as the later ones recorded by Eusebius, one would have to conclude that "an inhabited Egypt and some sort of kingdom came into existence with the very first men."[29] The material from Plato seemed to him to confirm this conjecture—especially if one reduced the supposed solar years to lunar ones. After all, Mercator reflected, the period before the Flood had been unequivocally evil, characterized by the spread of crimes and wars of all sorts. Very likely the terrible Atlantean war described by Plato had taken place then. "It seems plausible," he concluded, multiplying qualifiers to indicate the hesitance with which he stated his position,

> that the Egyptians, who always boasted so loudly of their antiquity, and who, more than anyone else, were blessed both with traditions handed

down by their earliest ancestors and by a form of writing that came into being in very ancient times, before all human memory, could have done their best to preserve a sort of general memory of events in their realm from the creation onwards.[30]

Panvinio, though clearly gratified by Mercator's insistence that the years of Egyptian tradition were not solar ones, was equally clearly unhappy with the idea that Egyptian tradition might provide a separate access, independent of the Bible though also much more general, to antediluvian history.

But Panvinio had been formed in the cosmopolitan 1540s and 1550s—not the harsher age of religious censorship during which the *Chronologia* appeared. Catholic chronologers like the Jesuits James Gordon and Jacobus Salianus sharply criticized Mercator for failing to see that Egyptian traditions about the earliest times all rested on the lies typically uttered by pagans claiming that they were older than the Jews—just as their colleague Denys Petau would flail Joseph Scaliger, who made a similar argument, even more ambivalently, in 1606, after he rediscovered the shattered fragments of Manetho's Greek dynasty list.[31] Late in the seventeenth century, the one Jesuit chronologer who could rival Annius in sheer fertility of imagination, Athanasius Kircher, was ordered by Jesuit censors to retract his own speculations on the antediluvian history of Europe, which were based on Scaliger and the fifteenth-century Cairene historian al-Suyuti.[32] Panvinio's letter shows that Mercator's work emitted signals that were genuinely difficult to decode. No wonder that the Catholic authorities eventually placed the book on the Index of Prohibited Books.

If Mercator's book was not a full bore work of Catholic scholarship, what was it? Mercator's project had two clear goals. He wanted to lay out a visually effective map of time, in two forms: the particular sacred time of Jesus's mission on earth, as narrated by the four Evangelists, which he represented with a four-column "harmony," and the more general time of world history, which he represented with a complex table that brought together events from many different kingdoms. Mercator understood that an artful layout could tell a technical story more effectively than words. He explained to his readers that he had tried to devise a visual language that could convey "a first sketch and rough image of the past, in which the major parts are laid out in accordance with their size, position, and proportion."[33] Historians would eventually polish this image, adding the details that would give it life. For the purposes of chronology, though, Mercator explained that he considered it most important to lay events out on a single, uniform scale, even at the cost of leaving large empty spaces:

I did not act randomly when I maintained the same amount of space even for the years in which I had nothing to record. For that uniformity is extremely appropriate, and proves very handy when a reader wishes to compute a particular number of years forwards or backwards without the trouble of addition or subtraction. True, this makes the work grow in size and price, but the handiness of the format makes up for that, and I wish this layout to remain unchanged in the future.[34]

It is tempting to see Mercator here as wedding cartography with chronology, applying principles first used for mapping space to the representation of time. In fact, though, his ancient predecessor Eusebius, who had drawn up his tabular chronology in the years around AD 300, had been fully aware of the ways in which an artful mise-en-page could make a text more instructive: he used parallel columns to show how all kingdoms rose and fell, until none remained except the Romans and the Jews, and then the Romans alone, just in time to unite the world to receive the message of Jesus.[35] Mercator's self-consciously visual approach to chronology emerged as much from the historical as from the scientific tradition.

The principle of uniform division of space that Mercator stated, as a kind of counterpart to his cartographical projection, enabled him to carry out one remarkable philological feat. He devised a revisionist theory of the chronology of the life of Jesus, according to which the Messiah's ministry on earth had lasted not for three years but for four, stretching across five distinct Passovers. To prove this theory, he laid out the passages in the four gospels that described Jesus's actions and travels in parallel columns. Mercator skillfully used mise-en-page and typography to clarify both his views on Christ's life and the relations of all its component stories to the Gospels they came from. Every time he quoted the beginning of a chapter, he started the verse on the extreme left of its line of text; other passages and summaries were indented. Brief summaries of material treated at greater length elsewhere he laid out as columns of single words, making their character evident at a glance. So far as possible, Mercator placed parallel passages next to one another.[36] The reader could see at once how each of the evangelists had narrated each major story.

A problem inevitably arose—one that had worried Christian scholars since ancient times. The evangelists do not tell their tales in the same order. To put all of the parallel passages at the same place in a Gospel harmony, one must destroy the original order of the Gospels themselves. Many scholars— for example, the Lutheran chronologer and theologian Andreas Osiander— compiled their harmonies to show that the Gospels actually never diverged, much less contradicted one another. When two Gospels described similar events in different places, they actually described two distinct events, an assumption that led to much multiplication of Jesus's actions.[37]

Mercator, by contrast, accepted that two Gospels might tell the same story at different points in their narratives and assumed that each Gospel had its own textual integrity. He still transformed some parallel passages into multiple events to eliminate apparent divergences. Thus, he posited two cleansings of the Temple in Jerusalem and two anointings of Jesus by a woman. For the most part, however, Mercator used his tabular presentation of the material to advance, through an ocular demonstration, a novel thesis: that the Gospels must be respected as independent texts before they could be used as sources, and that their authors, like classical writers, had freely arranged their material.[38] Mercator's Gospel harmony gave one area of chronological scholarship a newly objective and scientific flavor.

The real proof of Mercator's rich pudding, however, lay in the larger tables in which he laid out the history of the world from the creation to AD 1577 (like Hartmann Schedel, who gave readers of his *Nuremberg Chronicle* three pages in which to record the events that intervened between its publication in 1493 and the end of time, Mercator wanted readers to be able to insert events that followed the book's appearance in the proper place). These tables had certain qualities that distinguished them from their many competitors. To maintain their cartographic character—to give a visual sense of the major divisions of history and their relations to one another—Mercator did not crowd his pages with as many notices as many rivals did. Neat series of numbers located each year with respect to multiple eras, which served as chronological baselines: the interval from the Creation to the Flood, for example, located the earlier years of postdiluvian history in universal time, and the distances from the era of the Olympic Games, the era of Nabonassar, the era of Solomon's Temple, and the era of Rome's foundation, as well as the Creation, clearly identified each year in the first several centuries BC.

Most important, Mercator laid out these different systems of dating as individual columns, carefully designed to show that each set of years began and ended at a different point. The reader could see immediately that years *Ab urbe condita* began in the spring and trace them back to the foundation of the city in April of Olympiad 6.4 (752 BC).[39] Olympic years, by contrast, were counted from the summer solstice, when the games were held.[40] Years of Nabonassar were computed not in Julian years of 365 ¼ days but in the Egyptian years used by astronomers, which were 365 days long. They began on the Julian date February 26, 747 BC, but the first of Thoth, their new year's day, moved forward in the Julian calendar by one day every four years. Mercator even took care to indicate the passage of 1 Thoth through the Julian year. It was a masterly exercise in the visual presentation of complex information. A reader did not have to be an astronomer or an expert to collate years dated in any of the varied systems that Mercator worked with.

Antiquaries like Panvinio and Matal had shown how to write what contemporaries agreed was a new sort of history: a history that, instead of narrating the deeds of kings and generals, recreated the institutions and customs of past societies, relying wherever possible on the monuments that actually survived from the period in question.[41] Mercator also used ingenious and novel methods to connect individual histories, in a way the eye took in at once, to a larger, uniform framework. At the same time, he used dateable celestial events—above all the eclipses mentioned by ancient sources—to fix this whole framework, rigorously, by the movements of the sun and the moon. This world history genuinely was built of sterner stuff than most of its predecessors.

From the early eighth century BC onward, Mercator mentioned and described many of the eclipses mentioned by Ptolemy in the *Almagest* and by other ancient writers.[42] Famous celestial events like the supposed solar eclipse of Thales and the solar eclipse that took place at the beginning of the summer in the eighth year of the Peloponnesian war made their appearances at the proper point in Mercator's tables.[43] He even had a bash at solving some of the complex chronological problems that attend famous eclipse reports: for example, the mind-bendingly complex (in fact, insoluble) ones that accompanied the ancient descriptions of Thales's eclipse, which Herodotus, Pliny, and Cicero placed variously in the reigns of Cyaxares, Alyattes, and Astyages.[44]

Most impressive of all was Mercator's attack on a difficult problem first raised by an older astronomer, Petrus Apianus, in 1540. In his magnificent *Astronomicum Caesareum*, Apianus had argued, more than a generation before Mercator, that historians needed to learn astronomy. From the beginning of the *Astronomicum*, Apianus insisted that his work, with its colorful volvelles designed to enable readers to find the positions of the planets and the dates of lunar and solar eclipses, offered the basis for a radical transformation of historical scholarship and its practices as well as of astronomy. Historians, he explained, traditionally worked by simple addition—a process that necessarily introduced gross errors into their work.[45] If they would only attend to the dateable eclipses mentioned in their sources, he argued, they could attain a far higher level of precision, one that scholarship had never reached before:

> To show you the extraordinary power, excellence and utility that come with knowledge of eclipses, I found myself compelled to offer some examples from [the time] before Christ. These will reveal the utility of eclipses more certainly than later examples. Everyone knows that historians go wildly wrong when they state periods of years from the creation and the founding of Rome. The learned can judge how much the obscurity, ignorance, doubt and confusion introduced by this mistake have damaged all efforts to assess and understand history. For the result is that if they inform the Christian reader about some event that either preceded or fell not long after the founding of Rome, they cannot give a consistent account. Only knowledge of eclipses can correct this great evil and improve matters. For eclipses can make it possible to fix all events to particular years, before Christ just as much as after him. Once the historian has obtained the unquestionable date for some event, he can infer the other intervals, both those before and those after it.[46]

As a case in point, Apianus computed the date of the lunar eclipse that preceded the battle of Gaugamela at which Alexander defeated Darius and world history, as recounted by most Latin Christians, reached a dramatic turning point: the point where the second world monarchy of the Persians gave way to the third of the Macedonians. The event was well chosen. Though historians gradually abandoned the schema of the four monarchies in the late sixteenth and seventeenth centuries, Gaugamela remains a great dividing point in world history, and a lynchpin of chronology, to this day. Eusebius, Apianus noted, had dated the battle, using his annalistic methods, to 328 BC. But after checking every possible full moon, Apianus reckoned that the eclipse before the battle must have taken place on June 26, 326 BC.

Accepting Apianus's dating would have required scholars to make multiple changes in a fairly well documented historical record. It is a measure of Mercator's open mind and sound judgment that he found it possible to reject Apianus's problematic solution, while accepting his general thesis that historians needed the help astronomy could provide. Instead of assuming that astronomy necessarily undermined accepted datings, however, Mercator realized that it might also confirm them. When he computed the date of the Gaugamela eclipse, he found that it took place in Olympiad 112.2, in modern terms 331–330 BC.[47] This date meshed well with the other ones in Eusebius and other historians—a fact that should surprise no one, as it was the correct year.

Sadly, the astronomical data and the practices that Mercator grafted onto his history did not smoothly metamorphosize into organic parts of what he had hoped would be a coherent whole. In theory, he had claimed it was essential to make his tables absolutely uniform in structure. In practice, however, Mercator found himself dealing with particular years for which he had precise details of an astronomical or calendrical kind. Noah entered the ark 1,656 years after the Creation, and the biblical account followed the progress of the Flood and Noah's actions, month by month. A little less than 1,600 years later, Romulus founded Rome. Thirty years after that, lunar eclipses were observed in Babylon. For each of these events, and others as well, Mercator could not restrain himself from tailoring his tables to display and clarify the relations among these rich details.

At these junctures Mercator devised a way to slow the virtual passage of historical time. He substituted year-long columns, divided into months by the passage of the sun through the signs of the zodiac, for the shorter, undivided spaces with which he normally indicated whole years.[48] The table, thus adapted, dramatized certain historical moments as well as the new standard of precision that went with astronomical method. But it did so at the expense of abandoning uniform intervals. One could not reckon intervals between events by simply counting the pages between them in the *Chronologia*, as Mercator had suggested might be possible.

The disruption caused by astronomy took place on a deeper level as well. Mercator relied on Ptolemy's *Almagest*—rather than the more recent *Prutenic Tables* of Erasmus Reinhold, published in 1551—to compute the dates and times of individual eclipses. When he set out to identify an eclipse for which Ptolemy did not give a date and time—for example, that of Gaugamela—he would start from the nearest previous eclipse. In this case, he began from an eclipse observed at Babylon and reported by Ptolemy, on 16 Thoth Nabonassar 367 or December 12, 382 BC.[49] Since both eclipses had been lunar, and accordingly took place at full moon, Mercator knew that a whole number of lunar months must separate them. He worked out that the interval in this case amounted to 636 "periods of opposition," which he converted to 51 Egyptian years, 166 days, and 11 hours. By adding the motions of the various parts of the lunar and solar model for this period to their positions at the earlier eclipse described by Ptolemy, Mercator fixed a time for the Gaugamela eclipse. It reached full intensity at 3:40 p.m. at Alexandria and 4:30 at Babylon.[50] Mercator knew that this result looked very odd. Lunar eclipses normally take place at night, since at the point in the moon's cycle when it is in opposition to the sun, it rises at sunset and sets at dawn. As Mercator himself noted, moreover, Pliny had actually set the eclipse "in the second hour of the night" at Gaugamela and at moonrise in Sicily.[51] Accordingly, Mercator wound up his treatment by accepting that his own proposed time was wrong. Despite the discrepancy "there can be no doubt that Pliny assigned the correct time to this eclipse."[52]

By this point Mercator had traveled several exits past the limits of his own proficiency. And he knew he was at sea.[53] Although he claimed that he and Pliny more or less agreed on the time of the eclipse, unlike Apianus, he did not venture to offer an Egyptian month date—much less a Julian one—for it. This was just as well. In a further comment, Mercator noted that according to the *Varia historia* (*Miscellany*) of the sophist and polymath Aelian—a writer with no special claim to authority on astronomy, who included this information in a list of lucky and unlucky days—the eclipse had taken place in the Athenian month Thargelion, the eleventh month of the Attic year.[54] Thargelion, Mercator inferred, corresponded to the period when the sun was in Taurus, which yielded a date in Julian April or May 330 BC rather than in September 331 BC, when the eclipse actually took place.[55] For all Mercator's protests of his meticulous concern for details, this inconsistency was fairly typical of his work. He did not systematically identify and date all the eclipses that the Greek historians mentioned. Instead of using Erasmus Reinhold's reasonably accurate *Prutenic Tables* of 1551 to compute their dates, he took each eclipse in turn as a baseline from which to compute the date and time of the next, working with Ptolemy's methods and tables—a process that inevitably involved compound errors.

Not surprisingly, as Mercator totted up the intervals between eclipses and computed the dates of those not dated by Ptolemy using his own, rough, and ready methods, he found numerous discrepancies like the one that appeared in his treatment of the Gaugamela eclipse. As these mounted up, he decided that he had discovered an "anomaly, hitherto undetected by anyone" in the motion of the moon, and he laid out the evidence for this in a detailed chapter of his introduction.[56] What had begun as a chronology based on eclipses had mutated into something like its opposite: a call for a reform of models for the motion of the sun and the moon, based on the historical evidence. Yet Mercator insisted that the scale of error he had detected was not serious enough to undermine the historical structure he had built, and boldly moved the birth of Christ—the date of which had been one of the central objects of chronological research for centuries—to 2 BC.[57] For all its richness and ingenuity, the *Chronologia* irritated some astronomically trained readers: Bünting, for example, made fun of Mercator's ignorance of astronomical procedures even as he showed that his predecessor had tried to fix the date of the death of Augustus with reference to a solar eclipse that could not have been observed in Roman territory.[58] And no wonder. Mercator's splendid-looking but flawed book was at once a bold demonstration that history could be refounded on an adamantine new foundation of astronomical data and a reluctant admission that the foundation showed signs of erosion over time.

Accordingly, Mercator's successors treated him and his work rather as he had treated Apianus: they accepted his principles but replaced his methods. Paulus Crusius, writing in the 1570s and Heinrich Bünting, working in the 1580s, used the *Prutenic Tables* to establish a canon of dateable eclipses that has served, down to the present, to underpin the chronology of ancient history. In this respect Joseph Scaliger, who did more than any single other scholar to revolutionize the field, did no more than follow Crusius's lead, and he never became so proficient as Bünting in establishing the exact dates, durations, and visibilities of eclipses. Like Mercator, these scholars saw that historians could add the astronomer's practices to their own. Unlike him, they mastered these, in their most up-to-date form, and found them fully capable of solving most of the technical problems that historical eclipses posed them—a belief not seriously challenged until recent years, when the Russian topologist Anatoly Fomenko and others have tried to tear down the entire traditional structure of chronology, arguing, among other things, that close study of historical eclipses confirms that changes have taken place in the motion of the earth and the moon.[59]

Most of Mercator's work remains to be explored and contextualized. Yet it seems clear that his enterprise puzzled some of its early readers as much as it puzzles us now. The book was reprinted only once, and the edition was a strange, botched hybrid, carried out in Basel. The editor tried unsuccessfully to meld Mercator's cosmopolitan and speculative world history with the radically Biblicist chronology of the Genevan scholar Mathieu Béroalde, who ignored astronomy and denied that pagan historians like Herodotus, who were roasting in hell as they deserved, could command credence or reward study.[60] Particularly unusual passages like Mercator's meditations on early Egyptian history were silently omitted, making the book still less coherent.

The *Chronologia*, in the end, was only one among many tentative efforts to rebuild history on an astronomical basis. It awaits a full study, as well as a systematic comparison with the other texts with which Bodin grouped it—all of which have their place in a much longer story that actually began in antiquity, when scholars first tried to date past events by connecting them to celestial phenomena, and culminated, in the early modern world, in the great, contrasting historical works of Francesco Bianchini and Giambattista Vico.[61] It is devoutly to be hoped that some great scholar, master of both technical details and wide perspectives, will trace the great broken arc of this episode in the history of historical thought.

Notes

1. See in general Anthony Grafton, *What Was History? The Art of History in Early Modern Europe* (Cambridge: Cambridge University Press, 2007), esp. Chapter 3.

2. Princeton University Library, Ex PA6452. A2 1555, note on the dedicatory letter to Glareanus's chronologia, P verso: "*Quanta fides Eusebio tribuenda: consulendi Neochronologi: praesertim Funccius, Crusius, Mercator. Nam de authentico Synchronismo multi adhuc scrupuli.*"

3. Ibid.: "*Quanquam mihi valde profuit cum duobus peritissimis Gallis, Joanne Bodino et Petro Barone, viva collatio. Qui plus industriae certitudinisque tribuunt Glareano, Funccio, Mercatori, Crusio, quam ulli veterum Chronologorum. Salvo tamen cuiusque classici auctoris iure.*"

4. For a general map of the field see Anthony Grafton, *Joseph Scaliger: A Study in the History of Classical Scholarship* (Oxford: Clarendon Press, 1983–1993), II, pt. 1, and Benjamin Steiner, *Die Ordnung der Geschichte: Historische Tabellenwerke in der Frühen Neuzeit* (Cologne: Böhlau, 2008). On eclipse dating in this period, and its fortunes from Bacon through the sixteenth century, see C. Philipp E. Nothaft, *Dating the Passion: The Life of Jesus and the Emergence of Scientific Chronology (200-1600)* (Leiden and Boston: Brill, 160-171, 238-240,261-282.

5. Jean Bodin, *Methodus ad facilem historiarum cognitionem*, in J. Wolf, ed., *Artis historicae penus*, 2 vols. (Basel: Perna, 1579), vol. I, 332–332.

6. Mark Pattison, "Joseph Scaliger," in Henry Nettleship, ed., *Essays*, 2 vols. (Oxford: Clarendon, 1889), I, 162–163.

7. Bodin, *Methodus*, I, 334–335: "*erroris etiam non minima causa videri potest, quod Graeci annos ab aestivo solstitio incoeperunt: Latini ab hyberno: caeteri qui ad occasum, a verna sectione: Arabes ab ingressu Solis in Leonem, ut Solinus et Firmicus lib. iii. tradit. Orientales vero, id est Caldaei, Persae, Indi, Aegyptii, Hebraei, ab autumno: quo tempore mundus est a Deo creatus, ut Iosippus lib.1. antiquitatum, cap. iii. scribit, et Rabi Eleazar in Genesim:* בראשית *hoc est, initio,* בתשרי *id est, mense Septembri arcanum mensis eadem literarum conversione complecti putat. id etiam confirmat Rabi Abraham, in cap. viii Danielis: tametsi Moses Nisan, qui Xanticus Macedonibus, Latinis vero Aprilis dicitur, ordine primum posuerit, quod Deus eo mense populum liberasset, nihil tamen in reliquis immutavit, ut Iosippus testatur. atque id intelligi potest ex eo quod iubet Moses Exod. cap. xxiii. diem festum coles, inquit, in exitu anni, cum fruges in horrea congregaveris. quare fallit Garcaeus, qui Hebraeos a sectione verna coepisse tradit, quod mundum eo tempore creatum putarent, ut Rabi Iosue, quem refellit Rabi Eleazar. Plutarchus in symposiacis quaestionem illam disputat, sed omissa disputatione, si Veris initio fruges inchoatae non maturae fuissent, eodem modo animantia imperfecta aut lactentia Deus fecisset, ac nutrices adhibuisset. Non minus labitur Mercator, qui mundo nascente Solem in Leone fuisse putat, eoque fundamento male iacto, caetera quae de syderum motu ad historiae fidem retulit, ruinam minantur. omitto leves oleae coniecturas post eluviones: et caetera id genus quae leviora sunt, quam ut refelli mereant.*"

8. The arguments of these authorities were recorded in Babylonian Talmud, Rosh Hashanah 10b and summarized in Genesis Rabbah 22:3 to Genesis 4:3, but Bodin probably knew them through an intermediary source. They are summed up, without some of the details Bodin cited, in a Latin work first published in 1567, Gilbert Génébrard, *Chronographiae libri quatuor* (Lyon: Pillehotte, 1599), 2–3.

9. Bodin, *Methodus*, I, 335: "*Non minus labitur Mercator, qui mundo nascente Solem in Leone fuisse putat, eoque fundamento male iacto, caetera quae de syderum motu ad historiae fidem retulit, ruinam minantur. omitto leves oleae coniecturas post eluviones: et caetera id genus quae leviora sunt, quam ut refelli mereantur.*" This passage does not appear in the first edition of Bodin's *Methodus* (Paris: le Jeune, 1566), 291, and clearly reflects his response to Mercator's book. In 1575, when Ortelius called Mercator's attention to Bodin's critique, he dismissed it: *Correspondance Mercatorienne*, ed. M. van Durme (Antwerp: De Nederlandsche Boekhandel, 1959), 113–114.

10. G. Mercator, *Chronologia* (Cologne: Birckmann, 1569), a recto.

11. Ibid., a verso: *"Accedit aliud primi mensis argumentum ex Hebraeorum cabala, iuxta quam Aleph litera primum et aeternum principium Deum signat, Beth litera principium creationis, ut etiam literarum mysterio Ab principium creationis a principio aeterno factum, hoc est primum tempus conditi a Deo mundi significet."*

12. For an excellent review of Mercator's life, work, and posthumous reception see Mark Monmonier, *Rhumb Lines and Map Wars: A Social History of the Mercator Projection* (Chicago; London: University of Chicago Press, 2004).

13. See William Stenhouse, *Reading Inscriptions and Writing Ancient History: Historical Scholarship in the Late Renaissance* (London: Institute of Classical Studies, 2005) for an introduction to both men and their world. Also essential are Jean-Louis Ferrary, ed., *Correspondance de Lelio Torelli avec Antonio Agustín et Jean Matal (1542–1553)* (Como: New Press, 1992); Ferrary, *Onofrio Panvinio et les antiquités romaines* (Rome: Ecole française de Rome, 1996); and Ingo Herklotz, *Cassiano Dal Pozzo und die Archäologie des 17. Jahrhunderts* (Munich: Hirmer, 1999).

14. For Panvinio's chronological scholarship, see Nothaft, esp. 254–259.

15. See Roberto Bizzocchi, *Genealogie incredibili: scritti di storia nell'Europa moderna*, 2nd ed. (Bologna: Il Mulino, 2009).

16. Mercator, *Chronologia*, recto.

17. Ibid.: *"Quae [sc. Chronology] qui iactant se ex Hebraeorum Cabala restituere . . ."*

18. Ibid.: *"In Aegyptiis tantum, nonnihil mihi dubii relinquitur: ac tamen, ut Mercator tuus, brevi futurus quoque noster, ad calculum, eos annos, ex Solis Lunaeque motu cursuque revoco."*

19. See Gerald Verbrugghe and John Wickersham, ed. and tr., *Berossos and Manetho: Native Traditions in Ancient Mesopotamia and Egypt* (Ann Arbor : University of Michigan Press, 1996).

20. For a brief account see Anthony Grafton and Megan Williams, *Christianity and the Transformation of the Book* (Cambridge, Mass.: Harvard University Press, 2006).

21. Eusebius/Jerome, *Chronici canones*, ed. J. K. Fotheringham (London: Humphrey Milford, 1923), 17: *"Porro apud Aegyptios XVI potestas erat, quam vocant dynastiam, quo tempore regnabant Thebaei an. cxc."*

22. See most recently Walter Stephens, "When Pope Noah Ruled the Etruscans: Annius of Viterbo and His Forged Antiquities," *MLN* 119, no. 1, supp. (January 2004), S201–S223.

23. Annius of Viterbo, ed., *Berosi sacerdotis Chaldaici Antiquitatum libri quinque* (Antwerp: Steelsius, 1545), 105 verso: *"Dynastiae Aegyptiorum primae a diluvio usque ad xliii. annum Nini xv. fuere, ut Eusebius de temporibus scribit."*

24. See e.g., Giovanni Maria Tolosani [writing under the pseudonym Joannes Lucidus Samotheus], *Opusculum de emendationibus temporum* (Venice: Giunta, 1537), quoting Annius without naming him on 40 recto (*"Aegyptiorum Dynastiae primae a centesimo trigesimoprimo anno post diluvium, quo coeperunt, usque ad pene completum quadragesimum tertium annum Nini regis Assyriorum, fuerunt quindecim, ut Eusebius in libro de temporibus pandit"*) and building his table of Egyptian kings on Annius's account (80 verso–83 recto); Johann Funck, *Chronologia* (Nuremberg: Wachter and Jacob, 1545), starting at A5 recto; Theodor Bibliander, *Temporum a condito mundo usque ad ultimam ipsius aetatem supputatio* (Basel: Oporinus, 1558), 32, 47, 59.

25. Mercator, *Chronologia*, (†)5 recto-verso: *"Autores qui in opere citantur."* See also his comment on his method of citation, (†)4 verso.

26. Ibid., (†)4 recto: *"Haec admonere volui, quod magni apud multos videam hos autores Annianos fieri, et somnia illorum de amplis statim a diluvio per omnes terras regnis pro veris haberi, cum merae sint fabulae, ut nihil ad nostram hanc exactam temporum dispositionem conferre potuerint."*

27. Anthony Grafton, "Tradition and Technique in Historical Chronology," in Michael Crawford and C. R. Ligota, ed., *Ancient History and the Antiquarian: Essays in Memory of Arnaldo Momigliano* (London: Warburg Institute, 1995), 15–31.

28. Mercator, *Chronologia*, 5: "*quas exiguo annorum spacio concludere non convenit, non enim leviter et subito regnorum potestas ex uno stemmate unave natione ad aliam transit.*"

29. Ibid.: "*omnino apparebit cum primis fere hominibus Aegypti habitationem et qualecunque regnum natum esse.*"

30. Ibid.: "*credibile est Aegyptios, qui tantopere de sua antiquitate semper gloriati sunt, et tum proavorum per manus traditione, tum antiquissima et ante hominum memoriam nata scribendi ratione foelices prae caeteris fuerunt, potuisse ac studuisse generalem quandam rerum apud se gestarum memoriam inde a rerum primordiis conservare.*"

31. James Gordon, *Chronologia annorum seriem, regnorum mutationes, & rerum memorabilium sedem; annumque ab orbe condito ad nostra usque tempora complectens* (Bordeaux: Milanges, 1611); Jacobus Salianus, *Annales ecclesiastici Veteris Testamenti*, 6 vols. (Paris: Heuqueville, 1641), vol. I, 198–199; Denys Petau, *De doctrina temporum*, 3 vols. (Venice: Albrizzi and Baronchelli, 1757), vol. II, 19.

32. Anthony Grafton, "Kircher's Chronology," in Paula Findlen, ed., *Athanasius Kircher: The Last Man Who Knew Everything* (New York and London: Routledge, 2004).

33. Mercator, *Chronologia*, (†)4 recto-verso: "*cum Chronologia sit veluti prima historiae adumbratio et rudis quaedam imago, in qua primariae partes iuxta magnitudinem, situm et proportionem disponuntur.*"

34. Ibid., (†)4 verso: "*Annorum spacia non temere eadem magnitudine continuavi, etiam ubi nihil erat quod asscriberem, valde enim illa aequalitas idonea et commoda est cum quis certum aliquem annorum numerum in consequentia vel antecedentia sine additionis aut subtractionis molestia computare volet. crescit quidem inde opus et pretium, sed compensat utilitas. proinde hanc dispositionem inviolatam manere opto.*"

35. Grafton and Williams, *Christianity and the Transformation of the Book*; Daniel Rosenberg and Anthony Grafton, *Cartographies of Time* (New York: Princeton Architectural Press, 2010).

36. Mercator explains his procedures clearly, *Chronologia*, b verso.

37. Dietrich Wünsch, *Evangelienharmonien im Reformationszeitaltar: ein Beitrag zur Geschichte der Leben-Jesu-Darstellungen* (Berlin and New York : W. de Gruyter, 1983).

38. For a more detailed examination see H. J. de Jonge. "Sixteenth-Century Gospel Harmonies: Chemnitz and Mercator," in *Théorie et pratique de l'exégèse* (Geneva: Droz, 1990), 155–166.

39. Mercator, *Chronologia*, 54.

40. Ibid., 52.

41. See e.g. Arnaldo Momigliano, "Ancient History and the Antiquarian," *Journal of the Warburg and Courtauld Institutes*, 13,3/4 (1950), 285–315; Lionel Gossman, *Medievalism and the Ideologies of the Enlightenment: The World and Work of LaCurne de Sainte-Palaye* (Baltimore: Johns Hopkins Press, 1968); Roberto Weiss, *The Renaissance Discovery of Classical Antiquity*, 2d ed. (Oxford: Blackwell, 1988); Riccardo Fubini, *Storiografia dell'umanesimo in Italia da Leonardo Bruni ad Annio da Viterbo* (Rome: Edizioni di Storia e Letteratura, 2003); Ingrid Rowland, *The Scarith of Scornello: A Tale of Renaissance Forgery* (Chicago: University of Chicago Press, 2004); Graham Parry, *The Trophies of Time: English Antiquarianism in the Seventeenth Century* (Oxford: Oxford University Press, 2007); Peter Miller, ed., *Momigliano and Antiquarianism: Foundations of the Modern Cultural Sciences* (Toronto and Buffalo: University of Toronto Press, 2007); Peter Miller and François Louis, ed., *Antiquarianism and Intellectual Life in Europe and China, 1500–1800* (Ann Arbor: University of Michigan Press, 2012)

42. For eclipses from Ptolemy see e.g. 57, 64, and 89, where Mercator inserts lunar eclipses reported by Ptolemy in books 4 and 5 of the *Almagest*.

43. Mercator, *Chronologia*, 66, 83.

44. Ibid., 66.

45. Apianus's version of traditional history nicely adumbrates Collingwood's idea that all traditional histories were compiled with scissors and paste.

46. Petrus Apianus, *Astronomicum Casareum* (Ingolstadt: Apianus, 1540), J III verso: "*Quo videas penitus quidnam ecleipsium cognitio virtutis, praestantiae, utilitatisque in se complectatur, aliquot ante Christum exempla, unde certius quam ex prioribus commodum conspici quaeat, adiungere coactus sum. Nemo ignorat, rerum historiarumque traditores in enumerandis tum mundi, tum urbis conditae annis, tantum non, ut dicitur, coelo ipso excidere. Qui quidem error quantum tenebrarum, ignorantiae, dubii, confusionisque in omni alia historia iudicanda intelligendaque importet, doctorum esto iudicium. Inde enim fit, ut si cuiuspiam rei, memorabilis saltem, lectorem christianum commonefaciant, quae urbem conditam aut praecesserit aut non ita multo post, subsequuta sit, ipsis constare nullo modo possint. Quod quidem tam grande malum, sola ecleipsium cognitio emendare et in melius vertere potest. Per ecleipses enim omnia certos in annos reduci possunt, Christum praecedentes, non minus, quam sequentes. Historicus autem certum semel tempus rei gestae nactus, reliqua per se, tam ante quam retro elapsa gestorum spacia, colligere valet. Quae omnia exempla uberrime patebunt.*"

47. Mercator, *Chronologia*, 96.

48. Ibid., 4, 54–55, 57.

49. Ptolemy, *Almagest*, 4.10.

50. Mercator, *Chronologia*, 96.

51. Pliny, *Natural History*, 2.72.180: "*Ideo defectus solis ac lunae vespertinos orientis incolae non sentiunt nec matutinos ad occasum habitantes, meridianos vero serius nobis illi. apud Arbilam Magni Alexandri victoria luna defecisse noctis secunda hora est prodita eademque in Sicilia exoriens.*"

52. Mercator, Chronologia, 96: "*recte hujus eclypsis tempus a Plinio assignatum esse, dubitari non potest.*" Cf. ibid., e ij recto: "*418 a Nabon. demonstravimus eclipsim Lunae, quae contigit Alexandro ad Arbelam cum copiis Darii confligente, horis 5.30 circiter serius factam quam calculus Ptolemei ex eclipsi hic proxime praecedente colligat, quod apte admodum cum praecedentium syzygiarum defectu consonat.*"

53. His problems were sometimes exacerbated by the fact that he shifted basic eras, including that of the foundation of Rome and of Nabonassar. See Nothaft, 265–266, who observes that "the impressive visual façade of Mercator\s chronology concealed serious cracks in the technical basement, which threatened to bring the entire house down" (265).

54. Aelian, *Varia historia*, 2.25.

55. Mercator, *Chronologia*, 96: "*Inventa autem eclypsis haec ea in parte anni optime quadrat cum Aeliani testimonio, mense Thargelionis factam dicentis. Thargelion enim primo post intercalarem anno, in quem haec eclipsis incidit, Tauri signo respondet.*" The eclipse was observed with great care in Babylon, though there are some discrepancies between the account in a Babylonian astronomical diary, to which, of course, Mercator and his contemporaries had no access, and those in the Greek sources. See V. F. Polcaro, G. B. Valsecchi, and L. Verderame, "The Gaugamela Battle Eclipse: An Archaeoastronomical Analysis," *Mediterranean Archaeology and Archaeometry*, 8 (2008): 55–64.

56. Mercator, *Chronologia*, a iiii verso: "*Etsi autem eclipses hae a nobis demonstratae, non in idem quod invenimus forte tempus exacte incidant, propter inaequalitatem quandam et anomaliam in motu Lunae sive eius a Sole recessu, hactenus a nemine animadversam, ut tertio capite indicabimus, tamen non ita longe Ptolomei calculus a vero aberrat, ut in anni assignatione falli queamus.*"

57. Nothaft, 267.

58. Ibid., 269-270.

59. Florian Diacu, *The Lost Millennium: History's Timetables under Siege* (Toronto: Knopf Canada, 2005).

60. Matthieu Béroalde and Gerardus Mercator, *Chronologia, hoc est, Supputatio temporum, ab initio mundi ex eclipsibus & obseruationibus astronomicis & sacrae scripturae firmissimis testimonijs demonstrata* (Basel: Guarinus, 1577). The first edition of Béroalde's work had appeared as *Chronicum, Scripturæ Sacræ autoritate constitutum* (Geneva: Chuppin, 1575). Isaac Casaubon's copy, its margins thick with mordant notes, is in British Library C.79.e.12.(1.).

61. For the ancient precedents, well known to early modern scholars, see Anthony Grafton and Noel Swerdlow, "Greek Chronography in Roman Epic: The Calendrical Date of the Fall of Troy in the *Aeneid*," *Classical Quarterly*, n.s., 36 (1986); 212–218 and Grafton and Swerdlow, "Technical Chronology and Astrological History in Varro, Censorinus and Others," *Classical Quarterly*, n.s., 35 (1985): 454–465. For Bianchini see John Heilbron, "Bianchini as an Astronomer," in Valentin Kockel and Brigitte Sölch, ed., *Francesco Bianchini (1662–1729) und die europäische gelehrte Welt um 1700* (Berlin: Akademie Verlag, 2005) and Tamara Griggs, "Universal History from Counter-Reformation to Enlightenment," *Modern Intellectual History* 4 (2007): 221–228.

10
Rethinking 1633: Writing about Galileo after the Trial*

Paula Findlen

In 1632 Galileo Galilei was one of the most celebrated astronomers and mathematicians in Europe who had finally completed one of his long advertised and eagerly anticipated books, the *Dialogue Concerning the Two Chief World Systems*. One year later, he was a penitent Catholic, vehemently suspected of promulgating the idea of heliocentrism and openly advocating his preference for Copernican astronomy. He was famously forced to repent these beliefs. These events changed fundamentally how and why the life of Galileo would be written. To some degree or another, every biographer of Galileo would have to confront the matter of the trial.[1]

This essay explores the problems of writing the life of Galileo during the first half-century after the trial. It potentially provides a more general object lesson in confronting the problems of writing about the life of a scientist who had his enemies as well as his friends and whose understanding of the natural world came into conflict with the most important institution of the day, the Roman Catholic Church. How did Galileo's early modern biographers describe his accomplishments? What did they highlight and what did they omit? What solutions did they devise for dealing with the most controversial aspects of his career? Although referring to Vincenzo Viviani's well-known *Racconto istorico della vita del Signor Galileo Galilei*, written and rewritten by his last, living disciple starting in 1654 and unpublished until 1717, the most famous and authoritative life of Galileo is not the focal point of my research.[2] Instead, this chapter concentrates on a far less celebrated but equally revealing episode to create an authorized Jesuit biography of Galileo in the 1670s. Examining earlier and later biographies of Galileo—and of course Viviani's own celebrated account—we see how immediate and successive generations of scholars understood the significance of writing about a man whose list of scientific accomplishments was as impressive as the Catholic Church's harsh response to the religious implications of his astronomy.

Galileo's trial in Rome—culminating in his public abjuration in June 1633, the publication of his condemnation by the Holy Office, the placement of his *Dialogue* on the Index of Prohibited Books, and his subsequent silence on Copernican astronomy during the nine-year interval between these tragic events and his death at his villa in Arcetri in 1642—has made his biography one of the most inherently dramatic lives for historians of science to reconstruct. His earliest biographers, who could not distance themselves from these events, or the individuals and institutions which set them in motion, had an entirely different perspective on the difficulties of writing about a controversial life. To write the life of Galileo was neither an act of scholarly dispassion, if it has ever been, nor was this the criteria that shaped "life writing" in the seventeenth century whose goal was to offer exemplary tales of what a life ought to be. With far less evidence at hand than any modern historian but with far more tacit knowledge of what was at stake, Galileo's early modern biographers attempted to reconstruct his life to make it meaningful for their own times. How did the trial inform these early assessments of Italy's most famous mathematician and astronomer?

Creating silence

The first life of Galileo, completed in the midst of the trial, appeared in the Greek scholar and bibliographer Leone Allacci's *Apes Urbanae*, published in Rome in 1633. Allacci, who eventually became the Vatican librarian, had a difficult problem to confront. He was writing an account of the flourishing of arts and letters in Rome under the patronage of Urban VIII's nephew, Cardinal Francesco Barberini, and implicitly the pope. When he began his project, Galileo was one of the leading examples of the glory of Barberini patronage. Initially Allacci did not know a lot about Galileo's career but close associates supplied the missing details. Allacci updated the initial draft of his entry to include the recently published *Dialogue*. He even took the initiative of reading one of Galileo's unpublished treatises circulating in Rome, which he vaguely described as *Discourse on the Motion of the Earth*.[3]

Allacci's entirely positive evaluation of Galileo's contributions to knowledge rested on a critical piece of evidence: Urban VIII's unadulterated admiration for his fellow Tuscan. Cardinal Maffeo Barberini's 1620 poem in praise of Galileo's astronomical discoveries concluded the version of Galileo's biography that Allacci sent to press, probably around April 1633, with the approval of the master of the Sacred Palace, the infamous "Father Monster" Niccolò Riccardi whose unfortunate role in the labyrinthine process by which Galileo's *Dialogue* had been approved for publication is well known from the history of the trial.[4]

In the midst of publication, Allacci unexpectedly found himself the author of a highly controversial biography, the only one written during Galileo's lifetime. His unequivocal praise of Galileo no longer seemed appropriate.

His reference to the condemned *Dialogue* without any mention of its censure was highly inappropriate, and his allusion to reading a manuscript by
the author on the same subject might lead the Holy Office to suspect *him*
of harboring the same heretical beliefs for which Galileo was condemned.
Finally, Allacci's liberal quotation from the pope's poem about Galileo might
thwart his efforts to gain the favor of the Barberini. Grabbing these pages
from the hands of his printer Ludovico Grignano, Allacci furiously expunged
the offending passages in light of the June 22, 1633 condemnation and abjuration of Galileo.[5] The new version, now containing the pages reprinted by
Grignano, told an entirely different story of Galileo's life. Allacci's final entry
simply omitted the *Dialogue* from the list of Galileo's publications.[6] It was
as if the book and the trial it engendered never existed. Allacci's account of
Galileo's life simply stopped before 1632.

Having created a hermetically sealed version of Galileo's life, which ended
before the trial, Allacci was now free to treat the controversy over the *Dialogue*
as an entirely separate affair. The biography of one of Galileo's leading adversaries, the Jesuit astronomer Christoph Scheiner, proved to be the perfect
venue for raising the subject. Discussing their debate on sunspots and the
ensuing publications, including Scheiner's *Rosa Ursina* (1630), which was
highly critical of Galileo, Allacci referenced Galileo's discussion of sunspots
in the *Dialogue*. He indicated that Scheiner was at work on a treatise called
the *Forerunner in Favor of the Stability of the Earth against the Same Dialogue* "in
which the errors of Galileo in logic, physics, mathematics, ethics, theology,
and sacred matters are asserted at length."[7] In this devastating assessment
of the serious and comprehensive nature of Galileo's crime, Allacci proved
himself worthy of his eventual appointment to a plum position within the
Barberini household. By neatly separating Galileo's celebrated accomplishments from his recent disgrace, Allacci managed to do the one thing that
mattered most of all in seventeenth-century Rome: save appearances.

At least one sympathetic observer, the French libertine writer and Roman
transplant Jean-Jacques Bouchard, was so outraged by Roman efforts to malign
Galileo that he conceived of the idea of writing the life of Galileo as an anti-
dote to the efforts of the Holy Office to besmirch Galileo's reputation. In the
winter of 1638 Bouchard found his oration on the French savant Nicolas-
Claude Fabri de Peiresc brutally subjected to the censor's revisions because
it contained some favorable remarks about Galileo. The French scholar was
especially incensed that Riccardi, upon reading his account of Galileo's relationship with Peiresc, scrupulously "crossed out all that I had said in praise
of him." Bouchard shared his outrage with one of Galileo's sympathizers, the
Marchese Vincenzio Capponi, in Florence. "I am so disdainful of this barbar-
ity employed against poor Galileo in particular that I have decided to employ
whatever free time I have to write his life."[8]

For more than a year Bouchard enlisted Capponi in the project of collecting
"details of the life of Signor Galileo Galilei" from his Florentine friends and
disciples. Responding to questions about his plans for publication, Bouchard
reassured his informants that he had absolutely no plans to publish the life of

Galileo as long as the subject lived. Given how "things are today," he felt that there was no point in attempting such a risky publication.[9] The last we hear of this project was in April 1639. Bouchard collected his materials but never completed the work in his life; when he died in Rome in August 1641, Galileo was still alive. No longer able to read, he nonetheless possessed the censored and uncensored copies of Bouchard's life of Peiresc.[10]

Other early biographers found it equally convenient to simply omit the events of 1632–1633 from their accounts of Galileo's career.[11] Two 1647 biographies by Vittorio Siri and Girolamo Ghilini—the first to my knowledge to appear after Galileo's death in 1642 when Urban VIII curtailed efforts to bury Galileo in Santa Croce because he had caused "the greatest scandal in Christendom"[12]—adhered to this principle. Ghilini was only somewhat braver than Allacci. He mentioned the *Dialogue* in his list of Galileo's publications without alluding to its censure. Yet he then retreated into utter silence, indeed verging on misrepresentation, when discussing the final decade of Galileo's life. Ghilini described Galileo's retreat to Arcetri as a personal decision tied to his return to Tuscany rather than the product of his unhappy circumstances in 1633. The Venetian biographer suggested that Cosimo II's appointment of Galileo as court philosopher and mathematician ultimately gave him the leisure "to enjoy the happy repose of a private life until the age of 73, after the effort of teaching for many years."[13] Readers of Ghilini's account were left with the benign impression that the culmination of Galileo's lengthy career was not a disgraceful and troubling episode but a noble philosophical retirement to the countryside.

The historian Siri reinforced this understanding by being equally vague about the end of Galileo's life. Both mid-seventeenth-century biographers had a tangible accomplishment to which they could refer that deflected attention from the subject on which they were so silent: Galileo's *Two New Sciences* (1638), his great work of mechanics written and published after the trial. Siri briefly considered Galileo's final years in light of this important publication that cemented Galileo's reputation as mathematical physicist. The significance of the *Two New Sciences* allowed Siri to conclude his biography by declaring Galileo to be "the most perfect mathematician of the present century."[14]

The appearance of these two biographies may have stimulated another thwarted effort to write the life of Galileo. In November 1647 Carlo Dati—an admirer of Galileo, eventual participant in the experimental activities of the Accademia del Cimento, and one the great literary lights of his generation—approached the 77-year-old Fulgenzio Micanzio, theologian to the Republic of Venice and a loyal friend to Galileo even after the trial. Dati informed Micanzio that he had been repeatedly urged by Galileo's friends and admirers "to write the life, customs, accidents, opinions, sayings, and celebrated works of this sovereign philosopher and mathematician, of whom our age and my country are proud." He sought out Micanzio as one of the few remaining "contemporary friends and confidants of Signor Galileo who can give me some good information."[15] The Venetian theologian was an invaluable resource for filling in the missing details of Galileo's life and work in Padua, but he was also known to be an impassioned advocate for the preservation of Galileo's reputation. Dati's biography might have offered a richly textured account of Galileo's life combining the perspectives of both his Florentine and Venetian disciples. The biography never appeared though hints of what it might have been are scattered through his work. Interestingly, Dati specifically requested information on "the accidents, both good and bad" that befell Galileo.[16] Was this his way of saying that he was soliciting Micanzio's view of the events leading up to the trial?

One of the reasons to be silent about the trial was not only sympathy for Galileo but also understandable concern about the consequences of discussing it. Allacci's caution was shared by Abbé Ghilini who confronted the problem of writing a book that included the lives of "Cardinals Maffeo, today Our Father Urban VIII, and Francesco Barberini" adjacent to his account of Galileo.[17] To write about the trial was implicitly to offer an opinion about the reasons it had occurred. It is hardly surprising that in the first two decades after 1633 absolute silence was the most prudent choice. All three publications that I have mentioned were nothing more than short notices of Galileo's life and work, highly selective accounts of his accomplishments and publications containing many other omissions.

To my knowledge, only one biographer of Galileo discussed the trial in print before 1660. The Roman humanist Giovan Vittorio de' Rossi—friend of many of Galileo's Roman associates including Giovani Ciampoli, Cassiano dal Pozzo, and Gabriel Naudé, and enjoying the favor of Cardinal Francesco Barberini—had a great deal to say about it.[18] Given his intellectual network, he was unusually well-informed so his comments reflect a Roman critique of the trial that had already begun to take shape immediately after 1633. Discussing the emergence of the position that heliocentrism contravened "the testimony of Sacred Literature, the consensus of the Holy Fathers, and the truth of Catholic Faith," he underscored the fact that "many others" before Galileo had participated in this heresy. Implicitly he raised the question of why Galileo had been singled out. Regarding the publication of the *Dialogue* he discussed why its contents precipitated the judgment of the Holy Office. Unlike Ghilini, Rossi accurately described Galileo's confinement to Arcetri as a result of his sentence. He offered his opinion on this verdict, writing that it was to the detriment of "wisdom itself, from which the world has been constituted from its beginning."[19]

No one should be surprised to hear that Rossi published his *Pinacotheca*—a popular and opinionated account of the lives of modern scholars that competed openly with Allacci's *Apes Urbanae* as a counternarrative of scholarship in the age of Urban VIII—in Cologne under his academic pseudonym "Janus Nicius Erythraeus." He accepted his good friend Monsignor Fabio Chigi's—the future Alexander VII—generous offer to find a German publisher for this book, which contained the fullest and frankest life of Galileo to date. The pope reportedly laughed so hard reading the book that he took his glasses off to avoid breaking them, though one wonders how closely he read the life of

Galileo. By contrast, Allacci seems to have understood that Rossi neither liked nor respected him; he later took his revenge unmasking Rossi as the author of a vicious satire of Barberini Rome.[20] The reception of Rossi's well-known book highlights the paradoxes surrounding Galileo's trial. His positive assessment of Galileo would not have earned him an *imprimatur* in Rome, and yet he was published in Catholic Germany, with the assistance of the future pope Chigi, read by Urban VIII, and supported by the pope's nephew. Viviani expended a great deal of energy furiously correcting the one glaring error perpetrated by Rossi in Galileo's biography: his claim that Galileo was illegitimate.[21] No one in Florence would have made this mistake.

Tentative explanations

By contrast, the Neapolitan writer Lorenzo Crassi's life of Galileo in his *Eulogies of Learned Men* (1666) developed the kind of ambiguous formulation of the meaning of the trial that more accurately reflected mid-seventeenth-century views of an event that was increasingly in the past and did indeed deserve some commentary. He attributed Galileo's decision to offer an opinion "on matters of earth and heaven, on stability and motion against the establishment of the Roman Church" to his "overly speculative and subtle genius." In Crassi's account, Galileo had let his mind wander too freely where it should not have gone. Offering a psychological explanation, he suggested that accidental heresy was the potential flaw of unbounded genius. Crassi reminded his readers that Galileo suffered the consequences, though he never specified exactly what those consequences were. In his vague formulation, Galileo became another Icarus who fell to earth after flying too close to the sun, yet readers of Crassi's life of Galileo would have had absolutely no understanding of the concept of heliocentrism, why the Catholic Church had condemned it, and how this decision affected Galileo, if this was the only thing they read. Crassi loyally suggested that Galileo might have suffered even more, were it not for good advice and "help of great men."[22] But who were those great men and what exactly did they do for Galileo? Like a seventeenth-century cryptogram, you needed to know the cipher to figure it out.

The appearance of Crassi's biography was another reminder to the Florentine community devoted to preserving Galileo's legacy that they had yet to produce their own version of his life. Crassi's exercise in dissimulation pleased nobody, least of all Galileo's admirers and defenders who had been patiently working toward some sort of accord with the church that might lead to a partial, if not complete, rehabilitation.[23] From Rome, one of the key participants in this venture, the mathematician and theologian Michelangelo Ricci (1619–1682), offered an extensive critique of the deficiencies of this recent biography. Ricci told Leopoldo de' Medici (1617–1675) that he hoped it would finally spur Galileo's last disciple Vincenzo Viviani (1622–1703) to publish the long-awaited life of his master. Observing that "many students and friends of Signor Galilei are still alive," Ricci felt that they could provide Viviani with plenty of material for a more accurate and multidimensional life of Galileo that would truly honor his memory.[24]

Capturing the living memory of Galileo before it disappeared was a project with some urgency. With each passing year, the number of scholars who had been personally associated with Galileo dwindled. Ricci expressed his regret that Galileo's Roman disciple Raffaello Magiotti was no longer alive to participate in this project.[25] In the very same letter he informed his Medici patron that the pope had appointed him to a position in the Holy Office. Ricci, who ended his life a cardinal, did not see his official duties as being in any way incompatible with his unequivocal support for the cause of promoting Galileo. This juxtaposition suggests the landscape in which we need to consider what it meant to write the life of Galileo in mid-seventeenth-century Italy. The production of a good life of Galileo was perceived to be a vehicle through which to broker a better understanding of the relationship between science and faith, doing justice to Galileo's merits while also confronting, with the greatest tact and diplomacy and with full recognition for the authority of the institutions that had passed judgment upon Galileo, the implications of his trial.

To fully understand Ricci's frustration in 1666 with the absence of a life of Galileo that told the story of his accomplishments well, we need to explore the circumstances in which writing the life of Galileo became the collective preoccupation of a community of individuals either directly involved in the Accademia del Cimento or closely associated with it. Leopoldo de' Medici's idea of including a life of Galileo in the first edition of Galileo's works published in Italy after the trial—Carlo Manolessi's two-volume edition of the *Works of Galileo Galilei* published in Bologna in 1656—was the primary catalyst. As is well-known, in 1654 Leopoldo asked two people who had known Galileo well in the final years of his life to compose their version of his life. One was Viviani and the other was Niccolò Gherardini, prior of Santa Margherita a Montici, who had first met Galileo during the trial and left his legal studies in Rome to accompany Galileo back to Tuscany. Gherardini responded to Leopoldo's request with a letter that truly summed up his personal recollections of Galileo: how they had met and what he remembered. It was a heartfelt account of an encounter with a great scientist at the moment of his greatest moral anguish that changed both of their lives. To my knowledge, Gherardini never attempted to revise his account nor did anyone encourage him to publish it. Viviani found almost every element of this memoir objectionable, annotating it with a censorious pen that would have impressed even that painstakingly sincere censor of all Roman books, Father Riccardi.

Viviani spent the rest of his own life editing, polishing, and expanding his life of Galileo, which he initially based on Siri's brief biography of 1644. Responding to Ricci's entreaties, Viviani wrote in 1668 that it was "approaching perfection," yet he was still investigating the particulars of Galileo's life in 1692.[26] The *Racconto istorico* would remain unpublished until Salvino Salvini's edition of 1717. We do not know whether Leopoldo specifically asked the most distinguished literary member of the Accademia del Cimento, Carlo Dati, to also provide him with a biography, but Dati seems to have collected materials for a similar project.[27]

None of this material made it into Manolessi's 1656 edition of Galileo's works, which conspicuously omitted the *Dialogue* and other writings deemed potentially controversial, describing them ambiguously as works "suppressed with good cause" that the publisher would immediately print "when their publication is authorized."[28] Michael Segre has rightfully argued that Leopoldo and his collaborators ultimately withdrew from this imperfect homage to Galileo in hope that they might broker an agreement with the Roman Catholic Church that would allow them to create an official bilingual edition of all of Galileo's works in Italian and Latin, thereby superseding the partial edition in Italian published in the second city of the Papal States and the Latin translations of his Italian works published in northern Europe.[29] Initially Viviani held onto his life in the hope that it would appear in this grand edition of Galileo's works. Although he may not have shown the actual manuscript to many people, the scholarly community was certainly aware of its existence. Given the list of foreigners who signed Viviani's *album amicorum*, some of them actually may have seen the *Racconto istorico*. Robert Southwell was already trying to wrest it from Viviani's hands with the promise of making a Latin translation in 1662.[30] He did not succeed nor did Henry Oldenburg who dearly wanted to present it to the Royal Society.

It was that British gadfly of the Tuscan scene, Sir John Finch, who offered the most concrete account of the atmosphere surrounding the project of writing the life of Galileo in the 1660s. Responding to Galileo's English biographer and translator Thomas Salusbury, who expressed frustration at the problems of writing "the Life of a person that lived at such a distance, and dyed under a Cloud," Finch explained why it was so difficult for anyone to get good information about Galileo from Viviani whom Salusbury had probably written. "For the truth is that il Serenissimo Prencipe Leopoldo having a design to Print all Galileos workes prohibited and not prohibited in two Volumes in folio in Latin and Italian, He is resolved to Praefixe his life to his workes." Finch explained that the timing of this publication was entirely Leopoldo's decision. Viviani awaited his patron's wishes.

Finch also alluded to the extent to which this project was enmeshed in delicate negotiations between Tuscany and the Papal State. "Nor indeed can the Transactions concerning him at Rome be discoursed of with any certainty unless that the Grand Duke is pleasd to give way they should be made publique."[31] The end result was that even the most banal details of Galileo's life were uncertain since no one could discuss Galileo without bringing their opinion of him (and implicitly his trial) into the conversation. Salusbury's biography, recently rediscovered in a private collection and brilliantly analyzed by Nick Wilding, was indeed riddled with errors of fact. At the same time, freed from the constraints of writing about the trial in Catholic Europe, Salusbury offered by far the most detailed and sophisticated account of the events precipitating this event interwoven with a quintessentially British view of the dangers of the Jesuits as the architects of Galileo's fall.[32] Had all but one mangled copy survived the Great Fire of London, Salusbury's life would have surely engendered a lively discussion between England and Italy about foreign perceptions of Galileo's accomplishments and the reasons for his trial.

Riccioli's account and Viviani's response

It was instead a different, far more local, account of Galileo that was very much on the minds of the Florentines. In 1651 the Jesuit astronomer Giambattista Riccioli (1598–1671) published, for the first time in Italy since 1633, the Holy Office's 1616 decree against Copernican astronomy, the 1620 expurgation of Copernicus's *On the Revolutions of the Heavenly Spheres* (1543), the 1633 condemnation of Galileo, Galileo's abjuration, and Cardinal Antonio Barberini's unprecedented letter of July 1633 to inquisitors in the Italian peninsula and papal nuncios throughout Catholic Europe instructing them to read aloud to communities of mathematicians and natural philosophers, and to publicize in other ways the news of the trial. Riccioli included this material in his *New Almagest* (1651), a vast compendium of observational astronomy that weighed the evidence for and against all the different cosmological systems and also included a detailed history of astronomy since antiquity including brief lives of important astronomers. Although firmly supporting the official Jesuit position that some version of a Tychonic system incorporated all new observations without running the risk of contradicting scripture, Riccioli nonetheless wrote glowingly of heliocentrism as a hypothesis and professed his great admiration for Copernicus, Kepler, and Galileo, whom he described as "a mathematician of immense power wonderfully skilled in astronomy." At the same time, Riccioli considered it his duty to outline the Church's official position on Copernican astronomy by including its key documents and observing that Galileo "would have been greater still if he had put forward the opinion of Copernicus as a mere hypothesis."[33]

From Riccioli's perspective, he had done two things. He scrupulously defended the Roman Catholic Church's official position on astronomy, reminding his readers that "Catholics are bound in prudence and obedience" not to contradict the church's decision, which provided a concrete explanation of Galileo's error.[34] At the same time, Riccioli drew to his reader's attention the fact that heliocentrism was only contrary to scripture but not formally heretical, a subject that was the source of many misconceptions since some people argued that heliocentrism had been declared a heresy, not understanding that only the pope, and not a special commission, could pronounce this judgment; instead Galileo had been vehemently suspected of a heresy whose exact nature remained unspecified. Riccioli wished to remind his readers that geocentrism was not yet an absolute article of faith. When taken together with his admiring comments about Copernican astronomers, a sympathetic reader might conclude that although Riccioli was not trying to evade the unpleasant facts of the trial he was nonetheless pointing out that Galileo's error had not been a matter of doctrinal heresy, *de fide*, but an unfortunate transgression of scriptural interpretation. His explanation managed to make the trial less serious than a reading of the documents alone would have suggested; it clarified an important procedural and doctrinal issue without delving into the ambiguities and contradictions of the trial documents—the pope's attitude toward Galileo and the degree to which devising hypothetical accounts of new world systems was easier said than done.[35]

Florentine readers of the *New Almagest* did not see it that way. The appearance of Riccioli's book provoked outrage in Tuscany. Twenty years after the publication of the *New Almagest* when news reached Florence of Riccioli's death, Viviani was still fuming about the way in which the Jesuit astronomer demonstrated his animus toward "our Galileo" by publishing "unnecessarily and quite off the subject . . . that sentence of the abjuration that even the Holy Office in Rome did not deem necessary to publish then." He recalled how Grand Duke Ferdinando II had been "nauseated" and "greatly offended" by Riccioli's book. In Viviani's opinion, there was simply no need for Riccioli to neither invoke "the authority of the Supreme Tribunal of the Holy Office" nor mention Galileo's trial in discussing the condemnation of heliocentrism. "It was enough to say that the other opinion was prohibited."[36]

No doubt Viviani took some small satisfaction in hearing of Riccioli's subsequent troubles with the inquisition over his interpretation of the Immaculate Conception and the doctrine of papal infallibility. Unfortunately, it had the unintended effect of convincing Riccioli that he had equivocated too much. By 1669, even as he softened his scientific critique of Copernicanism, Riccioli revised his earlier description of the church's condemnation, writing that the censure of Copernicanism and Galileo was "absolute, and not only provisional" because it contravened scripture.[37] Surely it was Riccioli's final word on the trial that most angered Viviani, since he knew very well that Urban VIII had insisted that the text of Galileo's condemnation and abjuration be read aloud by inquisitors and papal nuncios, making it the most well-publicized outcome of a heresy trial in seventeenth-century Italy. The Cologne nuncio who preceded Fabio Chigi was the first to print a Latin summary of the trial that famously led Descartes to alter his publication plans of his book on *The World*, in part, because it mistakenly presented heliocentrism not simply as contrary to scripture but a full-fledged heresy. By the time Riccioli published his *New Almagest* many manuscript copies of the sentence and abjuration had been in circulation since 1633, and there were printed versions in Italian, Latin, and French.[38]

Grappling with the issue of Galileo's reputation was indeed a central question for Catholic scholars in the decades after the trial. The fact that a Jesuit astronomer had republished the key documents of this trial and continued to sharpen his interpretation of them over two decades added further ammunition to a longstanding perception that Galileo's fiercest Jesuit critics had been behind the decision to treat Galileo harshly.[39] Viviani eloquently summed up his view of Riccioli's perceived betrayal of Galileo's legacy when he said that it was "unnecessary and even unbecoming the condition of an old man in religious orders, who was otherwise venerated, to demonstrate a greater aversion to the man than to his assertions."[40] But was this really the case? Viviani considered Riccioli's publication to be a personal attack on Galileo more than a critique of Copernican astronomy. It was partly in response to reading Riccioli that Viviani began to develop his own account of the trial, critiquing Galileo's prominent Jesuit adversaries Christoph Scheiner and especially Orazio Grassi whom he called the "Mathematician of the Roman College" to suggest, in essence, that anyone who held this position was Galileo's enemy. Viviani described how the debate over the comet of 1618 led to Grassi's "eternal persecution" of Galileo.[41] These were precisely the sort of inflammatory comments that Ricci encouraged Viviani to edit out of any eventual publication because this explanation of the trial transferred all blame from Galileo to the Jesuits.

Much as Viviani might want to follow the lead of Galileo's earliest biographers, who had simply erased the trial from his life, he could not do it. As his faithful disciple, he committed himself to writing a complete history of the life of Galileo. His goal was to find a way to include these events without making them overwhelm the more glorious episodes that had justly earned Galileo his fame. He also needed to convince readers of Galileo's piety rather than his potential heresy (or at minimum disobedience) so as to convey the impression that, from the beginning to the end of his life, Galileo had been the model Catholic astronomer. Viviani achieved the effect his sought by presenting the trial as a sign of Galileo's necessary humanity:

> But given that Signor Galileo had been elevated all the way to the heavens with immortal fame for his other admirable speculations and many novelties which made him appear almost divine among men, Eternal Providence permitted him to demonstrate his humanity through error. Thus, in his discussion of the two systems he demonstrated himself to be more adherent to the Copernican hypothesis, already condemned by the Holy Church as repugnant to Divine Scripture.[42]

Describing the Roman inquisition's decision to command Galileo to come to Rome to answer its questions, Viviani underscored Galileo's absolute compliance with their decree. "He was arrested and in brief (having publicly recognized his error) he retracted as a true Catholic this opinion of his; but in penalty his *Dialogue* was prohibited."[43]

Viviani reinforced this perception of Galileo as a penitent and obedient Catholic by portraying his master as "greatly mortified" at the appearance of the Martin Bernegger's Latin translation of his *Dialogues* and Elia Diodati's Latin translation of the *Letter to the Grand Duchess Christina*. He wrote that Galileo realized that it would be an "impossibility ever to suppress them." The continued circulation of his earlier, erroneous opinions was part of his penance, even though he had "by the authority of the Roman censure, Catholicly abandoned" the idea of heliocentrism.[44] Since this account of the circumstances of the publication of Galileo's works after the trial in no way conforms to the actual record of Galileo's wheeling and dealing with the Elzevirs to try to encourage them to publish a complete edition of his works as well as the *Two New Sciences,* it is one of many instances in which Viviani put aside the documentary record, namely Galileo's correspondence from 1633 to 1642, to write the history that he wished to write. His Galileo had "rendered his soul to his Creator with philosophical and Christian constancy" on January 8, 1642—immortal for his works and pious in all his acts.[45]

A new Jesuit life of Galileo

However beautifully rendered Viviani's life of Galileo was, including his explanation of the trial as a necessary ingredient to instill some Christian humility into a man who otherwise ran the risk of being his own idol, it remained unpublished and unread. In the midst of his endless revision of this biography, the prospect of another Jesuit biography of Galileo came to his attention. It was the spring of 1678. Leopoldo de' Medici was no longer alive to pursue his dream of a complete edition of Galileo's works, and the activities of the Cimento had become an artifact of historic memory. One day Viviani received a letter from a young Jesuit mathematician named Antonio Baldigiani announcing his plan to include the lives of Galileo, Torricelli, Marchetti, and Viviani inside a book he was in the process of editing. Baldigiani was not the author of the book that bore the name of the most prolific scientific writer in the Society of Jesus: the German polymath Athanasius Kircher (1602–1680). Kircher briefly held the position of professor of mathematics at the Roman College until celebrity relieved him of teaching duties to devote greater time to showing visitors the marvelous machines, natural curiosities, Egyptian antiquities, and missionary artifacts of the Roman College museum he curated while churning out a seemingly infinite quantity of large encyclopedias on just about everything anyone wanted to know.[46]

For almost two decades Kircher had been writing a book he originally called the *Etruscan Journey.* Inspired by his brief stay in Florence at the Jesuit College of San Giovanni in 1659, Kircher hoped to gain Medici's patronage. He initially envisioned it as both a history and a topography of the ancient province of Etruria, which included the northernmost parts of the Papal States as well as the Grand Duchy of Tuscany. An early version of the manuscript earned an explosively negative review from a Tuscan Jesuit assigned to read it. In 1660 Father Domenico Ottolini of Lucca considered it so riddled with errors and gaping in its omissions that he wrote: "The book of Father Athanasius Kircher entitled *Iter Hetruscam* in my judgment not only fails to surpass mediocrity; it seems to me not even to reach that standard."[47] Despite this devastating judgment, Kircher persisted with the project, consulting with Tuscan scholars about some of the details he could not personally know (especially concerning Ottolini's hometown) and gradually transforming the book from a hastily composed history and chorography into an encyclopedia of Tuscany, now titled *Etruria Illustrated,* that included some discussion of Tuscany's contributions to science.

By the time Baldigiani wrote Viviani, he had been editing the *Etruria Illustrata* for several years. He was himself a Florentine and was closely connected to the world of the Cimento.[48] He was also a self-avowed Galilean who had been actively struggling to find the right way to reconcile his scientific beliefs with his faith. In short, he was the Jesuit of Viviani's dreams. Baldigiani saw the task of editing Kircher's book as his opportunity to create an officially sanctioned account of Tuscan science. He had already inserted an account of Redi's natural history that, Baldigiani assured his friend, would do justice to its merits without revisiting the longstanding debate between Kircher and Redi on the spontaneous generation of insects.[49] Emboldened by this success, Baldigiani expanded the section on Tuscan science, rewriting an earlier entry on Galileo, and either editing or adding biographies of Galileo's two most important disciples (Torricelli and Viviani) and the figure who most represented the revival of atomism in Tuscany (Marchetti). This was a bold move indeed for a Jesuit in Rome.

Writing to Viviani in late May, Baldigiani described the circumstances under which he had been allowed such freedom of revision. "Today the said Father is rather weakened by old age and is largely unable to pursue his studies," he informed Viviani:

> It is my responsibility to take over most of the task of fixing some parts of it, not otherwise having the time to recast it all. I managed to insert here the eulogies of scholarly men, among which there is that one of Galileo...and I am certain that no one until now has written so magnificently of him.[50]

Implicitly challenging Viviani's decision to write but not to publish the definitive life of Galileo, Baldigiani promised to send copies of the eulogies of Galileo, Torricelli, Marchetti, and of course Viviani. In return, he asked Viviani for the wording of the inscription commemorating Galileo in Santa Croce and a list of all the published and unpublished works by these four scientists.

Viviani was intrigued. Having spent several decades collecting biographical materials and defending Galileo's reputation, he had a vested interest in any published account of his master. He praised Baldigiani's knowledge of Tuscan science and his love of Galileo, "our compatriot and I will say also our common master," and was encouraged by the idea that a Florentine Jesuit writing about Galileo and his disciples would solicit his input. By mid-June Viviani had read Baldigiani's drafts. He made no changes to Baldigiani's account of Marchetti, a sworn adversary, and suggested only minor revisions to the entry on Torricelli. With his own biography, after giving Baldigiani a list of unpublished projects he hoped to see in print, Viviani reminded Baldigiani that readers should know that "he had the fortune to be the last disciple of Galileo."[51]

Viviani now began to edit Baldigiani's life of Galileo. He provided Baldigiani with his description of Galileo's position to appear just after his name: "Florentine Patrician and Chief Philosopher and Mathematician to the Most Serene, Magnificent, and Excellent Lords Cosimo II and Ferdinand II." He encouraged Baldigiani to say more about Galileo's celestial discoveries, namely the moons of Jupiter and the sunspots. Finally, he advised Baldigiani to remove any mention of the reasons for Galileo's trial, citing one passage in the biography that especially bothered him: "Who if he were more cautious in a number of things, etc." (*qui si in nonnullis cautior fuisset, etc.*).[52] Viviani's Galileo was a pious Catholic who had neglected to mention that Copernicanism was nothing but an interesting hypothesis. No opinion of why he had been brought to trial, and whether it could have been avoided, should enter into the discussion.

From the Villa Pamphili at Frascati, Baldigiani informed him that the book had now cleared the revisors and was en route to Amsterdam. He warned Viviani that the result was not exactly what he had hoped for. "Father Athanasius himself cancelled from the Eulogy of Galileo that paragraph that pleased me more than the others." Despite reports of Kircher's feebleness, Baldigiani was not able to act quite as autonomously as he might have liked—and there were the mechanisms of Jesuit and papal censorship to consider as well, which could not entirely be circumvented. He described how other Jesuits present during Kircher's editing of Baldigiani's life of Galileo barely prevented Kircher from removing the entire entry. Although he sympathized with Viviani's perspective, Baldigiani felt that it could not reflect only a Tuscan interpretation of Galileo's trial. He ended his description of the negotiations then underway about this especially difficult section of the *Etruria Illustrated* by lamenting the fact that he could not do more. "If I had to write it and if the book were mine," he told Viviani, "it would have turned out much better."[53] Rather than addressing the question of how to treat the trial in this letter, Baldigiani promised to respond to this concern separately.

Both Viviani and Baldigiani were extremely concerned about the sensitive nature of their correspondence. Repeated requests to return letters suggest that they did not wish them to fall into anyone else's hands. Baldigiani did not want Viviani's correspondence arriving directly at the Roman College. Instead it went initially to Father Giovanni Martini, superior of the Congregation of the Missions, in Montecitorio, and ultimately to Baldigiani's brother Niccolò. Viviani seems to have responded through Magalotti and possibly through the other Jesuit in the family, Giovanni Maria. Both sought to create a space in which they could talk freely and confidentially about the editing of this Jesuit life of Galileo.

Baldigiani also had to contend with the prickly ego of Galileo's last disciple. When he asked how soon Viviani's edited collections of the published and unpublished works of Galileo and Torricelli might appear, let alone the many unrealized projects of Viviani himself, the temperature of their discussion rose a few degrees. When Viviani found out that Baldigiani probably would not have an opportunity to make the changes he had requested to his own entry, he tartly requested that he remove it entirely.[54] The further their correspondence proceeded, the more he insisted on this avenue—to the great discomfort of Baldigiani who seems to have personally persuaded Kircher to include his eulogies of these four Tuscan scientists.

Kircher's editing of Galileo's life distressed Viviani—he described the final version as "enervated"—but he agreed with Baldigiani that since the rest of the learned world would praise Galileo more fully, he should not insist on a full accounting of his science in this particular book. However, he failed to understand why he had been asked to work so hard to improve a biography that ultimately was so unsatisfactory. He had assumed that Baldigiani seriously wanted his input.[55]

Baldigiani had not forgotten about Viviani's desire to alter the interpretation of Galileo's trial. In fact, he had probably thought of little else since he began this venture. On July 18, 1678, he composed a letter that gave even him a pause. Insisting that Viviani share this letter with no one but Magalotti, Baldigiani explained his own reaction to Viviani's request. He began by recounting the history of the manuscript:

> I tell you that when Father Kircher's work on *Tuscany Illustrated* fell into my hands, it had been viewed and reviewed by many people over here, and yet they had left aside the abovementioned words with other words that were even more offensive and sharp. I had more than a little difficulty inducing that Father to take his pen to that passage and to content himself with the addition that I made of a few pages, and when I saw it had been reduced to *Si in nonnullis etc.*, it seemed to me that I had done the impossible.[56]

The consummate editor of other people's prose, Baldigiani reminded Viviani how hard it was to get authors to make changes.

Baldigiani was highly sensible to the fact that Kircher's interpretation of the condemnation was far closer to Riccioli's than his own, perhaps even harsher, since he "treated him as a Heretic." In his response to Viviani, we have some insight into the delicate nature of the conversation Baldigiani had been having with Kircher, his fellow Jesuits, and key Tuscans in Rome. On the one hand, he hoped to create the most sympathetic portrait of Galileo that was permissible. On the other hand, it had to be approved by Kircher, the Jesuit revisors, and the Master of the Sacred Palace. Citing Kircher's age and ill health, Baldigiani suggested that he "easily became irritated and suspicious."[57] He implored Viviani to discuss the matter with Magalotti who studied at the Roman College in his youth and recently returned from Vienna where he had been the Tuscan ambassador at the Habsburg court.[58] Magalotti perhaps might explain why things that might be done in Florence simply could not occur in Rome.

Finally, Baldigiani got to the heart of the matter. The life of Galileo that he was crafting ran the risk of censure by the Master of the Sacred Palace, who had final approval on the publication of all books in the city of Rome and the Papal States. Baldigiani had taken certain liberties because of how this system of censorship treated Kircher's publications. "I am certain, indeed most certain that if the Master of the Sacred Palace, who signs off on Kircher's books without seeing them, had seen that eulogy, it would not have approved it at all or even a little."[59] Baldigiani invoked the difficulties Francesco Nazari experienced when he tried to cite far less controversial but prohibited authors in the *Giornale de' letterati* as an example of the vigilance normally practiced in Rome. Rather than lamenting the limits of Kircher's account of Galileo, Baldigiani celebrated it as an opportunity to expand the discussion with the official approval of the church.

Turning to the most controversial issue, Baldigiani explained his own approach to the trial of Galileo. By making Galileo guilty of imprudence rather than doctrinal error, Baldigiani could present the trial as more of a "civil" than a "criminal" case. Galileo had simply been rash and repented his error as opposed to advocating a system of the world in literal contradiction with scripture and church authority, as Riccioli and others argued. Baldigiani proudly wrote that "those words" that Viviani so disliked "are the most honorable to Galileo in that entire eulogy." The argument he had chosen reflected a sympathetic view of Galileo among the Jesuits—first presented by the Cardinal Inquisitor Robert Bellarmine, then reiterated in Bartoli's biography of Bellarmine, and now presented by Baldigiani in the guise of Kircher.[60] It took Riccioli's basic point—that the Roman Inquisition could not declare Copernicanism a heresy because only the pope could decide this issue[61]—and expanded it to encompass Galileo's crime, transforming the vehement suspicion of heresy at the heart of his 1633 trial into something far more benign.

To go further was to risk everything. Baldigiani reminded Viviani of the harsh realities of June 1633. "An entire Congregation declared him a heretic, rash, and a contradictor of the Scriptures, etc. Whoever having signed such a judgment would then write: *qui si in nonnullis cautior fuisset*? No Catholic speaks thus about someone he considers a heretic."[62] He had indeed revised the meaning of the trial.

Baldigiani's passionate conviction comes through in every sentence of this remarkable letter. He had thought long and hard about Galileo's condemnation and come to his own conclusions about how to reconcile the fact of the trial with his admiration for the scientist. Unlike Viviani, he had no desire to gloss over the events and also could not envision fighting the institutions of the church to prove that Galileo should not have been condemned. Of Galileo he said:

> He was summoned, interrogated, and condemned: what could one say? Should I not say what it was but what it should have been? That he was completely innocent, that an entire Congregation erred, that the most holy tribunal was unjust? Who would ever speak in this way, even if he might believe it? And even if he were to speak thus, whom would he persuade? Isn't it better to have said that he was mortified, and with some reason since he provided some cause, that he should have been able to comport himself with a little more prudence, that he injured Urban VIII and the Barberini, and gave them cause to be justifiably resentful?[63]

Given these events, it simply was not possible to expunge the entire record of Galileo's trial.

Baldigiani's willingness to divulge his personal opinion suggests just how much he wanted Viviani to understand the magnitude of the task he had undertaken. His carefully worded letter intimated his personal disagreement with the Roman Catholic Church's treatment of Galileo. But it was as hypothetical as any account of the cosmos then taught in the Jesuit classrooms. What purpose would it serve to decry the system? What was done was done, and Baldigiani was in search of the most productive way to reinterpret Galileo's trial, almost a half-century after it had occurred. He felt confident that Galileo himself would have appreciated his efforts. "As for myself, I have always spoken in this way [about the trial] and I felt that I was doing him a great service, and I believe that if he were alive, he would thank me for it."[64]

Thwarted initiatives

Baldigiani and Viviani never did resolve their differences. The question quickly became moot. By August 1678, Baldigiani was fully immersed in his theological studies in preparation for being examined publicly in this final field of study. He soon tired of Viviani's incessant requests to edit Viviani's life in the *Etruria Illustrated*. "And for this alone you want me to bother Father Kircher...enter into negotiations with his companion, his scribe, our revisors, and Jansson?" Ultimately he disclaimed responsibility for the book's content. "If you don't believe me, you can easily find out, since this is not my affair and has passed into the hands of others, since Kircher is still alive as is his companion, his scribe, Jansson and his agent in Venice."[65] He washed his hands off this beleaguered affair and left Rome the following year to teach philosophy at the Jesuit College in Fermo where he successfully completed his four vows and ordination on February 2, 1681.[66]

During Baldigiani's absence from Rome, on November 27, 1680, Kircher finally did pass to a better life. In his last letter to Cosimo III, personally delivered by Baldigiani in the fall of 1678, Kircher had suggested that the *Etruria Illustrated* "will perhaps be the last of my endeavors."[67] This was not to be. The book, so long anticipated and so painstakingly edited, never appeared. The death of the author and subsequently the publisher placed it in limbo. By the time Jansson's heirs rediscovered the manuscript in 1688, even Magliabechi could not tell them why they should publish it.[68] It has simply vanished—far more definitively than its editor Baldigiani and his discussion with Viviani about how to write the life of Galileo.[69]

Or so everyone thought until I took a closer look at the archives. The entire book continues to be untraceable in Amsterdam, Florence, or Rome. But page 276 of the *Etruria Illustrated* still survives. Worried that his carefully crafted biographies of Tuscan scientists might never appear, in June 1678 Baldigiani painstakingly copied them out and sent them to Viviani. There were "new difficulties" impeding the publication of Kircher's book. The Jesuit revisors had gotten wind of the additions to the manuscript and wanted to review it again. As Baldigiani told Magliabechi, they could make changes entirely at their discretion. He was especially worried about his "eulogy of Galileo," though he hoped it was simply his own anxiety. "However, can you do me the favor of preserving this copy with you, given all that might occur? But don't talk about it with the others." We do not know whether Baldigiani's life of Galileo survived the final edit. But thanks to his correspondence with Viviani, including this crucial letter that Antonio Favaro curiously did not publish in his 1882 edition of the Baldigiani-Viviani correspondence, we know exactly what Baldigiani intended to do by writing a life of Galileo under Kircher's name. As he told Viviani in their secret correspondence, he had deliberately written it *alla Romana* for a non-Florentine audience. "The book will be read promiscuously by everyone," he reminded Viviani, "and I can assure you that it will bring some glory to the Most Serene House [of Medici] and our common fatherland, both because the things it says are true and great and because they are said by with complete dispassion by a foreigner, and one whom the Northern lands are accustomed to believe with their eyes closed."[70] Such was his assessment of Father Kircher's reputation and the reach of his books.

Baldigiani's forgotten life of Galileo helps us to understand more clearly the gradual process by which Italian admirers of Galileo, themselves good Catholics, came to terms with the meaning of the trial and actively negotiated ways to minimize its more deleterious effects on the reputation of Catholic science. Neither Baldigiani nor Viviani ever returned to this particular project but they did not forget to look out for each other's interests. From a distance, Viviani observed the evolution of Baldigiani's lengthy career as a professor of mathematics at the Roman College where he would ultimately teach Roger Boscovich's own mentor, Orazio Borgondio. He must have also observed how Baldigiani's prudent diplomacy earned him the confidence of key figures of great authority within the Roman Catholic Church. As Maurizio Finocchiaro and Domenico Bertoloni Meli have highlighted in their own research, Viviani's discussions with Leibniz, who met and admired Baldigiani in Rome in 1689, stimulated a new correspondence by Viviani in August 1690 regarding the longstanding desire to create a corrected edition of Galileo's *Dialogue* equivalent to the one done for Copernicus's *On the Revolutions*.

Praising Baldigiani's knowledge and authority, Viviani also hoped to entice him with the prospect of rehabilitating singlehandedly the reputation of the Society of Jesus in the Galileo affair through this magnanimous act of heroic Christian piety. He reminded Baldigiani that it had been Cardinal Leopoldo's fondest desire. He recalled the Florentine Jesuit's affection for both Galileo and Tuscany. Suggesting that Baldigiani was the most overly Galilean Jesuit he had had the pleasure of knowing, Viviani concluded: "If the desired end is not achieved now through you, one does not hope that it will ever be achieved by anyone else."[71]

Baldigiani's response to this letter, if there was any, does not survive. In May 1691 he was appointed consultor to the Congregation of the Index. It is indeed possible that Viviani knew, from Leibniz or someone else, that he had been reading books for the Congregation for several years prior to this appointment.[72] Revision of Galileo's *Dialogue* was neither approved during this period nor would it be until the 1744 edition published in Padua. In 1693 Viviani finally gave the world a taste of his life of Galileo by inscribing it over the door of his palace (today on Via S. Antonino, 11) and publishing the inscriptions in a work of geometry in 1701.[73] Others would edit and publish the *Racconto istorico* posthumously while Baldigiani's effort lay dormant in his archive of correspondence.

It is tempting to think that, as Baldigiani ascended the ecclesiastic hierarchy in Rome—ending his life as a confidant of Innocent XII, pious reformer of many of the practices associated with the Barberini papacy, and as a functionary of the Holy Office as of 1710—he might have had the opportunity to examine the dossier of Galileo's trial. Whether he did or what he might have thought of these crucial documents, we simply do not know.

Notes

*Versions of this essay were presented at the History of Science Society 2009 meeting, the University of British Columbia, NYU, and the Davis Center at Princeton. It has greatly benefited from comments from these audiences, and especially Jessica Riskin, Mario Biagioli, Roger Hahn, Eileen Reeves, Daniel Stolzenberg, and Nick Wilding. A special thanks to John Heilbron for generously sharing his knowledge of science and religion in post-Galilean Italy and for providing an example of how to write about the inhabitants of this world.

1. John L. Heilbron, *Galileo* (Oxford: Oxford University Press, 2010) and David Wootton, *Galileo: Watcher of the Skies* (New Haven: Yale University Press, 2010).
2. Vincenzo Viviani, *Vita di Galileo*, ed. Luciana Borsetto (Bergamo: Moretti & Vitali, 1992); Michael Segre, "Viviani's Life of Galileo," *Isis* 80 (1989): 207–231.
3. Thomas Cerbu and Michel-Pierre Lerner, "La disgrace de Galilée dans les *Apes Urbanae*: sur la fabrique du texte de Leone Allacci," *Nuncius* 15 (2000): 589–610.
4. Heilbron, *Galileo*, 288–309, 320.
5. Cerbu and Lerner, "La disgrace de Galilée" and Heilbron, *Galileo*, 299–300.
6. Leone Allacci, *Apes Urbanae, sive, de viris illustribus, qui ab anno MDCXXX per totum MDCXXXII Romae adfuerunt, ac typis aliquid evulgarunt* (Rome, 1633), 118–119.
7. Ibid., 70 This project was posthumously printed as *Prodromus pro Sole Mobili et Terra Stabili contra Galilaeum a Galileis* (Prague, 1651).
8. Stefano Caroti, *Nel segno di Galileo: erudizione, filosofia e scienza a Firenze nel secolo XVII. I "Trattati Accademici" di Vincenzio Capponi* (Florence: Edizioni SPES, 1993), 8–9 and Stéphane Garcia, *Élie Diodati et Galilée. Naissance d'un réseau scientifique dans l'Europe du XVIIᵉ siècle* (Florence: Olschki, 2004), 323–324.
9. Caroti, *Nel segno di Galileo*, 7.
10. Ibid., 9.
11. By contrast, the Tuscan diplomat Giovanfrancesco Buonamici, a relative of Galileo's by marriage who was in Rome during the trial, wrote an unpublished account of the events leading up to the trial that circulated strategically throughout Europe as a letter written in July 1633. He presented Urban VIII as a defender of Copernican astronomy and an advocate of Galileo's until the commissary-general of the Holy Office, Father Vincenzo Maculano da Firenzuola, persuaded the pope to allow Galileo's enemies to humiliate him. Buonamici prefaced his letter by saying that Galileo was "too well known to the world for me to have to give an account of his personal life on the occasion of relating the long molestation he suffered because of the system of Nicolaus Copernicus." As quoted in Maurice

Finocchiaro, *Retrying Galileo 1633–1992* (Berkeley: University of California Press, 2005), 33.

12. Maurice Finocchiaro, *Defending Copernicus and Galileo: Critical Reasoning in the Two Affairs* (Dordrecht: Springer, 2010), 95.

13. Girolamo Ghilini, *Teatro di uomini letterati* (Venice, 1647), 69.

14. Vittorio Siri, *Del Mercurio overo Historia de' correnti tempi* (Casale, 1647), vol. 2, 1720–1722. See Heilbron, *Galileo*, 330–349 for the significance of the *Two New Sciences*; also Paolo Palmieri, *Re-enacting Galileo's Experiments: Rediscovering the Techniques of Seventeenth-Century Science* (Lewiston, NY: Edwin Mellen Press, 2008); and Stillman Drake, *Galileo at Work* (Chicago: University of Chicago Press, 1978).

15. Francesco Fontani, *Elogio di Carlo Roberto Dati* (Florence, 1794), 74 (Dati to Micanzio, Florence, November 9, 1647).

16. Ibid., 75.

17. Ghilini, *Teatro*, 160.

18. Luigi Gerboni, *Un umanista nel Seicento. Gian Nicio Eritreo* (Città di Castello: S. Lapi Tipografo-Editore, 1890). See also Eraldo Bellini, *Umanisti e Lincei. Letteratura e scienza a Roma nell'età di Galileo* (Padua: Antenore, 1997); and Bellini, *Stili di pensiero nel Seicento italiano: Galileo, i Lincei, i Barberini* (Pisa: ETS, 2009).

19. Janus Nicius Erythraeus [Giovanni Vittorio de' Rossi], *Pinacotheca imaginum illustrium doctrinae vel ingenii laude virorum, qui auctore superstite diem suum obierunt* (Cologne, 1643), 279, 281.

20. Gerboni, *Un umanista*, 34, 36–37, 75, 83–84, 131. The history of Rossi's Roman satire, *Eudemia* (1637) can be found on pp. 109–134.

21. Finocchiaro, *Retrying Galileo*, 82.

22. Lorenzo Crassi, *Elogi d'huomini letterati* (Venice, 1666), 245.

23. Finocchiaro, *Retrying Galileo*, 86–107.

24. Angelo Fabroni, *Lettere inediti di uomini illustri per servire d'Appendice all'Opera intitolata Vitae Italorum doctrina eccellentium* (Florence, 1773–1775), vol. 2, 142–143 (Michelangelo Ricci to Leopoldo de' Medici, Rome, November 14, 1666); p. 144 (Rome, December 20, 1666). Quote on p. 144.

25. Maurizio Torrini, "Due galileiani a Roma: Raffaelo Magiotti e Antonio Nardi," in G. Arrighi et.al., *La scuola galileiana. Prospettive di ricerca* (Florence: La Nuova Italia, 1979), 53–88.

26. As quoted in Michael Segre, *In the Wake of Galileo* (New Brunswick: Rutgers University Press, 1991), 109; Segre, "Viviani's Life of Galileo." On the mid-seventeenth century Galileans, see Maurizio Torrini, *Dopo Galileo. Una polemica scientifica 1684–1711* (Florence: Olschki, 1979) and Luciano Boschiero, *Experiment and Natural Philosophy in Seventeenth-Century Tuscany: The History of the Accademia del Cimento* (Dordrecht: Springer, 2007).

27. Vincenzo Antinori, "Notizie istoriche all'Accademia del Cimento," in Antinori, ed., *Saggi di naturali esperienze fatte nell'Accademia del Cimento* (Florence: Tipografia Galileiana, 1841), 76.

28. Carlo Manolessi, "A' Discreti, e Virtuosi Lettori," in *Opere di Galileo Galilei Nobile Fiorentino* (Bologna, 1656), vol. 1, sig. +2r.

29. Segre, "Viviani's Life of Galileo," 211.

30. Nick Wilding, "The Return of Thomas Salusbury's *Life of Galileo* (1664)," *British Journal for the History of Science* 41 (2008): 253.

31. John Finch to Thomas Salusbury, April 17, 1664, as quoted in Wilding, "The Return of Thomas Salusbury," 257.

32. Thomas Salusbury, *Galilaeus Galilaeus His Life: In Five Books* (London, 1664), as quoted in Wilding, "The Return of Thomas Salusbury," 260.

33. Giambattista Riccioli, *Almagestum novum* (Frankfurt, 1651), Part 1, Chapter 1, xxiv, in John L. Heilbron, "Censorship of Astronomy in Italy after Galileo," in Ernst McMullin, ed., *The Church and Galileo* (Notre Dame: University of Notre Dame Press, 2005), 288 and Heilbron, *Galileo*, 359–360.

34. Riccioli, *Almagestum novum*, Part 1, Chapter 2, 52. See Alfredo Dinis, "Was Riccioli a Secret Copernican?" in *Giambattista Riccioli e il merito scientific dei Gesuiti nell'età barocca* (Florence: Olschki, 2002), esp. 58–60; Finocchiaro, *Retrying Galileo*, 82–83; and Heilbron, "Censorship of Astronomy."

35. Francesco Beretta, *Galilée devant le tribunal de l'inquisition: une relecture des sources* (Friburg: Université of Friburg, 1998); Richard Blackwell, *Behind the Scenes at Galileo's Trial: Including the First English Translation of Melchior Inchofer's Tractatus Syllepticus* (Notre Dame: University of Notre Dame Press, 2006); Jules Speller, *Galileo's Inquisition Trial Revisited* (Frankfurt: Peter Lang, 2008); and Finocchiaro, *Defending Copernicus and Galileo*.

36. Biblioteca Laurenziana, Florence, *Cod. Ashb.* 1811, cc. 29v–30r (Vincenzo Viviani to Geminaro Montanari, Florence, September 26, 1671); Paolo Galluzzi, "Galileo contro Copernico," *Annali dell'Istituto e Museo di Storia della Scienza* 2 (1977): 94n23.

37. Riccioli, *Apologia*, 104. See Heilbron, "Censorship of Astronomy," 289–290 and Heilbron, *Galileo*, 360.

38. Finocchiaro, *Retrying Galileo*, 26–42, 79–85.

39. Richard Blackwell, *Galileo, Bellarmine, and the Bible* (Notre Dame: University of Notre Dame Press, 1991); Blackwell, *Behind the Scenes*; Rivka Feldhay, *Galileo and the Church: Political Inquisition or Critical Dialogue?* (Cambridge, UK: Cambridge University Press, 1995); and John L. Heilbron, *The Sun in the Church: Cathedrals as Solar Observatories* (Cambridge, MA: Harvard University Press, 1999).

40. Galluzzi, "Galileo contro Copernico," 94n23.

41. Vincenzo Viviani, *Racconto istorico di Vincenzio Viviani*, in Galileo Galilei, *Le Opere di Galileo Galilei*, ed. Antonio Favaro (Florence: Giunti Barbera, 1890–1909), vol. 19, 615–616.

42. Viviani, *Racconto istorico*, 617.

43. Ibid., 617.

44. Ibid., 618.

45. Ibid., 623. On Galileo's attitude towards publication after the trial, see Tara Nummedal and Paula Findlen, "Scientific Publishing in the Seventeenth Century," in *Scientific Books, Libraries and Readers* (London: Scolar Press, 1999), 164–215.

46. Ingrid Rowland, *Ecstatic Journey: Athanasius Kircher in Baroque Rome* (Chicago: University of Chicago Library, 2000); Daniel Stolzenberg, ed., *The Great Art of Knowing: The Baroque Encyclopedia of Athanasius Kircher* (Stanford: Stanford University Libraries, 2001); Paula Findlen, ed., *Athanasius Kircher: The Last Man Who Knew Everything* (New York: Routledge, 2004); and Joscelyn Godwin, *Athanasius Kircher's Theatre of the World* (London: Thames & Hudson, 2009).

47. Archivum Romanum Societatis Iesu, *Fondo Gesuitico* 663, c. 314r. See Ingrid Rowland, "The Lost *Iter Hetruscam* of Athanasius Kircher (1665–78)," in Sinclair Bell and Helen Nagy, eds., *New Perspectives on Etruria and Early Rome: In Honor of Richard Daniel De Puma* (Madison: University of Wisconsin Press, 2009), 274–289 (quote on p. 276).

48. Paula Findlen, "Living in the Shadow of Galileo: Antonio Baldigiani (1647–1711), a Jesuit Scientist in Seventeenth-Century Rome," in Maria Pia Donato and Jill Kraye, eds., *Conflicting Duties: Science, Medicine, and Religion in Rome, 1550–1750*, Warburg Institute Colloquia 15 (London: The Warburg Institute and Nino Aragno Editore, 2010), 211–254.

49. Biblioteca Laurenziana, Florence, *Redi* 219, f. 179r (Baldigiani to Redi, n.d. but probably 1675).
50. Favaro, "Miscellanea galileiana," 822–823 (Baldigiani to Viviani, Rome, May 26, 1678). These events are briefly discussed in Finocchiaro, *Retrying Galileo*, 89–92.
51. Ibid., 823 (Viviani to Baldigiani, Florence, June 7, 1678) and 830 (Florence, June 14, 1678).
52. Ibid., 829 (Viviani to Baldigiani, Florence, June 14, 1678).
53. Ibid., 832 (Baldigiani to Viviani, Rome, June 18, 1678).
54. Ibid., 836 (Viviani to Baldigiani, Florence, July 12, 1678).
55. Ibid, 835.
56. Ibid., 837 (Baldigiani to Viviani, Rome, July 18, 1678).
57. Favaro, "Miscellanea galileiana," 837 (Baldigiani to Viviani, Rome, July 18, 1678).
58. Eric Cochrane, *Florence in the Forgotten Centuries 1527–1800* (Chicago: University of Chicago Press, 1976), 274.
59. Favaro, "Miscellanea galileiana," 837 (Baldigiani to Viviani, Rome, July 18, 1678).
60. Blackwell, *Galileo, Bellarmine, and the Bible* and Feldhay, *Galileo and the Church*.
61. Heilbron, *Galileo*, 360.
62. Favaro, "Miscellanea galileiana," 838.
63. Ibid. My translation differs slightly from the one in Paolo Galluzzi's excellent article on the posthumous reputation of Galileo; see his "The Sepulchers of Galileo: The 'Living' Remains of a Hero of Science," in Peter Machamer, ed., *The Cambridge Companion to Galileo* (Cambridge, UK: Cambridge University Press, 1998), 421–422.
64. Favaro, "Miscellanea galileiana," 838.
65. Ibid., 842–843 (Baldigiani to Viviani, Rome, September 3, 1678).
66. Archivum Romanum Societatis Iesu, *Rom.*, 82, f. 142v; *Rom.* 88, f. 20; *Ital.* 18, 62–63.
67. Alfonso Mirto, "Le lettere di Athanasius Kircher dell'Archivio di Stato," *Atti e Memorie dell'Accademia Toscana di Scienze e Lettere La Colombaria* 65, n.s., 51 (2000): 236 (Kircher to Cosimo III, Rome, October 30, 1678).
68. Biblioteca Nazionale Centrale, Florence (hereafter BNCF), *Magl.* VIII. 1223, letter 28, f. 53r (Jean and Gilles Jansson to Magliabechi, July 1, 1688).
69. Finocchiaro, *Retrying Galileo*, 89: "The final results are unknown because Kircher's book apparently was never published and the manuscript was lost."
70. BNCF, Gal. 257, cc. 182r–183v (*Ex Etruria Illustrata P. Atanasii Ki[r]keri pag. 276 in auctoris Manuscripto in quo Character est D. Eli[a]e Loreti Amanuensis P.A.K.*); quotes on cc. 183r–v.
71. Finocchiaro, *Retrying Galileo*, 92.
72. Marta Fattori, "Le censure di Antonio Baldigiani alla rivista di Pierre Bayle," *Nouvelles de la Republique des Lettres* 26 (2006): 105–121.
73. Vincenzo Viviani, *De locis solidis secunda divinatio geometrica: in quinque libros iniuria temporum amissos Aristæi senioris geometræ / autore Vincentio Viviani... opus conicum continens elementa tractatuum ejusdem Viviani, quibus tunc ipse multa, maxima, & abdita in mathesi theoremata demonstrare cogitaverat* (Florence, 1701), esp. 120–128. See Robert Lunardi and Oretta Sabbatini, eds., *Il rimembrar delle passate cose. Una casa per memoria: Galileo e Vincenzo Viviani* (Florence: Edizioni Polistampa, 2009).

Part IV
Things

11
Machines in the Garden*

Jessica Riskin

Ideas are inseparable from things and vice versa. Intellectual history and the history of material culture, accordingly, are not discrete endeavors: they are conjoined. Here is a guiding principle of our volume and the focus of the current section, "Things." This chapter demonstrates these propositions for what is arguably the hardest case, the idea most reputed to have detached modern philosophy from the objects of daily experience: Descartes's philosophical revolution cleaving mental self from mechanical body. A particular kind of machine, proliferating across the landscape of late medieval and early modern Europe, closely informed this philosophical revolution. Moreover, to approach the revolutionary philosophy in terms of the devices that informed it is to arrive at a kind of instability, a fault-line running through its very core.

Descartes's idea of the mechanical body took on an array of meanings— such as passivity, unresponsiveness, and even lifelessness—that it did not initially hold. Indeed, in the first instance, Descartes's animal-machine idea meant something like the opposite: responsiveness, feeling, and vitality. Looking at the actual lifelike machines to which Descartes referred reveals this fundamental instability in his idea of living machinery. In so doing, such an investigation reopens an older, eclipsed set of possibilities for what it can mean to be both mechanical and alive.

The machines in question were close cousins, sometimes appendages, of clocks and organs. They were moving mechanical images of living creatures: people, angels, devils, and animals. By the early 1630s, when Descartes framed his argument that animals and humans, apart from their capacity to reason, were automata, European towns and villages were positively humming with mechanical vitality. Mechanical images of living creatures had been ubiquitous for several centuries. Descartes and other seventeenth-century mechanists were therefore able to invoke a plethora of animal- and human-like machines. These machines fell into two main categories: the great many devices to be found in churches and cathedrals, and the automatic hydraulic amusements on the grounds of palaces and wealthy estates.

229

The contraptions of neither category signified, in the first instance, what machine metaphors for living creatures later came to signify: passivity, rigidity, regularity, constraint, rote behavior, or soullessness. Rather, the machines that informed the emergence of the early modern notion of the human-machine held a strikingly unfamiliar array of cultural and philosophical implications, notably the tendencies to act unexpectedly, playfully, willfully, surprisingly, and responsively.

Moreover, neither the idea nor the ubiquitous images of human-machinery ran counter to Christian practice or doctrine. Quite the contrary: not only did automata appear first and most commonly in churches and cathedrals, but the idea as well as the technology of human-machinery was also indigenously Catholic. The church was a primary sponsor of the literature that accompanied the technology of lifelike machines, and the body-machine was also a recurrent motif in Scholastic writing.[1]

Automata were therefore theologically and culturally familiar, things with which one could be on easy terms. They were funny, sometimes bawdy, and they were everywhere. To understand what Descartes and the other seventeenth-century mechanists did with the idea of animal- and human-machinery, one needs to take into account its preexisting familiarity and older meanings. During the early to mid-seventeenth century, at the hands of mechanist philosophers, matter and its mechanical combinations were divested first of soul and then of life. This essay tours a mechanical culture that flourished before that development in which machines represented precisely the capacities that the mechanists would later deny them: divinity and vitality.

Deus qua machina

A mechanical Christ on a crucifix, known as the Rood of Grace, drew flocks of pilgrims to Boxley Abbey in Kent during the fifteenth century. This Jesus, which operated at Easter and the Ascension, "was made to move the eyes and lipps by stringes of haire."[2] Moreover, the Rood was able

> to bow down and lifte up it selfe, to shake and stirre the handes and feete, to nod the head, to rolle the eies, to wag the chaps, to bende the browes, and finally to represent to the eie, both the proper motion of each member of the body, and also a lively, expresse, and significant shew of a well contented or displeased minde: byting the lippe, and gathering a frowning, forward, and disdainful face, when it would pretend offence: and shewing a most milde, amiable, and smyling cheere and countenaunce, when it woulde seeme to be well pleased.[3]

Even before approaching the Rood for benediction, one had to undergo a test of purity administered by a remote-controlled saint:

> Sainct Rumwald was the picture of a pretie Boy sainct of stone...of it selfe short, and not seeming to be heavie: but for as much as it was wrought out

of a great and weightie stone...it was hardly to be lifted by the handes of the strongest man. Nevertheless (such was the conveighance) by the helpe of an engine fixed to the backe thereof, it was easily prised up with the foote of him that was the keeper, and therefore, of no moment at all in the handes of such as had offered frankly: and contrariwise, by the meane of a pinne, running into a post...it was, to such as offered faintly, so fast and unmoveable, that no force of hande might once stirre it.[4]

Having proven your "cleane life and innocencie" at the hands of the rigged Saint Rumwald, you could proceed to the mechanized Jesus. Automaton Christs—muttering, blinking, and grimacing on the cross—were especially popular.[5] One, a sixteenth-century Breton Jesus, rolled his eyes and moved his lips while blood flowed from a wound in his side. At his feet, the Virgin and three attendant women gesticulated, while at the top of the cross, a head symbolizing the Trinity glanced shiftily from side to side.[6]

Mechanical devils were also rife. Poised in sacristies, they made horrible faces, howled, and stuck out their tongues. The Satan-machines rolled their eyes, and flailed their arms and wings; some even had moveable horns and crowns.[7]

And then there were the automaton angels. A host of these, in one Florentine festival, carried the soul of Saint Cecilia up to heaven.[8] For the feast of the Annunciation at San Felice, the fifteenth-century Florentine architect Filippo Brunelleschi sent the archangel Gabriel in the reverse direction by mechanical conveyance. Brunelleschi mechanized heaven too. His mechanical paradise was "truly marvellous...for on high a Heaven full of living and moving figures could be seen as well as countless lights, flashing on and off like lightning."[9] The heavenly machinery was balanced beneath by elaborately engineered hells. One mechanical inferno's moving gates, when they gaped ajar amid rumbling thunder and flashes of lightning, spewed forth writhing automaton serpents and dragons.[10]

A menagerie of mechanical beasts played parts in religious theater. Daniel's lions gnashed their teeth[11] and more lions knelt before Saint Denis.[12] Balaam's ass balked and swerved before the angel of the Lord.[13] The serpent twined itself round the trunk of the Tree of Knowledge to offer its apple to Eve.[14] A wild boar tracked by hunters; a leopard that sniffed Saint André; a dromedary that wagged its head, moved its lips, and stuck out its tongue; a host of dog- and wolf-shaped devils surging up from the underworld; and serpents and dragons spewing flames from their mouths, noses, eyes, and ears rewarded the devoted spectators at the 40-day performance of the *Mystère des actes des apôtres* in Bourges in 1537.[15] The machines were commissioned from local artisans, usually clockmakers.[16]

Mechanical enactments of biblical events spread across the European landscape, reaching a crescendo during the late fifteenth and early sixteenth centuries.[17] And the holy machinery was not solely to be found in cities. In May of 1501, an engineer in the village of Rabastens, near Toulouse, was engaged to build an endless screw that could propel the Assumption of the Virgin.

He did his job: the following August, the Virgin rose heavenward, attended by rotating angels.[18] Another mechanical Ascension of the Virgin took place annually in Toulouse.[19] Children built small replicas of the Virgin-ascender at home.[20]

Many automata were connected with church clocks—outgrowths of the church's drive to improve timekeeping for the better prediction of feast-days.[21] These often enacted biblical scenes but there were also figures of people and animals, among which mechanical roosters were popular. The renowned Rooster of Strasbourg Cathedral, for nearly five centuries, cocked its head, flapped its wings, and crowed on the hour.[22]

The other prime spot for mechanical figures was church organs.[23] These housed entire choirs and orchestras of mechanical angels.[24] Saint Peter towered atop an organ of the late fourteenth or early fifteenth century at the cathedral in Beauvais and blessed the congregation on his feast-day by nodding his head and moving his eyes.[25] Strasbourg Cathedral was hectic with mechanical activity, having not only the Rooster and clock-automata but also three moving figures attached to the strings of the organ: Samson opening and closing the jaws of a lion; the Héraut de la ville, lifting his trumpet to his lips; and the Bretzelmann in a red and black cape. The Bretzelmann had long hair and a shaggy beard, an aquiline nose, and an evil aspect. When set in motion, he seemed to speak with great emphasis, opening and shutting his mouth while shaking his head and gesticulating.[26] At Pentecost, throughout the service, the Bretzelmann mocked the priest, laughing, hurling insults and coarse jokes, and singing nasty songs.[27]

Other organs sported disembodied heads that frowned, contorted their faces, rolled their eyes, and stuck out their tongues as the music played.[28] In the organ gallery of the cathedral in Barcelona, the head of a moor hung by its turban, made mild expressions when the music played softly and, when the strains grew louder, rolled its eyes and grimaced.[29] And in the *Cloître des Augustins* in Montoire, in the Loire valley, a mechanical head on the organ gallery gnashed its teeth with a noisy clatter.[30]

Early modern Europe, then, was alive with mechanical beings and the Catholic Church was their main sponsor. As J. L. Heilbron has noted, the Catholic Church was the leading patron of early modern natural sciences. As an example of this patronage and its importance, Heilbron has detailed the transformation of churches into astronomical observatories. Here we can see the same to be true of mechanical innovation: churches and cathedrals constituted the original, primary locus of clockwork and automatic machinery.

The church was also a primary sponsor, between the late fifteenth and late sixteenth centuries, of the translation and printing of a small flood of ancient texts on mechanical and hydraulic automata, which then informed the construction of such devices throughout the Renaissance. The first printed edition of Vitruvius's *De Architectura*, for example—containing descriptions of the third century BC engineer Ctesibius's water organ and other automata—appeared in 1486 as a key part of the Renaissance popes' project to build a Christian Rome.[31]

Not only the technology but also the idea of the animal-machine was at home in the Catholic tradition. Aquinas himself proposed that animals might be regarded as machines some four centuries before Descartes. Like clocks and other "engines put together by the art of man," Aquinas argued, animals were moved by reason although they themselves lacked reason. Just as a clock exhibited regular actions because it had been built by an intelligent maker, so an irrational animal behaved in an orderly fashion, not through will or reason, but because it had been devised by "the Supreme art."[32]

A Franciscan monk built around 1560, now at the Smithsonian, offers a final example of the early modern mechanization of faith.[33] The monk paces, raises a crucifix and rosary, turns it eyes and head to look at the cross, moves its mouth in prayer, strikes its breast, and lifts the cross to its lips.[34] Its eerie, riveting performance embodies the power of an image, especially a moving image, and even more so, a moving devotional image. Mechanization is often taken as an index of modernization. But automaton icons had a medieval impetus in a tradition of imagery in which the tangible, visible, and earthly representations of Christian lore and doctrine were pushed ever farther.[35] The icons were representations in motion, inspirited statues: they were mechanical *and* divine. Rolling their eyes, moving their lips, gesturing, and grimacing, these automata dramatized the intimate, corporeal relation between representation and divinity, icon and saint. As this relation became increasingly fraught, the machinery took on new meanings. Reformism and clockmaking developed side by side from Augsburg to Strasbourg to Geneva. The flood of mechanized religious images coincided both in time and, centrally, in place with the heating-up of the question of whether and how religious images blurred the boundary between image and deity.

The Reformation cast a partial hush over the humming, groaning, chirping, whistling, and chattering ecclesiastical machinery. The uncouth Bretzelmann of Strasbourg Cathedral was silenced along with many of his fellow organ automata and, indeed, with many of the church organs themselves, which became emblematic of Catholic ritual.[36] Henry VIII banned mechanical statues from English churches.[37] The grimacing Rood of Boxley Abbey gave its last performance in 1538, after being snatched from Boxley by Geoffrey Chamber as part of his commissioned defacement of the abbey. Chamber wrote to Thomas Cromwell[38] that he had found in the Rood

> certain engines and old wire, with old rotten sticks in the back, which caused the eyes to move and stir in the head thereof, "like unto a lively thing," and also, "the nether lip likewise to move as though it should speak," which was not a little strange to him and others present.[39]

But can it have been any surprise that the Rood was made of wood and wire? It, and its many cousins, had been built by local artisans—clockmakers, carpenters—and treated by its local beholders with great familiarity, inspiring at least as much laughter as awe. The Bretzelmann of Strasbourg Cathedral was obviously funny. So was the lever-and-pulley-operated Saint Rumwald.

According to one contemporary, "many times it mooved more laughter than devotion, to beholde a great lubber to lifte at that in vaine, which a young boy (or wench) had easily taken up before him."[40]

That mechanical icons were mechanical cannot have been big news. But Chamber and his fellow iconoclasts introduced the idea that, by virtue of being mechanical, such icons were deceptions. The destruction of mechanized icons represented only small swells inside the larger surges of iconoclasm that spread across Europe during the middle decades of the sixteenth century.[41] But the demolition of the Rood and its ilk reveal that one core logic of iconoclasm—the rigorous distinction between the divine and the artifactual—brought with it a fundamentally transformed view of the ontology of machines.

Chamber removed the Rood to Maidstone, where he displayed it in the public market and instilled in the townspeople a "wondrous detestation and hatred" of it. The Rood was then transported to London where John Hilsey, bishop of Rochester, exhibited it during a sermon at Saint Paul's Cross, after which it was torn apart and burned. The chronicler Charles Wriothesley described the events as follows:

> Allso the sayde roode was sett in the market place first at Maydstone, and there shewed openlye to the people the craft of movinge the eyes and lipps, that all the people there might see the illusion that had bene used in the sayde image by the monckes of the saide plaace of manye yeares tyme out of mynde, whereby they had gotten great riches in deceiving the people thinckinge that the sayde image had so moved by the power of God, which now playnlye appeared to the contrarye.
> ... [T]he image of the roode ... was brought to Poules Crosse, and there, at the sermon made by the Bishopp of Rochester, the abuses of the ... engines, used in old tyme in the said image, was declared, which image was made of paper and cloutes from the legges upward; ech legges and armes were of timber; and so the people had bene deluded and caused to doe great adolatrie by the said image.[42]

As with other reformist initiatives, both sides of the confessional divide participated in this partial rejection of mechanized religious images. By the mid-seventeenth century, certain Catholic monarchs had developed a distaste for automaton angels and mechanical Ascensions. In 1647, Louis XIV and the Queen Mother came to view the automaton angels of Dieppe and found them not to their liking; that was the end of the angels.[43] An interdiction of 1666 put an end to the Virgin's annual mechanical Ascension in Toulouse on the grounds that it distracted the congregation and caused "irreverent reflections."[44]

But mechanized devotional objects did not disappear; on the contrary, they survived and flourished. Thus, during the late sixteenth and seventeenth centuries, the proliferating and elaborating machines coexisted with proliferating and elaborating theological and philosophical suspicions of them.

The Council of Trent, in its 1563 decree on the use of sacred images, placed a ban on "unusual" images except when they were approved by a bishop.[45] Rather than eliminating mechanical icons, this ban helped to motivate a thematic shift. Mechanical nativity scenes, for example, became popular, especially at the hands of the Jesuits,[46] who made automata a central tool in their promulgation of Christianity. The Jesuit ambassador to the Chinese Mission sent an elaborate mechanized nativity scene in 1618, with the Magi bowing; the Holy Virgin gesturing; Joseph rocking the cradle; the Holy Father making a benediction; and angels, shepherds, and barn animals all moving about.[47] The Jesuits also included worldly themes in their automatic offerings, such as a spring-driven android knight that marched about with a drawn sword for a quarter of an hour.[48] In this and other senses, religious automata brought secular ones in their wake.[49]

Waterworks on the grounds of estates constituted the main secular tradition in automata. The wealthy and powerful found in lifelike machinery an endless source of comedy, and of the most bawdily uproarious, knee-slapping variety. The first part of this essay has traced the predominantly Christian origins of androids and other mechanical creatures and described an early intimacy between machinery and divinity. The second part takes up the relation of machinery to a vitality represented by a remarkably vivacious vulgarity. From the sublime, onward to the ridiculous.

Waterworks

"Frolicsome engines" [*engiens d'esbattement*] were to be found as early as the late thirteenth century at the chateau of Hesdin (in what is now Pas-de-Calais), seat of the comtes d'Artois. The machines are mentioned in the account-books of Mathilde de Brabant (known as Mahaut), comtesse d'Artois, in 1299. The next year, the family appointed a castle "Master of Engines."[50] After that, the engines make regular appearances in the accounts. They included mechanical monkeys, an elephant, a goat, and a boar.[51] The comtesse Mahaut's descendent, Philippe le Bon, Duke of Burgundy from 1419 until his death in 1467, left in his own account-books a catalogue of the mechanized tricks he inflicted on visitors:

> Painting of 3 personages that spout water and wet people at will...a machine for wetting ladies when they step on it...an "engien" which, when its knobs are touched, strikes in the face those who are underneath and covers them with black or white...another machine by which all who pass through will be struck and beaten by sound cuffs on their head and shoulders...a wooden hermit who speaks to people who come to that room...6 personages more than there were before, which wet people in various ways...eight pipes for wetting ladies from below and three pipes by which, when people stop in front of them, they are all whitened and covered with flour...a window where, when people wish to open it, a personage in front of it wets people and closes the window again in spite of

them...a lectern on which there is a book of ballades, and, when they try to read it, people are all covered with black, and, as soon as they look inside, they are all wet with water...[a] mirror where people are sent to look at themselves when they are besmirched, and, when they look into it, they are once more all covered with flour, and all whitened...a personage of wood that appears above a bench in the middle of the gallery and fools [people] and speaks by a trick and cries out on behalf of *Monsieur le Duc* that everyone should go out of the gallery, and those who go because of that summons will be beaten by tall personages...who will apply the rods aforesaid, or they will have to fall into the water at the entrance to the bridge, and those who do not want to leave will be so wetted that they will not know where to go to escape from the water.[52]

The Hesdin engines, in all their malicious glory, inspired many imitations.[53]

By the time Montaigne went traveling in 1580 and 1581, hydraulic automata had grown so commonplace that he eventually got bored of them. Outside Augsburg, at the summer palace of a rich banking family, Montaigne saw sprays of water from hidden brass jets activated by springs. "While the ladies are amused to see the fish play, one simply releases some spring: suddenly all these jets spurt thin, hard streams of water to the height of a man's head, and fill the petticoats and thighs of the ladies with this coolness." Elsewhere, hidden jets could be triggered to gush directly into the face of a visitor who stopped to admire a particular fountain.[54] The palace also had an automaton lion in one room that sprang forward when a door was opened.[55]

At Pratolino, a palace of Francesco I de' Medici, Grand Duke of Tuscany, Montaigne marveled at a "miraculous" grotto where he saw singing birds and other automaton animals moving to music. The garden housed an automaton lady who emerged from behind a door to fill a cup with water. In one part of the garden, "all the seats gush water on your buttocks."[56] At another of the Grand Duke's residences, Montaigne recorded miniature, hydraulically driven "water mills and windmills, little church bells, soldiers of the guard, animals, hunts, and a thousand such things."[57]

Arriving at the already famous Villa d'Este in Tivoli, Montaigne was unimpressed. The Tivoli palace and gardens had been built by Cardinal Ippolito II d'Este, the then governor of Tivoli, as consolation after a failed bid for the papacy. The grottoes were done in 1572 and were already old news by 1580. Montaigne wrote with a yawn that the "gushing of an infinity of spouts of water checked and launched by a single spring that one can work from far away, I had seen elsewhere on my trip." He then gave a meticulous yet jaded account of the water organ, with its many moving figures of birds and other things, concluding, "All these inventions...I have seen elsewhere."[58]

Twenty years later, Tomaso Francini, engineer to Ferdinando I de' Medici, then grand duke of Tuscany, was enticed away by Henri IV to give him some waterworks.[59] At Saint Germain en Laye, Francini built grottoes devoted to Neptune, Mercury, Orpheus, Hercules, Bacchus, Perseus, and Andromeda, all standard features of garden hydraulic amusements.

Figure 11.1   Neptune grotto of Salomon De Caus.
Source: *Les raisons des forces mouvantes* (Francfort: J. Norton, 1615), p. 35r.

An automaton Neptune with a streaming blue beard rode the waves at Saint Germain en Laye; Mercury posed by a window with one foot carelessly propped and loudly sounding a trumpet; Orpheus played his lyre for an audience of animals who craned eagerly toward him;[60] Perseus freed Andromeda from her dragon. There were automaton blacksmiths, weavers, millers, carpenters, knife-grinders, fishermen, and farriers, "their faces blackened with grime and sweat," who hammered iron and stealthily conducted the obligatory watery attacks on spectators.[61]

What was it like to live amid such machines? We happen to have a daily record of the life of a child who grew up with the hydraulic grottoes of Saint Germain en Laye in his garden (a record that includes every lisping pronouncement, the numbers of plums and grapes consumed at each meal, and careful descriptions of each bowel movement). He was the future Louis XIII, the son of Henri IV and Maria de' Medici, born just when Francini was getting to work on his father's fountains. The Dauphin spent his childhood mostly at Saint Germain en Laye and surrounded by machines.

According to the journal of his doctor and caretaker, Jean Hérouard, the Dauphin as a toddler watched the workers from his windows[62] and, from age three, began visiting the grottoes several times a week.[63] We overhear him in bed one morning instructing a chambermaid, "Pretend dat I am Ofus [Orpheus] and you are da fountainee [foutaineer], you sing da canaries."[64] Soon afterward, he was working the grotto faucets, spraying himself and everyone else with water.[65] The prince plagued Francini with visits to his workshop, demanding names and explanations of each instrument.[66] At home, he pretended to be Francini, building and working the fountains. He played fountains in bed, in his gilt washbasin, and under the dining table—*"fssss"* and *"dss."*[67] The passion contained more than a hint of childish eroticism. Hérouard dutifully recorded: "Says he has a faucet in his ass and another in his willie: *'fs fs'."*[68] As a child king, following his father's assassination, Louis XIII continued to visit Francini, amusing himself for hours at a time by forging, soldering, and filing fountain pipes.[69]

Louis XIII liked clockwork as well as hydraulic automata. In Hérouard's journal, we see him at age four, beating his spoon against his plate and announcing to his governess: "Maman ga I am ringing da hour *dan, dan,* it rings like da jackamart who beats on da anvil."[70] Here he is at six, shopping in Paris along the rue Saint Honoré, choosing a spring-driven toy carriage on offer for 15 *écus.*[71] Later in the same year, the Dauphin was given a cabinet of automata that enacted Christ's Passion and the taking of Jerusalem, driven by falling sand, which he demonstrated to everyone in the palace.[72]

You didn't need to be a king or a prince. The popes competed in the coin of hydraulic trickery. When Ippolito Aldobrandini became Pope Clement VIII in 1592, he had a magnificent villa built that included what Edith Wharton, on her tour of Italian villas, would describe wearily as "the inevitable *théâtre d'eau."*[73] The popes, their nephews and their grandnephews too: all the little cardinals and archbishops needed hydromechanical amusements to call their own.[74] Markus Sittikus von Hohenems, sovereign and Archbishop of Salzburg from 1612 until his death in 1619, installed waterworks at his Schloss Hellbrunn that remain in operation almost four centuries later.[75] When he was elected archbishop, Sittikus was already a connoisseur of automata. He had lived briefly at the Villa Aldobrandini; moreover, his uncle, Cardinal Marco Sittico Altemps, nephew of Pope Pius IV, had built the Villa Mondragone, which had a renowned Water Theater designed by the engineer Giovanni Fontana.[76] In Sittikus's garden, visitors are still invited to seat themselves around a stone table, on stone benches with hidden spouts that release jets of water on command, drenching the obedient from below. In the Neptune Grotto to which they proceed, dripping and uproarious, guests gape at the Germaul, a stone gargoyle that rolls its eyes menacingly and sticks out its tongue. Fleeing the Germaul, the visitors are again watered down from spring-triggered spouts concealed in the walls. Arriving remoistened in the Birdcall Grotto, they are surrounded by the hydraulically produced sound of chirping and twittering birds.

You might think the joke would wear thin. But you would be wrong. The sport proceeded right on through the seventeenth century. By the 1660s, when John Evelyn was writing his gardening manuals, he assumed it was essential to include automata:

We may...people our Rocks with *Fowle, Conies, Capricornes, Goates* {& rapitary beasts, with} *Hermites, Satyres,* {Masceras} *Shepheards,* {rustic workes river gods Antiqs etc} and with divers *Machines* or *Mills* made to move by the ingenious placing of wheels, painted & turned by some seacret pipes of waters;...By these motions, histories, {*Andromedas*} and *sceanes* may be represented.[77]

Evelyn described with malicious satisfaction the "wayes of contriving seacret pipes to lie so as may wett the {gazing} Spectators, underneath, behind, in front and at every side according as the Fontaneere is pleased to turne & governe these clandestine & prepostrous showers." He mentioned a "waggish invention" he had found in the garden of the pope's crossbearer: a chair with a lion's head on the back that would vomit water onto the neck of anyone who sat on it.[78] The victims continued to take their licks with surprise and delight. Anne-Louise d'Orléans, duchesse de Montpensier, the memoirist and wayward cousin of Louis XIV, cheerfully recorded her experience at the Essonnes estate of the master of finances for the royal household, which she visited with her friend, Madame de Lixein, in the summer of 1656:

> As I passed through a grotto, they released the fountains, which came out of the pavement. Everyone fled; Madame de Lixein fell and a thousand people fell on her...We saw her being led out by two people, her mask muddy, and her face the same; her handkerchief torn, her clothes, her oversleeves, in short, disconcerted in the funniest way in the world, and I cannot remember it without laughing. I laughed in her face and she started laughing too, finding that she was in a state to inspire it. She took this accident as a person of humor. She took no meal and went right to bed...Upon returning, I visited her: we laughed a lot again, she and I.[79]

Robert Darnton has suggested that historians take note of the mystifying jokes of the past, as these indicate "where to grasp a foreign system of meaning in order to unravel it."[80] To what exotic tapestry do these mischievous machines in their endless funniness connect? Bergson described the quintessential comic situation as "something mechanical encrusted on the living": the appearance of a human being as an automaton. We laugh, Bergson claimed, as a "corrective": to reassert the distance between machinery and life.[81] But, as Darnton's recommendation assumes, humor has a history[82], and the need to establish that human beings are not machines cannot have had the same urgency in 1500 or 1600 as it had in 1900. Rabelais's, not Chaplin's, was the sense of humor at play. The frolicsome engines catalogued in this essay represented something like the opposite of Bergson's scenario: not people as rote automata but machines as responsively alive. The machines' human targets, laughing at the machines' whimsical vitality, do not seem to me to have been reasserting their own transcendence of machinery. I think they were doing something more like delighting in a base corporeality that they took to anchor even the very highest of human lives in an actively material world.

Arriving, then, at the mid-seventeenth century, when the idea of the animal-machine began to flourish in philosophical discussion, we can see that actual mechanical images of living creatures were already there everywhere. They were familiar, not only to the nobility and the wealthy bourgeoisie but also to their servants, to the engineers and artisans who built the machines, as well as to the audiences who flocked to witness them, and to the literate who read about them. The culture of lifelike machinery comprised these devices and the surrounding conversation projected no central antithesis between machinery and either divinity or vitality. On the contrary, the automata represented spirit in every corporeal guise available and life at its very liveliest. Here, then, was the culture that gave rise to the seventeenth-century animal-machine. That comparatively confined being represented more a narrowing than an opening of intellectual and cultural possibilities. To make full sense of this development, we must consider the world of objects that preceded it. Before machines became mindless and rote, they were the very life of the party.

Notes

*This essay is adapted from the opening chapter of a forthcoming book on the history of the human-machine—philosophy, physiology, technology, culture and politics—entitled *The Restless Clock*.

1. Aquinas himself proposed that animals might be regarded as machines some four centuries before Descartes and in strikingly similar terms: Thomas Aquinas, *Summa Theologica*, Prima Secundæ Partis, Qu. 13, Art II, Reply obj. 3.
2. Charles Wriothesley, *A Chronicle of England During the Reign of the Tudors, 1485–1559*, ed. William Hamilton, 2 vols. (London: Printed for the Camden Society, 1875), vol. 1, 74. On the Boxley Abbey Rood of Grace, see also Alfred Chapuis and Edouard Gélis, *Le Monde des Automates*, 2 vols. (Paris: Blondel La Rougery, 1928), vol. 1, 95 and Michael Jones, "Theatrical History in the Croxton *Play of the Sacrament*," *ELH* 66, no. 2 (1999): 223–260, on pp. 243–244.
3. William Lambarde, *A Perambulation of Kent: Conteining the Description, Hystorie, and Customes of that Shire. Written in the Yeere 1570. by William Lambarde of Lincolns Inne Gent.*, ed. Richard Church (Bath: Adams and Dart, 1970), 205–206. See also Wriothesley, *Chronicle*, 1: 75; Edward, Lord Herbert of Cherbury, *Life and Reign of King Henry the Eighth, Together with Which Is Briefly Represented a General History of the Times* (London: Mary Clark, 1683), 494; John Stow, *Annales, or a Generale Chronicle of England*, ed. Edmund Howes (1631), 575.
4. Lambarde, *A Perambulation of Kent*, 209–210.
5. Later examples of automaton Christs include an eighteenth-century one in Dachau, Germany that has human hair. Hidden in its beard are strands that control the movements of the eyes, mouth, and head. Another, in Limpias, Spain, moves its lips, rolls its eyes, blinks, and grimaces. See Chapuis and Gélis, *Le Monde des Automates*, vol. 2, 95–96, 1: 104. On mechanized Passions, see also Johannes Tripps, *Der Handelnde Bildwerk in der Gotik* (Berlin, Gebr. Mann, 1998), 159–173; plates 10e and f, pp. 292–293; plates 42a and b, p. 325; and plate 43a, p. 326.
6. Alfred Chapuis and Edmond Droz, *Automata: A Historical and Technological Study*, tr. Alec Reid (Geneva: Editions du Griffon, 1958), 119–120.
7. Some mechanical devils are depicted in drawings by the fifteenth-century engineer, Giovanni Fontana, Bayerische Staatsbibliothek, Cod. Icon. 242 (MSS. mixt. 90), 59-vo—60vo, 63vo; printed in Eugenio Battisti and Giuseppa Saccaro Battisti, *Le macchine cifrate di Giovanni Fontana* (Milan: Arcadia, 1984), 134–135. On Fontana's automaton devils, see especially Anthony Grafton, "The Devil as Automaton," in

Jessica Riskin, ed., *Genesis Redux* (Chicago: The University of Chicago Press, 2007), Chapter 3. See also Chapuis and Gélis, *Le Monde des Automates*, vol. 2, 97–101. I am grateful to Paula Findlen for bringing the Milan devil to my attention.

8. Alessandro D'Ancona, *Origini del teatro in Italia* (Firenze: Successori le Monnier, 1877), vol. 1, 424; D'Ancona, *Sacre rappresentazioni dei secoli XIV, XV e XVI* (Firenze: Successori le Monnier, 1872), vol. 2, 321; and Philippe Monnier, *Le Quattrocento, essai sur l'histoire littéraire du XVè siècle italien* (Paris: Perrin et Cie, 1908), vol. 2, 204.

9. Giorgio Vasari, *Lives of the Most Eminent Painters, Sculptors & Architects*, tr. Gaston de Vere (London: Philip Lee Warner, 1912–1914), vol. 2, 229–232. On Brunelleschi's and other *ingegni*, see Frank D. Prager and Gustina Scaglia, *Brunelleschi: Studies of His Technology and Inventions* (Cambridge: MIT Press, 1970); Paolo Galluzzi, *Gli ingegneri del Rinascimento da Brunelleschi a Leonardo da Vinci* (Bologna: Istituto e Museo di storia della scienza, 1996); and Anthony Grafton, "From New Technologies to Fine Arts," in *Leon Battista Alberti: Master Builder of the Italian Renaissance* (Cambridge: Harvard University Press, 2000), Chapter 3. Jacob Burkhardt was scathing on the subject of the *ingegni*: the drama, he said, "suffered from this passion for display," in Burkhardt, *The Civilization of the Renaissance in Italy* (London: Penguin, 1990 [1860]), 260. The moving figures in Heaven that Vasari mentions were in fact little boys; nevertheless he considers them as components in an overall work of "machinery." The Paradise apparatus was so heavy that it pulled down the roof of the Carmine monastery where it was installed and necessitated the monks' departure. See Giorgio Vasari, *Lives of the Most Eminent Painters*, vol. 2, 229. On moving Annunciations, see also Tripps, *Der Handelnde Bildwerk*, 84–88.

10. D'Ancona, *Origini del teatro*, vol. 1, 424; D'Ancona, *Sacre rappresentazioni*, vol. 2, 135, 232; and Monnier, *Le Quattrocento* (1908), vol. 2, 204. These mechanized performances were elaborations of an older tradition of religious puppet-shows, or "motions." An individual puppet might also be called a "motion," and the puppeteer was known as the "motion-master." William Lambarde described a motion that took place in Witney, in Oxfordshire, in which the priests used articulated figures to enact the Resurrection. These puppets might have included some mechanized features. The figure of the waking watchman who saw Christ rise made a continuous clacking noise that earned him the nickname "Jack Snacker of Wytney." See Lambarde, *Dictionarium Angliæ topographicum & historicum. An alphabetical description of the chief places in England and Wales; with an account of the most memorable events which have distinguish'd them. By... William Lambarde... Now first publish'd from a manuscript under the author's own hand* (London: F. Gyles, 1730), 459; William Hone, *The Every-Day Book and Table Book* (London: William Tegg, 1825–1827), entry for September 5, 1825, titled "Visit to Bartholomew Fair."

11. Edélestand Du Méril, *Origines latines du théâtre moderne* (Paris: Franck, 1849), 153; Gustave Cohen, *Histoire de la mise en scène dans la théâtre religieux français du moyen age* (Paris: Champion, 1926), 31.

12. Arnoul and Simon Gréban, *Mystère des actes des apôtres, representé à Bourges en avril 1536, publié depuis le manuscrit original*, ed. Auguste-Théodore, Baron de Girardot (Paris: Didron, 1854); Cohen, *Histoire de la mise en scène*, 147. The *Actes des Apôtres* was an assemblage of dramatizations of stories from the Old Testament collected around 1450 and staged in Paris at the beginning of the sixteenth century.

13. A. Gasté, *Les drames liturgiques de la cathédrale de Rouen* (Evreux: Imprimerie de L'Eure, 1893), 75; Cohen, *Histoire de la mise en scène*, 31.

14. Anonymous, *Jeu d'Adam* [ca. 1150], Wolfgang van Emden, ed. (Edinburgh: British Rencensvals Publications, 1996) [original is in the Bibliothèque municipale de Tours, MS 927], 23 (l. 292); Cohen, *Histoire de la mise en scène*, 54.

15. Gréban and Gréban, *Mystère des actes des apôtres*, livre IV, fol. 67 v°, livre III, fols. 21 r° and 23 r°, livre II, fol. 9 r°; Cohen, *Histoire de la mise en scène*, 147.

16. Cohen, *Histoire de la mise en scène*, 143–144; Chapuis and Droz, *Automata*, 356–357.

17. Vasari described the height of Florentine festivals as having coincided with the career of the architect Francesco d'Angelo (known as Cecca), in the second half of the fifteenth century: Cecca "was much employed in such matters at that time, when the city was greatly given to holding festivals." These took place, according to Vasari, not only in churches "but also in the private houses of gentlemen." There were also four public spectacles, one for each quarter of the city. For example, the Carmine kept the feast of the Ascension of Our Lord and the Assumption of Our Lady. See Vasari, *Lives of the Most Eminent Painters*, vol. 3, 194.

18. See Auguste-Michel Vidal, *Notre-Dame du Montement à Rabastens: projet pour la construction d'un appareil destiné à figurer l'Assomption* (Paris: Imprimerie nationale, 1910)), ; Chapuis and Gélis, *Le Monde des Automate*, vol. 1, 102.

19. See Alphonse Auguste, "Gabriel de Ciron et Madame de Mondonville," in *Revue historique de Toulouse* (Toulouse: Saint-Cyprien, 1919), 20–69, on p. 26; Chapuis and Gélis, *Le Monde des Automates*, vol. 1, 103.

20. This practice was thriving in the mid-seventeenth century, when Madame de Mondonville (Anne-Jeanne Cassanea de Mondonville) recalls in her memoirs having built a Virgin-ascender with her brothers. She used her precious, glittering bits of quartz crystal to carve the Virgin. See Auguste, "Gabriel de Ciron et Madame de Mondonville," 26; Chapuis and Gélis, *Le Monde des Automates*, vol. 1, 103. On mechanical Assumptions, see also Tripps, *Der Handelnde Bildwerk*, 174–190.

21. See J. L. Heilbron, *The Sun in the Church: Cathedrals as Astronomical Observatories* (Cambridge, Mass.: Harvard University Press, 1999) and David Landes, *A Revolution in Time: Clocks and the Making of the Modern World* (Cambridge: Belknap, 2000), especially Chapter 3.

22. Chapuis and Gélis, *Le Monde des Automates*, vol. 1, 120–127; Derek J. De Solla Price, "Automata and the Origins of Mechanism and Mechanistic Philosophy," *Technology and Culture* V, no. 1 (winter 1964): 9–23, on pp. 18 and 22; Silvio A. Bedini, "The Role of Automata in the History of Technology," ibid.: 24–42, on p. 29 and figs. 2 and 3. The Clock of the Three Kings was originally built between 1352 and 1354 and refurbished by the clockmaker brothers Isaac and Josias Habrecht between 1540 and 1574. Beneath the Rooster——overshadowed by it, both literally and figuratively——the astrolabe turned and the Magi scene played out its familiar sequence. In the Habrecht version, the Rooster, Magi, Virgin, and Child were joined by a host of other automata: Roman gods, angels, a mechanical Christ, and so on.

23. On the connection between organs and automata in the medieval period, see Merriam Sherwood, "Magic and Mechanics in Medieval Fiction," *Studies in Philology* 44: 4 (October 1947): 567–592, on pp. 588–589.

24. Marie Pierre Hamel, "Notice historique abrégée pour l'histoire de l'orgue," in Dom Bedos de Celles, *Nouveau manuel complet du facteur d'orgues* (Paris, 1849), xxiv-cxxvi, see especially pp. l-li; Chapuis and Gélis, *Le Monde des Automates*, vol. 1, 106. In a bill from 1541: "*A Nicolas Quesnel, ymaginier, pour faire deux ymages des anges mouvantz, pour mettre sur l'amortissement des orgues.*" Léon Emmanuel Simon Laborde, *Notice des emaux, bijoux et autres objets divers, exposés dans les galeries du Musée du Louvre* (Paris: Vinchon, 1853), 11e Partie, Documents et glossaire, "Images mouvantes"; Chapuis and Gélis, *Le Monde des Automates*, vol. 1, 105.

25. Chapuis and Gélis, *Le Monde des Automates*, vol. 1, 106. Joseph Gallmayr built a Saint Cecilia in Munich powered by organ-peddles in the latter eighteenth century. Ibid.

26. "Rohraffen" meant "roaring apes" and referred to the grotesque, bellowing figure of the Bretzelmann. He had help, for the vocal part of his act, from a *Münsterknecht*, a servant of the Cathedral hidden in the pendentive that held the organ. But the physical motions were the Bretzelmann's own and were controlled by the organ strings. See *Les Orgues de la Cathédrale de Strasbourg à travers les siècles. Etude historique, ornée de gravures et de planches hors texte, à l'occasion de la bénédiction des Grandes Orgues Silbermann-Roethinger, le 7 juillet 1935* (Librissimo, Phénix Editions, 2002); Conseil Régional d'Alsace, *Orgues Silbermann d'Alsace* (Strasbourg: A.R.D.A.M., 1992); Chapuis and Gélis, *Le Monde des Automates*, vol. 2, 108–109; Sherwood, "Magic and Mechanics," 585–586.

27. Pierre Schott, ammeister of Strasbourg, 1490: *"Par des mouvements désordonnés, des cantiques profanes et inconvenants criés à haute voix, il trouble les hymnes des pèlerins arrivants et les couvre de ridicule. De cette façon, il tourne la dévotion des arrivants en distraction, leurs pieux soupirs en rire, mais il dérange aussi les clercs qui psalmodient le saint office, il est cause d'une abominable et exécrable perturbation pendant le sacrifice de la ste messe."*

28. Jean Bourdette, *Le monastère de Saint-Sabi de Labéda (Ou Saint-Savin de Lavedan) et la vie de Saint-Sabi, Ermite* (Saint-Savin: au Presbytère et Toulouse. Chez l'auteur, 1911) ; Chapuis and Gélis, *Le Monde des Automates*, vol. 1, 106–107.

29. Ibid., 1: 106.

30. Alexandre de Salies, "Lettre sur une tête automatique autrefois attachée à l'orgue des Augustins de Montoire," in *Bulletin de la Société archéologique, scientifique et littéraire du Vendômois* (1867), 97–188; Chapuis and Gélis, *Le Monde des Automates*, vol. 1, 106.

31. The text was edited by Giovanni Sulpizio da Veroli, a close collaborator of the young Cardinal Raffaele Riairo, to whom the edition was dedicated and who was a central actor in the Renaissance renovation of Rome. Most of Vitruvius's illustrations had been lost; the Veronese friar Fra Giovanni Giocondo reconstituted these in a 1511 edition printed under the auspices of Pope Julius II. Vitruvius's treatise continued to inform the great building projects of the Renaissance popes through the reigns of Leo X and Paul III, not least Saint Peter's Cathedral. See Ingrid D. Rowland's introduction to Rowland, ed., *Vitruvius: Ten Books on Architecture, the Corsini Incunubulum with the Annotations and Drawings of Giovanni Battista Da Sangallo* (Rome: Edizioni dell'Elefante, 2003), 1–31.

32. Thomas Aquinas, *Summa Theologica*, Prima Secundæ Partis, Qu. 13, Art II, Reply obj. 3.

33. My discussion of the monk is derived from Elizabeth King's beautiful work on it. See King, "Perpetual Devotion: A Sixteenth-Century Machine that Prays," in Riskin, ed., *Genesis Redux* (2007), Chapter 13; and King, "Clockwork Prayer: A Sixteenth-Century Mechanical Monk," *Blackbird: An Online Journal of Literature and the Arts* 1, no. 1 (Spring 2002). The monk now resides in storage at the Smithsonian Institution's National Museum of American History; it has a twin at the Deutsches Museum in Munich. On Turriano, see José A. García-Diego, *Los relojes y autómatas de Juanelo Turriano* (Madrid: Tempvs Fvgit, Monografías Españolas de Relojería, 1982); Garcia-Diego, *Juanelo Turriano, Charles V's Clockmaker* (Madrid: Editorial Castalia, 1986); King, "Clockwork Prayer: A Sixteenth-Century Mechanical Monk"; Silvio A. Bedini and Francis R. Maddison, *Mechanical Universe: The Astrarium of Giovanni De' Dondi* (Philadelphia: The American Philosophical Society, 1966), 56–58; Bedini, "The Role of Automata," 32; and Chapuis and Gélis, *Le Monde des Automates*, vol. 1, 90–91. Originally from Cremona, Turriano was also known as Giovanni Torriani and as Gianello della Torre.

34. King, "Perpetual Devotion," 264–266.

35. The art historian David Freedberg has described a certain kind of viewing that comes in response to powerful religious representations. The devout beholder

"reconstitutes" the thing being represented, turning its representation into a presence: "The slip from representation to presentation is crucial ... from seeing a token of the Virgin to seeing her there." David Freedberg, *The Power of Images: Studies in the History and Theory of Response* (Chicago: University of Chicago Press, 1989), 28. I am indebted here to King's discussion of Freedberg's observation about religious images as applied to the monk, in "Perpetual Devotion."

36. Leaders of the Reformed movement, notably Ulrich Zwingli, denounced the use of organs and other musical instruments in church. See Charles Garside, *Zwingli and the Arts* (New Haven: Yale University Press, 1966); Quentin Faulkner, *Wiser than Despair: The Evolution of Ideas in the Relation of Music and the Christian Church* (Greenwood Press, 1996), Chapter 9; and Diarmaid MacCulloch, *The Reformation: A History* (New York: Penguin, 2003), 146, 590.

37. Chapuis and Gélis, *Le Monde des Automates*, vol. 1, 104–105.

38. Thomas Cromwell was an advisor to Henry VIII, also chancellor of the exchequer, secretary of state, and master of the rolls. He helped the king become head of the Church of England. In 1535 Cromwell was appointed vicar-general.

39. Chamber to Cromwell, February 7, 1538, in J. S. Brewer, ed., *Letters and Papers, Foreign and Domestic, of the Reign of Henry VIII* (London: Longman and Co., 1862–1910), vol. 13, Part I: # 231, on p. 79.

40. Lambarde, *A Perambulation of Kent*, 210.

41. For synoptic discussions of this vast and complex subject, see Alain Besançon, *The Forbidden Image: An Intellectual History of Iconoclasm* (Chicago: University of Chicago Press, 2001), Chapters 5 and 6; Carlos N. M. Eire, *The War Against Idols: The Reformation of Worship from Erasmus to Calvin* (Cambridge: Cambridge University Press, 1989), Chapters 3, 4, 6, and 8; Freedberg, *The Power of Images*, Chapter 8; Sergiusz Michalski, *The Reformation and the Visual Arts: The Protestant Image Question in Western and Eastern Europe* (New York: Routledge, 1993); and Edward Muir, *Ritual in Early Modern Europe* (Cambridge: Cambridge University Press, 1997), Chapters 5 and 6. For a study of the relations of Reformation art to iconoclasm, see Joseph Leo Koerner, *The Reformation of the Image* (Chicago: University of Chicago Press, 2004).

42. William Page, ed., The *Victoria History of the County of Kent* (Rochester, NY: Boydell and Brewer, 1974), vol. 2, 154; Wriothesley, *A Chronicle of England*, vol. 1, 74–75.

43. Chapuis and Gélis, *Le Monde des Automates*, vol. 1, 104.

44. Auguste, "Gabriel de Ciron et Madame de Mondonville"; Chapuis and Gélis, *Le Monde des Automates*, vol. 1, 103.

45. James Waterworth, ed. and tr., *The Canons and Decrees of the Sacred and Oecumenical Council of Trent* (London: Dolman, 1848), Session the Twenty-Fifth, 235–236.

46. The sixteenth-century architect Bernardo Buontalenti built a clockwork *presepio* for his pupil, Francesco, son of Cosimo de' Medici, with opening and closing heavens, flying angels, and figures walking toward the manger. And the Augsburg clockmaker Hans Schlottheim built an elaborate mechanical crèche around 1589 for the Court of Saxony. The crèche, which is now in the Museum für Sächsische Volkskunst in Dresden, includes shepherds and kings that process past the manger while angels fly down from heaven, Joseph rocks the cradle, and an ox and an ass rise up to stand before the holy Infant. Nesta Robeck, *The Christmas Crib* (Milwaukee: The Bruce Publishing Company, 1956), Chapter X and fig. 39. On mechanical nativity scenes, see also Chapuis and Gélis, *Le Monde des Automates*, vol. 2, 200–202.

47. The mechanical nativity scene was sent by Nicholas Trigault and had been donated by Ferdinand of Bavaria, the elector of Cologne. It and the other automata that Trigault shipped in 1618 did not arrive directly at their destination since the

Emperor Wan-Li was even then expelling the Jesuits from China. But a couple of decades later, after the Ming dynasty gave way to the Manchurian, T'sing dynasty, many of the automata arrived belatedly as the Jesuits returned under Emperor Shun-che and set up clockmaking workshops. See Edmond Lamalle, "La propagande du P. Nicolas Trigault en faveur des missions de Chine (1616)," *Archivum Historicum Societatis Jesu* 9, no. 1 (1940): 49–120; Chapuis and Droz, *Automata*, 77–84; Jonathan D. Spence, *The Memory Palace of Matteo Ricci* (New York: Penguin, 1985), 180–184. On the Jesuits' clockwork gifts in China, see also Trigault, ed., *China in the Sixteenth Century: The Journals of Matthew Ricci*, tr. Louis J. Gallagher, SJ (New York: Random House, 1942 [1615]), Bk. 1, Chapter 4 and Bk. 4, Chapter 12.

48. The Jesuit priest Gabriel de Magalhaens, who arrived in China in 1640, presented the knight to the Emperor Kang'hi. Louis Pfister, SJ, *Notices biographiques et bibliographiques sur les Jésuites de l'ancienne mission de Chine (1552-1773)*, 2 vols. (Shanghai: Imprimerie de la Mission Catholique, 1932, 1934), notice 88; Chapuis and Droz, *Automata*, 315.

49. In the clockmaking region of southern Germany during the late sixteenth and early seventeenth centuries, mechanical animals such as Schlottheim's mechanical crayfish became popular: automaton spiders; Neptune astride a creeping bronze tortoise; and a life-sized bear, wearing real fur, beating on drum. Chapuis and Gélis, *Le Monde des Automates*, vol. 1, 192; vol. 2, 141–143, 152–153; Chapuis and Droz, *Automata*, 242. In the 1680s and 1690s, clockmakers began to fabricate animated paintings (*tableaux mécaniques*) depicting hunting parties and other rustic scenes. Principal builders of animated paintings during the last decades of the seventeenth century included the clockmakers Abraham and Christian-Théodore Danbeck and Christophe Leo, of Augsburg, and during the early eighteenth century, Jean Truchet (le père Sébastien). See Chapuis and Gélis, *Le Monde des Automates*, vol. 1, 319. From their very inception, automata with secular themes often represented relations of power, not only metaphorically but also in the most explicit and literal way possible. A prime and very early example is the clock that Charles IV commissioned for the Frauenkirche in Nuremberg to commemorate his Golden Bull, which established the constitutional structure of the Holy Roman Empire and set the number of electors at seven. On the clock, which was inaugurated in 1361, seven figures known collectively as the Männleinlaufen (parade of little men) emerge at noon to bow before the emperor. Chapuis and Gélis, *Le Monde des Automates*, vol. 1,165–166. The sovereign and his seven electors reappeared on the clock of the Marienkirche in Lubeck where they paraded before Christ; Chapuis and Gélis, *Le Monde des Automates*, vol. 1, 170.

50. *Maistre des engiens du chastel.*

51. See Jules-Marie Richard, *Une petite-nièce de saint Louis. Mahaut, comtesse d'Artois et de Bourgogne (1302-1329). Etude sur la vie privée, les Arts et l'Industrie, en Artois et à Paris au commencement du XIVe siècle* (Paris: Champion, 1887), which is a study of the comtesse Mahaut's account books and Sherwood, "Magic and Mechanics," 589–590; Archives departementales du Nord, série B.

52. V*ᵉ compte de Jehan Abonnel dit Legros, conseilleur et receveur général de toutes les finances de monseigneur le duc de Bourgoingne . . .*, in the *Recette générale des finances, Chambre des comptes de Lille*, Archives departementales du Nord, série B no. 1948 (Registre); Léon Emmanuel Simon Laborde, *Les Ducs de Bourgogne: études sur les lettres, les arts et l'industrie pendant le XVe siècle, et plus particulièrement dans les Pays-Bas et le duché de Bourgogne* (Paris: Plon frères, 1849–1852), Seconde Partie, Tome 1: 268–271; Sherwood, "Magic and Mechanics," 587–590 (the quoted passage is taken from the excerpt cited in translation in Sherwood); Price, "Automata and the Origins of Mechanism," 20–21; and Chapuis and Gélis, *Le Monde des Automates*, vol. 1, 72.

53. On the fame and influence of the Hesdin engines, see Sherwood, "Magic and Mechanics," 590 and Price, "Automata and the Origins of Mechanism," 21.
54. Michel de Montaigne, *Journal de voyage*, ed. Louis Lautrey (Paris: Hachette, 1909)), 125.
55. Chapuis and Gélis, *Le Monde des Automates*, vol. 1, 74.
56. Montaigne, *Journal de voyage*, 187 and Chapuis and Gélis, *Le Monde des Automates*, vol. 1, 75. On hydraulic entertainments in Italy, see Philippe Morel, *Les grottes maniéristes en Italie au XVIe siècle: Théâtre et alchimie de la nature* (Paris: Macula, 1998). The waterworks Montaigne described here were built by the sixteenth-century architect Bernardo Buontalenti. On the Pratolino grotto, see also Salomon De Caus, *Les raisons des forces mouvantes avec diverses machines tant utiles que plaisantes, auxquelles sont adjoints plusieurs dessings de grotes & fontaines* (Francfort: J. Norton, 1615).
57. Montaigne, *Journal de voyage*, 388.
58. Montaigne, *Journal de voyage*, 270. The owl-and-birds arrangement was taken from a much-imitated design by Hero of Alexandria. See *The Pneumatics of Hero of Alexandria*, tr. and ed. Bennet Woodcroft (London: Taylor Walton and Maberly, 1851), #15. On Hero's automaton, see Sylvia Berryman, "The Imitation of Life in Ancient Greek Philosophy," in Riskin, ed., *Genesis Redux* (2007), Chapter 2.
59. The primary source of information about the Francini fountains at Saint-Germain-en-Laye is a collection of engravings by Abraham Bosse, done from Francini's own designs. Tomaso Francini's brother, Alessandro, who collaborated with him on the fountains at Fontainebleau, is listed as the author of the collection: Alessandro Francini, *Recueil. Modèles de grottes et de fontaines. Dessins lavés.* BNF Estampes et photographie, Réserve Hd-100(A)-Pet Fol; also in the Archives nationales, O¹ 1598. There is also John Evelyn's description in his diary: Evelyn, *Diary and Correspondence*, ed. William Bray (London: George Bell and Sons, 1883), Vol. 1: entry for February 27, 1644. See also Evelyn, *Elysium Britannicum, or the Royal Gardens in Three Books* (Philadelphia: University of Pennsylvania Press, 2000)—Book II, Chapter 9, "Of Fountains, Cascades, Rivulets, Piscenas, Canales, and Water-works"; Book II, Chapter 12, "Of Artificial Echoe[']s, Automats, and Hydraulic motions"; Book III, Chapter 9, "Of the most famous Gardens in the World, Ancient and Modern." For other primary accounts of the hydraulic automata at Saint-Germain-en-Laye, see Georges Houdard, *Les châteaux royaux de Saint-Germain-en-Laye, 1124–1789: Etude historique d'après des documents inédits recueillis aux Archives Nationales et à la Bibliothèque Nationale* (Saint-Germain-en-Laye: Maurice Mirvault, 1910–1911), vol. 2. For secondary descriptions, see Albert Mousset, *Les Francine, créateurs des eaux de Versailles, intendants des eaux et fontaines de France de 1623–1784* (Paris: E. Champion, 1930), Chapter 1 and plates 2, 3, and 4; Lionel du Sorbiers de la Tourrasse, *Le Château-neuf de Saint-Germain-en-Laye, ses terrasses et ses grottes* (Paris : édition de la Gazette des beaux-arts, 1924); Alfred Marie, *Jardins français créés à la Renaissance* (Paris: Vincent Fréal & Cie., 1955); and Chapuis and Droz, *Automata*, 43–47. The fountains were abandoned after Louis XIV moved his court to Versailles in 1682 and virtually no trace remains. On the fountains at Versailles, however, as emblems of material power, see Chandra Mukerji, *Territorial Ambitions and the Gardens of Versailles* (Cambridge: Cambridge University Press, 1997), esp. 181–197.
60. Evelyn, *Diary*, entry for February 27, 1644; André du Chesne, *Les antiquités et recherches des villes, chasteaux, et places plus remarquables de toute la France* (Paris: Pierre Rocolet, 1637), 221–224.
61. Evelyn, *Diary*, entry for February 27, 1644; Mousset, *Les Francine*.
62. See e.g. the entry for June 6, 1605, Jean Hérouard, *Journal de Jean Hérouard*, 2 vols., ed. Madeleine Foisil (Paris: Fayard, 1989), Vol. 1: 676.

63. In April and May of 1605 alone, visits to the grottoes were recorded on: April 11, 13, 14, 17, and 29; May 2, 8, 9, 15, 27, and 29. Ibid., 1: 638, 639, 643, 653, 655, 657, 658, 660, 666, 668.

64. *"Faite semblan que je sui Ophée (Orphée) e vous le fontenié (-er), dite chante le (les) canarie."* Ibid., 1: 633 (entry for March 20, 1605).

65. See entries for April 13, 14, and 17, 1605: Ibid., 1: 638, 639, 643.

66. See entries for May 25; June 2; July 3 and 30; August 20 and 28; October 25, 1605; May 9, June 15, and July 8, 1606; Ibid., 1: 664–665, 672, 703, 722, 741, 759, 809, 943, 987, 1000.

67. See entries for April 16 and 18; June 1, 2, 6, 10, 13, 15, 22, 26, and 30; and August 3, 1605; Ibid., 1: 640, 643, 671, 672, 673, 676, 681, 684, 686, 692, 696, 698, 725.

68. "Dict qu'il y a ung robinet a son cul et ung autre sa guillery: *'fs fs'.*" Entry for April 18, 1605, Ibid., 1: 643. For some repetitions of the joke, see entries for April 2; June 1 and 10, 1605; Ibid, 1: 638, 671, 681. The eroticism that pervades Hérouard's journal has occasioned a good deal of analysis. See, for example, Philippe Ariès, *L'Enfant et la vie familiale sous l'Ancien Régime* (Paris: Plon, 1960).

69. See entries for July 31, 1611 and October 27, 1612, Hérouard, *Journal* (1989), 2: 1939, 2066.

70. *"Maman ga, je sone les heure, dan, dan, I (il) sone come le jaquemar qui frape su l'enclume."* Ibid., 1: 699 (entry for June 30, 1605). The prince referred to the clock at Fontainebleau. *Maman ga* was *Mme. de Montglat.*

71. Ibid., 1: 1396 (entry for March 5, 1608).

72. Ibid., 1: 1434 (entry for May 16, 1608).

73. The engineers were Orazio Olivieri and Giovanni Guglielmi. Edith Wharton, *Italian Villas and Their Gardens* (New York: Da Capo, 1977 [1904]), 154. Other spouts of water and water-powered jets of air played organ- and fife-music and produced eerie sounds—thunder, wind, rain, whistles, shrieks—while wooden globes danced magico-mechanically across the floor. On the Villa Aldobrandini (originally named the Villa Belvedere), see Carla Benocci, *Villa Aldobrandini a Roma* (Rome: Argos, 1992). Giovanni Battista Falda's engraving of the Stanza dei Venti is in in *Le fontane di Roma*, Part Two, *Le fontane delle ville di Frascati* (Rome: Giovanni Giacomo de' Rossi, 1691), plate 7.

74. Farther down the social hierarchy was Martin Löhner, a hydraulic engineer and the master of wells (*Brunnenmeister*) for Nüremberg, who established a much-visited host of automata at his own house. It included Vulcan laboring at his forge; Hercules bludgeoning his dragon; Acteon surprising Diana and her nymphs in their bath, whereupon Diana threw water at Acteon, who turned away, grew antlers on his head, and was attacked by his own dogs; Cerberus spitting fire at Hercules; a lion emerging from his cave to drink from a basin, then retiring; the nine muses, each engaged at their appointed art. On Martin Löhner and his automata, see Johann Gabriel Doppelmayr, *Historische nachricht von den nürnbergishcen mathematicis und künstlern* (Nürnberg: P.C. Monath, 1730), 306; Chapuis and Gélis, *Le Monde des Automates*, vol. 1, 76.

75. On Sittikus and the Schloss Hellbrunn, see Chapuis and Gélis, *Le Monde de Automates*, vol. 1, 76–77; Alfred Chapuis, "The Amazing Automata at Hellbrunn," *Horological Journal*, Vol. 96, no. 6 (June 1954): 388–389; Bedini, "The Role of Automata," 27; and Wilfried Schaber, *Hellbrunn: Schloss, Park und Wasserspiele* (Salzburg: Schlossverwaltung Hellbrunn, 2004). When he was elected as archbishop, Sittikus was already a connoisseur of automata. He had lived briefly at the Villa Aldobrandini; moreover, his uncle, Cardinal Marco Sittico Altemps, nephew of Pope Pius IV, had built the Villa Mondragone, which had a renowned water theater designed by the engineer Giovanni Fontana. See Falda's engraving of the semicircular water theater of the Villa Mondragone in *Le fontane della ville di Frascati* (1691), plate 18.

76. See Ibid.
77. John Evelyn, *Elysium Britannicum, or The Royal Gardens* (ca. 1660), ed. John E. Ingram (Philadelphia: The University of Pennsylvania Press, 2001), 231, 242, 191.
78. Evelyn, *Elysium Britannicum*,184, 439.
79. Anne-Louise d'Orléans, duchesse de Montpensier, *Mémoires de Mlle. de Montpensier*, ed. Bernard Quilliet (Paris: Mercure de France, 2005), Chapter XXIII (July–September, 1656). The owner of the estate, whom the writer identifies as "Esselin," was Louis Cauchon d'Hesselin, who served as Maître de la chambre aux deniers de la maison du roi. Madame de Lixein was Henriette de Lorraine, daughter of François de Lorraine, who had married the Prince of Lixen.
80. Robert Darnton, *The Great Cat Massacre and Other Episodes in French Cultural History* (New York: Vintage, 1985), 78.
81. Henri Bergson, *Laughter: An Essay on the Meaning of the Comic*, tr. Cloudesley Brereton and Fred Rothwell (Whitefish: Kessinger Publishing, 2004 [first published in French as *Le rire: essai sur la significance du comique* (Paris: Editions Alcan, 1900)]), 16, 81. For Freud's use of Bergson's theory of humor in his own very different account, see Sigmund Freud, *Jokes and Their Relation to the Unconscious*, tr. James Strachey (New York: W.W. Norton, 1963 [first published in German as *Der Witz und Seine Beziehung zum Unbewussten* (Leibzig and Vienna: Deuticke, 1905)]), Chapter 7, especially 259–260.
82. For an anthology of recent treatments of this subject, see Jan M. Bremmer and Herman Roodenburg, eds., *A Cultural History of Humour: From Antiquity to the Present Day* (Cambridge: Polity Press, 1997). Peter Burke, in Chapter 5 of the volume, "Frontiers of the Comic in Early Modern Italy, c. 1350–1750," mentions Renaissance palace waterworks in passing, p. 65. The history of humor is an integral part of the discussion in Norbert Elias's classic work, *The Civilizing Process: Sociogenetic and Psychogenetic Investigations* (London: Blackwell, 1994 [first published in German as *Über den Prozess der Zivilisation* (Basel: Haus zum Falken, 1939)]).

12
Cosmography and the Meaning of Sundials

Jim Bennett

The sundial conundrum

In the history of the use of geometry to regulate our engagement with the motions we observe in the heavens, we find in dialing the widest gap between the commitment and enthusiasm of historical practitioners, and the attentions of historians. Astrology now has a scholarly community to recover and communicate its theories, methods, influence, and social and cultural significance. Dialing—despite an enormous following in its Renaissance and early modern heyday, the penetration of its practice to almost all levels of society, and the complexity, subtlety, and originality of its development by leading geometrical astronomers—now languishes beyond the concerns of nearly all historians, its memory sustained by enthusiasts for instruments and by practitioners more engaged with the exercise of horological geometry than with its history. A recent exception has been John Heilbron's book *The Sun in the Church*, telling the story of some of the grandest of all sundials, where the image of the sun thrown by an aperture high in a vault or a façade of a great church marks and measures the annual solar cycle by a meridian line set into the floor.[1]

One reason for the more general neglect is surely a misapprehension of the role of dialing in the Renaissance and an impoverished appreciation of its purpose. In short, we too readily restrict its function to telling the time. Where sundials are functional today, from the monumental to the recreational, they do little more than simply tell the time and, even then, this is not a method of time-telling that anyone would rely on. We are touchingly surprised when a sundial we come across might work at all, forgetting for the moment that it represents the source and regulation of all our time-telling. The danger for historians is to project that impoverished functionality back into the sixteenth century, when dialing was pursued by the leading geometrical astronomers. The design at the center of this article, for example, was the work of no less an astronomer than Johannes Regiomontanus.

249

The Regiomontanus dial appeared initially in the astronomer's *Kalendarium*, first published in Venice in 1474.[2] Figure 12.1 is an expanded and more detailed representation, more typical of surviving instruments from the sixteenth century, taken from Oronce Finé's *De Solaribus Horologiis et Quadrantibus* of 1560. What assumptions can a modern author make about his historian readers, even if they are well-informed on sixteenth-century astronomy? How might they engage with an image such as this and what might it mean to them? If it concerned planetary theory, for example, comprising a deferent circle and an epicycle, he could be fairly confident in taking as understood an appreciation of its basic purpose and functionality. Here he cannot have that confidence,

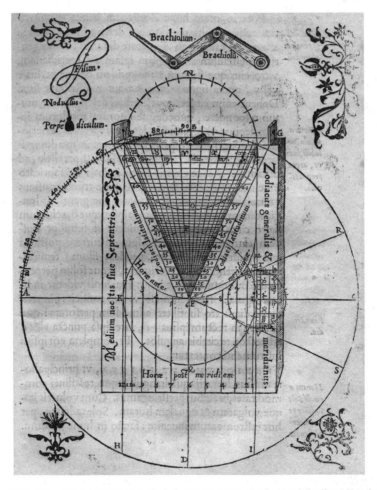

Figure 12.1 The Regiomontanus dial from Oronce Finé, *De Solaribus Horologiis et Quadrantibus* (Paris, 1560). All the figures are © Museum of the History of Science, Oxford.

even though an image such as figure 12.1 is a much more common occurrence in the astronomical geometry of the sixteenth century than is a theoric of planetary motion.

Telling the time "here and now" is of course a feature of the Regiomontanus dial, even if we may agree eventually that it is only part of the story. Time-telling is achieved by means of the altitude of the sun, but it might be misleading to say that the altitude is "measured," since it is not known at the end of the operation. What becomes known is the current local time, derived by the instrument from the solar altitude. There are two other variables to take account of in this operation: the solar altitude depending on the time of year and the latitude of the observer, as well as the time of day. Date and latitude are adjusted for by setting the point of suspension of a plumb-line (a weighted thread suspended from the tip of an articulated arm) to the intersection of the appropriate values for these variables on the triangular grid on the upper part of the dial, and then extending the thread from this point across the date (zodiac/solar declination) scale on one side of the dial (to the right in figure 12.1). A bead on the thread is slid into place at its intersection with the appropriate zodiacal position of the sun. Then, with the instrument held vertically, when the sights at the top are trained on the sun and the thread hangs freely, the bead will indicate the time on the vertical hour lines on the lower part of the dial. The central line indicates 6 o'clock, while the extreme lines to the left and right are for midnight and midday respectively. If the reader is wondering why a sundial should have a line indicating midnight, a doubt has been sown that the instrument is not simply for telling the time "here and now."

Sundials as mathematical instruments

We do not yet have the general history of Renaissance and early modern sundials that would extend their story beyond its current location in a geometrical and technical discourse and demonstrate its significance for other themes in social, intellectual, and cultural history. Dials could be objects of prestige and patronage, symbols of learning, or of devotion and piety. A monumental dial could be a project worthy of a prince. Utilizing the different edges and planes of a crucifix for casting and receiving shadows might create an object of devotion, even in some cases a reliquary to wear. Sundials were natural candidates for *memento mori*. Yet there was a range of products from instrument workshops that catered for a wide diversity of society.[3] In *As You Like It*, Shakespeare has Touchstone consult a dial he carries in his bag, an event undeserving of the moral soliloquy the "fool" then elaborates from this everyday gesture:

> And then he drew a dial from his poke,
> And looking on it, with lack–lustre eye,
> Says, very wisely, "It is ten o'clock."[4]

If the bawdy reading sometimes offered was intended, so much the more noteworthy that the theater audience was assumed to be familiar with portable sundials.

The notable development of dialing in the sixteenth century was one instance of what might be called a craft tradition in astronomical practice that was characteristic of the period. The work of a practitioner such as Gerard Mercator demonstrates an integration of learning and skill, as he combined the work of engraver, cartographer, cosmographer, printer, and instrument-maker. Other examples, among many, would be Johann Schöner, Georg Hartmann, Peter Apian, and Gemma Frisius. Innovative astronomy and cartography were disseminated through objects made by craftsmen in the leading workshops and print-shops. Some of these objects were themselves important inventions, such as the printed atlas or the printed globe.

If a rich history of sundials, a further instance of astronomical craft, is yet to be written, in recent years we have learnt much about the general class of mathematical instruments, to which they belong, and have taken greater care over how we characterize such instruments. In particular we are careful not to equate their functionality with that of the later class of *scientific* instruments. Instruments from the sixteenth century—"mathematical" instruments, as they were known—scarcely ever espoused pretensions to discover truths about the natural world. They were for solving problems susceptible to mathematical treatment, such as finding the time, a position at sea, or the range of an artillery target; or laying out a fortification or drawing a map.

This does not mean that they might not be technically sophisticated, far from it. Although they scarcely engaged with causal explanations in the manner of natural philosophy, their operations were grounded in the mathematical *science* of geometry and their output—designers and commentators insisted—was correspondingly reliable. Our next section follows the sixteenth-century geometrical discourse of the Regiomontanus dial, revealing something of the nature of this mathematical practice by working through a little of it, and revealing its constructive or crafted character.

The sixteenth-century discourse of the Regiomontanus dial: An exercise in practical geometry

If a sundial whose operation extends to midnight sits outside our usual characterization of such an object, where might we place the Regiomontanus dial in the broader context of sixteenth-century learning? One source of guidance would be to follow carefully the way it is presented in the texts of the period, when there is a vogue for printed descriptions of the construction and use of mathematical instruments. To this end we shall take the early account of Sebastian Münster, first published in his *Compositio Horologiorum* of 1531,[5] where, in keeping with the convention adopted in these texts, the reader is taken through the construction of the instrument as a practical geometrical exercise. This will allow us to imitate John Heilbron's expository technique in his *Geometry Civilized*,[6] where engaging with geometry becomes a route to historical insight, by taking the reader into the mathematical literacy of a time and a culture.

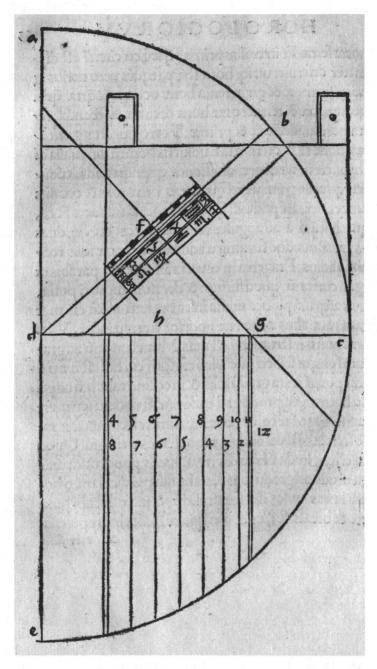

Figure 12.2 The construction of the rectilinear altitude dial sometimes called the "Capuchin dial," from Sebastian Münster, *Compositio Horologiorum* (Basel, 1531).

Here, however, we shall stick closely to the presentation and the limitations of the contemporary text, the only difference being that it will be helpful to introduce a simpler dial before moving to the instrument of Regiomontanus. This simpler altitude dial is not universal, but confined to a single latitude. The didactic technique of beginning with the simpler instrument can be found in modern treatments of the Regiomontanus dial but was not adopted by either Münster or Finé. Since Münster does deal with this dial at a later stage in his treatise, we can allow his account to present it also.[7]

Münster begins with instructions to draw a semicircle on a vertical diameter (see figure 12.2), divide it into two quadrants, and divide the upper quadrant arc into 90 degrees. The reader is then to mark off from the top of this quadrant the latitude where the dial is to be used, draw in the radius at this latitude, and on the line of the radius construct a zodiacal scale. The normal to this radius at its center f intersects the lower radial edge of the quadrant (dc) at the point g. Centered at g, arcs of 23 degrees 30 minutes are set out on either side of gf, the terminal radii being said to represent the tropics. Angles are then given for the other boundaries between the zodiacal signs, so that the reader can construct the complete zodiacal scale. A line from f parallel to the diameter of the semicircle will be the line for both the hours of 6 (morning and evening) and intersects dc at the point h. The reader is told to construct a semicircle below h with the radius hg and divide it into 12 equal parts. Parallels to the hour line for 6, drawn from the divisions on the semicircle, moving toward g, will be the lines for the pairs of hours 7 and 5, 8 and 4, 9 and 3, 10 and 2, 11 and 1, and ending with 12 midday at g. At this point Münster suggests that subdivisions of the hours may be added, so as to be useful when finding the length of day and times of sunrise and sunset through the year—an indication that the instrument is not just for finding the time here and now.

To find the time, however, two vanes, pierced for receiving the rays of the sun, are set above the zodiacal scale on a line at right angles to the diameter of the semicircle, and a slot is cut in the zodiacal scale on the line of the radius. In this slot moves a cursor and from it hangs a weighted thread with a pearl or some less precious index that can be moved, friction-tight, along the thread. When used for time-telling, as Münster then explains, the cursor is moved to the appropriate point on the zodiacal scale for the time of year, the thread stretched across the point g, and the index set to that point (g will be the bead's position for noon throughout the year). With the dial held vertical, the sights are then trained on the sun, and the position of the index among the hour lines will give the time. To find the length of the day at any time of the year, the cursor and the index are adjusted as before and, with the thread set perpendicular to dc, the hour lines will give the times of sunrise and sunset, since this is equivalent to sighting the sun on the horizon. Münster's dial has hour lines to the left of the line for 6, symmetrical to those on the right, up to 4 in the morning and 8 in the evening, the maximum length of day in the latitude for which the dial is constructed and coinciding with the perpendicular from the cursor's position at the summer solstice.

Münster offers his readers nothing further by way of geometrical proof or even plausible explanation for why this construction may be relied upon to find the time or yield the length of day throughout the year. If we consider the situation at the equinoxes, f may be regarded as the center of the celestial sphere and the path of the sun for the day would lie in the plane containing fg at right angles to the page. The circumference of the sphere is traced by the arc through g centered at f, that is, the path of the bead of the sundial. The hour lines as drawn are orthographic projections of the lines of equal altitude for the whole-hour positions of the sun, that is at 15-degree intervals in the circle, hour 6 being the horizon (altitude zero) and 12 being the meridian altitude, which at the equinox is the complement of the latitude. For other dates in the year the sun will be on a different circle on the celestial sphere and the orthographic projection of the lines of equal altitude will be different, which would be incompatible with a useful sundial. This dial accommodates that by changing the radius of the arc traced by the bead, while moving the point of suspension also accommodates the changing relationship between the daily path of the sun and the horizon. (The geometry is correct though no proof of this is given.[8])

Despite the simplicity of this dial in its construction, it is a relatively sophisticated piece of geometry, especially in the way the same pattern of lines serves for different positions of the projection of the daily solar motion on the celestial sphere. It works in only a single latitude, and it might be thought that adding different declination lines for different latitudes would rapidly make the instrument complicated and unmanageable. But Peter Apian offers solutions in quadrant designs included in his *Instrument Buch* of 1533.[9] In one quadrant there are zodiacal scales for every 2 degrees of latitude from 30 to 60 degrees and a scale along the meridian or 12 o'clock line for use in adjusting the bead for the corresponding latitude. The result is more an instrument for calculation than immediate time-telling. Among the multifunctional applications of a second instrument (figure 12.3) is a similar dial where zodiacal scales are provided for latitudes from the equator to the arctic circle, the scale for the equator being set vertical, in line with the hour line for 6 and indicating no variation in the times of sunrise and sunset throughout the year. (Ignore the curved lines for this purpose: the relevant hour lines are the straight, vertical ones.) We shall return to the clear geographical—or, more properly, cosmographical—meaning of such an instrument.

In fact a very satisfactory solution to extending the latitude range of the rectilinear altitude dial already existed in the Regiomontanus dial. We shall again follow the instructions given by Sebastian Münster, who provides the reader with two woodcuts—one is for following the details of the construction (figure 12.4), whereas the other is a finished instrument.[10] Münster begins with instructions to draw a circle and to divide one of its quadrants into degrees, beginning with 0 at the horizontal point a. The reader is then told to mark off the maximum declination of the sun, 23 degrees 30 minutes, on both sides of the 90-degree mark, b, giving h and f, and on either side of the point d, diametrically opposite to b, giving i and g. Münster says that the

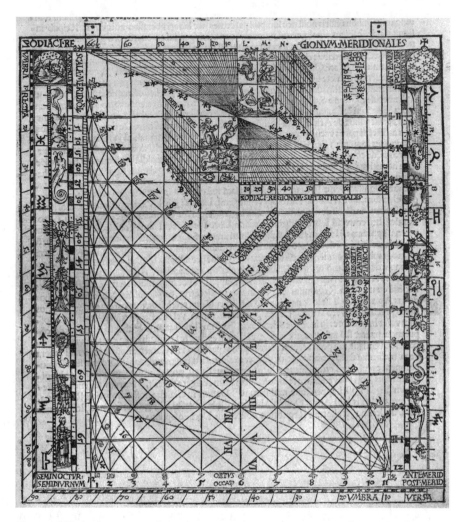

Figure 12.3   An horary quadrant from Peter Apian, *Instrument Buch* (Ingolstadt, 1533).

lines hi and fg represent 12 midnight and midday, and that the other hour lines will fall in the space between.

He then describes the construction of the triangular grid in the upper part of this space, beginning by drawing ef and eh to represent the tropics of Capricorn and Cancer. Dividing a circle constructed on hf as diameter into 12 equal parts gives the zodiacal divisions of the sun's annual motion, which are projected orthographically on to hf and joined by lines drawn to e. The latitude component of the grid is added by finding the intersections between the lines from the center e to the required latitudes and the line fm, and from these points of intersection lines are drawn, numbered appropriately,

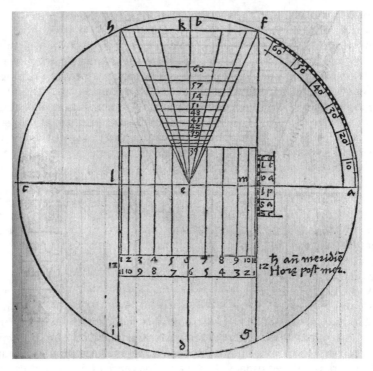

Figure 12.4   The construction of the Regiomontanus dial, from Sebastian Münster, *Compositio Horologiorum* (Basel, 1531).

parallel to ca and contained by the boundaries of the triangle, representing the tropics.

The reader is then instructed in drawing the hour lines by constructing a circle on lm and dividing it into 24 equal parts; the lines through these divisions parallel to bd each serve for morning and afternoon hours: 7 and 5, 8 and 4, 9 and 3, 10 and 2, and 11 and 1. Subdivisions of the hours are also possible. The reader is then told to transfer the zodiacal scale for the 45-degree latitude line to the line fg, centered at m, where the scale can be subdivided and marked with the signs of the zodiac.

Münster's instrument is completed by an arm that is adjustable over the triangular grid, so that a weighted thread with an adjustable bead can hang freely from any point in the grid, and by a pair of sights set above the transverse line for the most northerly latitude. For direct time-telling by the sun, as we have already seen, the arm is set to the appropriate intersection of latitude and zodiacal sign on the triangular grid, the thread stretched across the appropriate point of the zodiacal scale on the side of the hour lines, and the bead brought to this intersection. With the dial held vertical, the sights are trained on the sun, and the position of the freely hanging bead among the hour lines will indicate the time. However, as Münster points out, as the full cycle of 24 hours is present, this curious "sundial" extends at least some of its functionality to midnight.

From what we have seen already, we can have some intuition about why this construction might fit the bill. With the single-latitude rectilinear dial, the solar declination line, for adjusting the point of suspension through the year, is perpendicular to the equinoctial line for the given latitude. In each of the Apianus dials (one is illustrated in figure 12.3) there is a set of declination lines for a range of latitudes, but so as not to have a confusion of intersecting lines, they all share the same equinoctial center and intersect there, fanning out according to the latitude. The necessary adjustment for the different radii over the range of latitudes is by that latitude scale on the noon line, where the bead is set. With the Regiomontanus dial, the declination lines are set out quite differently: the fanning is gone; they are all parallel; and their position is found by the latitude construction explained by Münster. Since they are no longer at right angles to the equinoctial lines, a single noon point adjustment for each latitude will be insufficient, and an adjustment of the radius is given by a scale that introduces an additional zodiacal variable.

A little thought confirms that the adjustment to the radius operates in the right sense but a modern reader may want something much closer to a proof at this point in the explanation. Yet even this vague justification is more than can be found in Münster. (The construction is geometrically correct. [11]) Both he and Finé carefully tell their readers exactly what to do, step by step, and then how to use the result. They do not offer a geometrical proof for establishing the instrument's legitimacy. Typical of this genre of mathematics, at least in its published presence, legitimacy comes from the use of a set of geometrical techniques accepted within mathematical practice.

The Regiomontanus dial as an instrument of cosmography

What qualities are presented by this dial? Why take these steps to arrive at just the properties and characteristics it offers, with its rectilinear arrangement of hour lines and its distribution of parallel lines arranged by ascending latitude, each with an appropriate scale of solar declination? Its functionality can be viewed in the context of Münster's wider reputation, which lies, of course, in cosmography. Finé too, though a polymath in the geometrical arts and sciences, is mainly remembered as a cosmographer, although horology and cosmography form a ubiquitous conjunction in the sixteenth century.[12]

The discipline of cosmography deals with the presentation of the whole cosmos—the heavens *and* the earth—and in particular the relationship between them. As John Dee explained succinctly in his "Mathematicall Praeface" to Henry Billingsley's English translation of Euclid of 1570, "Cosmographie, is the whole and perfect description of the heauenly, and also elementall parte of the world, and their homologall application, and mutuall collation necessarie."[13] Cosmographical astronomy emphasized that the imaginary circles according to which we organize the spatial account of our stationary earth—equator, tropics, and lines of latitude—are equivalent to and originate in the corresponding circles in the movement of the heavens. Münster's book on sundials begins, in the way a contemporary cosmography might begin, with the celestial sphere and the motion of the sun. Cosmography has to deal with the starry heavens and their daily motion about the pole, their appearance from different parts of the earth marking out geographical location. It does not move on from there to tackle the motions of all the planets: Ptolemy's *Geographia* (or *Cosmographia*, to use the more common title from the sixteenth century) is restricted to the motions of the stars and the sun, the other planets being the business of his *Almagest*.

Dee offers a material and instrumental commentary on the way the circles of the heavens are applied to the earth, by referring to the contemporary practice of drawing them on a terrestrial globe. Although there were, as yet, no English globes, he was familiar with this practice from his association with Gemma Frisius and Gerard Mercator. The art of cosmography, Dee explains, "matcheth Heauen, and the Earth, in one frame, and aptly applieth parts Correspōdent: So, as, the Heauenly Globe, may (in practise) be duely described vpon the Geographicall, and Hydrographicall Globe." The word "describe" is used here in its original, literal sense, as it still is in geometry, where a circle might be "described." Dee's "frame" also has a literal sense: he adds to his explanation of the circles their accommodation in a stand with a horizon ring: "by an Horizon annexed, and reuolution of the earthly Globe (as the Heauen, is, by the Primouant, carried about in 24.æquall Houres) to learne the Risinges and Settinges of Sterres." Here Dee points very particularly to the curiosity of this instrument, where the heavenly circles are described on the terrestrial globe to create what he calls a "cosmographical globe": the globe of the earth is given a rotation on the poles of the heaven for the convenience of cosmography. "By the Reuolution, also, or mouing of the Globe Cosmographicall, the Rising and Setting of the Sonne: the Lengthes, of dayes and nightes: the Houres and times (both night and day) are knowne." This revolution does not have a Copernican meaning; it is "artificial," the globe being an instrument of the art of cosmography.

In the context of this discipline, time is the most immediate link between the heavens and the earth, and the astronomical component of cosmography therefore incorporates the motion of the sun, adding the ecliptic circle to the celestial equator and the tropics, and "describing" it on the cosmographical globe. Ptolemy refers to the successive parallels of latitude in terms of the lengths of the longest day in those places on the earth.

Münster and Finé are far from being unusual in the period for combining cosmographical work with a concern for dialing. Examples range from small portable dials to cathedrals with meridian lines, such as the meridian in Florence by the cosmographer Paolo dal Pozzo Toscanelli.[14] Other prominent practitioners were Peter Apian, Johannes Stoeffler, Gemma Frisius, Gerard Mercator, Erlard Etzlaub, Johannes Werner, and Johannes Stabius. Working in Florence on the great cosmographical project of Cosimo I de' Medici—now surviving as the geographical room in the Palazzo Vecchio—were Miniato Pitti, Egnatio Danti, and Stefano Buonsignori.[15] After his enforced move to

Bologna, Danti built a meridian line there, while a Bolognese example of the cosmographer and dialist would be Giovanni Antonio Magini. The northern dominance of cosmography toward the end of the sixteenth century brings in Willem Janszoon Blaeu and Michiel Coignet.

The need or desire to find the time here and now is a hopelessly inadequate motive for the development of sundials we see in the sixteenth century, evident in both published treatises and surviving instruments. A solution to this historical conundrum is to take the dials, or at least the more ambitious designs, out of simple time-telling, to see them instead as instruments of contemporary cosmography, and to link the enthusiasm for sundials—universal dials in particular—with the contemporary rise of cosmography.

An instrument like the Regiomontanus dial, or "general horological quadrant" as Münster calls it, is not adequately characterized in the way we think of our simple and impoverished sundials of today, so that we are obliged to imagine a traveler carrying his dial to different parts of the world where he uses it to tell the time. That is only part of the intended or pretended functionality. A different class of dial from the altitude quadrant, the universal equinoctial dial, also has a clear cosmographical context—being adjustable to latitude by bringing its hour circle parallel to the equator and its gnomon in line with the pole. This adjustment is surely something we might more easily think of as a cosmographical gesture for the generality of owners and users than as a resetting for an unfamiliar location, according to the extensive travels of the user. The tables of the latitudes of places, found on many dials with latitude adjustment, might be seen as an aide-memoire for the travelling user, but they are also reminiscent of the latitude tables in Ptolemy's *Cosmographia*.

If we look at the Regiomontanus dial in this context, we can begin to see its advantages and to understand the investment of geometrical work involved in its design. Think of it more as a kind of map than as a dial, but at the same time think of a map more as a geometrical "theoric" than a picture or a bird's eye image of the earth. A theoric was an encapsulation of information, secured by a systematic technique (usually a geometrical one), in a device that might be an instrument but could also be a diagram or a construction. Results could be obtained from the theoric that were not entered in its construction and that were extracted by applying the proper protocols by the knowing user. As the vehicle for an operative technique rather than a causal explanation, the theoric belongs in the mathematical arts and sciences rather than in natural philosophy. The example most familiar to historians is the theoric of planetary motion, but mathematical practice has many other examples in different disciplines and a map drawn to scale is such a device. At the level of the world map, the theoric can take a variety of forms, shaped by different geometrical projections, and these varieties can coexist to be deployed according to their suitability for different purposes. They have different properties and advantages. So it is with the instruments we call sundials.

The Regiomontanus dial sets out very effectively the relationship between the seasonal variation in the length of the day and the latitude, presented systematically by increasing latitude and the correspondingly lengthening declination scales, indicating the growing discrepancy between summer and winter days. We might look at this as a kind of map, with information set out in a projection, but here the information incorporates the variable of time in relation to latitude and date. Looking at figure 12.1, you can see that at the equinox the sun rises at 6 everywhere in the world; that this is the case at the equator throughout the year, the declination "scale" being reduced to a point for latitude zero; that at the equator the sun is at the zenith at noon at the equinoxes; that there is a range of latitudes where the sun can ever reach the zenith; and so on. And of course by positioning the suspension point appropriately, you can find the times of sunrise and sunset and the length of the day for any date at any latitude (or rather, up to 65 degrees north in this example). So a great many cosmographical operations can be performed, in addition—if there is a pair of sights—to finding the time here and now.

The Regiomontanus dial is not the only "sundial" whose meaning is enhanced and whose apparent incongruity is resolved by closer attention to its disciplinary location. A form of what today would be called a universal altitude dial is described by Finé under the name "horologium generale."[16] Though not in Münster's treatise, it became familiar in the sixteenth century by its appearance in the many editions of Apian's *Cosmographia* as a working paper *volvelle* with rotating parts and index threads. Apian tells his readers how to use the paper instrument on the page of their book to solve such problems as finding the latitude from the altitude of the sun (knowing the date); the time from the sun knowing the latitude and date; the altitude of the sun anywhere knowing the time, date, and latitude; the times of sunrise and sunset anywhere; the length of the day; and so on. As the instrument is intended to be used, time is one parameter in the complex of interdependent variables that belong to the business of cosmography. Time is integral to its functionality, but once again this is not an instrument just for finding the time here and now, in the manner of a sundial as generally understood.

An instrument in the collection at the Museum of the History of Science has on one face a brass version of the *horologium generale* and on the other a second *volvelle* from Apian's *Cosmographia*, his "speculum cosmographicum," translated from paper into brass.[17] Here the planispheric projection of a normal astrolabe is applied to the earth rather than the heavens, so the only plate is a terrestrial planisphere extending to the Tropic of Capricorn. Above this rotates a rete comprising only a zodiacal band with eight stars within (i.e. to the north of) the ecliptic. Above this in turn, pivoted at the center (i.e. the pole), is an index arm extending to a time scale beyond the planisphere. The user who had access to Apian, as any user surely did, would have known how to trace the daily and annual cycles of the sun in relation to the earth, how to find where the sun is overhead for the user's time for any date, the time differences between geographical locations, and so on. Although this instrument is generally, and perfectly reasonably, referred to as a "geographical astrolabe," the combination of the *speculum cosmographicum* and the *horologium generale* makes it a versatile instrument of cosmography. It was made by

Gillis Coignet of Antwerp, whose son Michiel continued the cosmographical tradition in books as well as in instruments. An instrument by Michiel
Coignet in the Oxford collection combines the *horologium generale* with the
nocturnal, which was also described by Apian in his *Cosmographia*.[18]

If we are unwilling to call the great meridian instruments built into
Renaissance cathedrals mere "sundials," then that reluctance should apply
to many small, portable instruments as well. Or, we could instead avoid the
perils of projecting the impoverished functionality of modern dials back on
to the sixteenth century. Either way, we might encourage historians to pay
more attention to a geometrical discipline that its many practitioners found
engaging, satisfying, and meaningful.

Notes

1. J. L. Heilbron, *The Sun in the Church: Cathedrals as Solar Observatories* (Cambridge,
   Mass. and London: Harvard University Press, 1999).
2. I have used the 1476 edition of Johannes Regiomontanus, *Aureun hic Liber est: Non
   est Preciosior ulla Gemma Kalendario* (Venice, 1476).
3. See, for example, the range of ivory diptych dials, Penelope Gouk, *The Ivory
   Sundials of Nuremberg 1500–1700* (Cambridge: Whipple Museum of the History of
   Science, 1988).
4. William Shakespeare, *As You Like It*, 2nd series, ed. Agnes Latham (London: Arden
   Shakespeare, 1975), 2.7.20–2.7.23.
5. Sebastian Münster, *Compositio Horologiorum* (Basel: Heinrich Petri, 1531), 25–31;
   see also Sebastian Münster, *Horologiographia* (Basel: Heinrich Petri, 1533), 35–43.
6. J. L. Heilbron, *Geometry Civilized: History, Culture and Technique* (Oxford: Oxford
   University Press, 1998).
7. Münster, *Compositio*, 151–154; Münster, *Horologiographia*, 250–254.
8. F. W. Cousins, *Sundials: A Simplified Approach by Means of the Equatorial Dial*
   (London: John Baker, 1969), 168–174.
9. Peter Apian, *Instrument Buch durch Petrum Apianum erst von new beschriben*
   (Ingolstadt, 1533).
10. Münster, *Compositio*, 29–30; Münster, *Horologiographia*, 40–41.
11. For a modern proof, see A. W. Fuller, "Universal Rectilinear Dials," *The Mathematical
    Gazette* 41 (1957): 9–24.
12. I have offered a more general account of the connection between dialing and
    cosmography in "Sundials and the Rise and Decline of Cosmography in the Long
    Sixteenth Century," *Bulletin of the Scientific Instrument Society*, no. 101 (2009): 4–9.
13. John Dee, "Mathematicall Praeface," in Euclid, *The Elements of Geometrie*, tr. H.
    Billingsley (London, 1570), sig. b. iii.
14. Heilbron, *The Sun in the Church*, 70–71.
15. Jim Bennett, "Cosimo's Cosmography: The Palazzo Vecchio and the History of
    Museums," in M. Beretta, P. Galluzzi, and C. Triarico, eds., *Musa Musaei: Studies on
    Scientific Instruments and Collections in Honour of Mara Miniati* (Florence: Olschki,
    2003), 191–197.
16. Oronce Finé, *De Solaribus Horologiis et Quadrantibus* (Paris, 1560), 155–163.
17. Museum of the History of Science, inventory no. 53211.
18. Ibid., inventory no. 44721.

13
The Web of Knowing, Doing, and Patenting: William Thomson's Apparatus Room and the History of Electricity

Giuliano Pancaldi

Introduction

The periodization of the history of electricity seems to have posed no major problems to historians of science. Scholars agree that around 1800 there was a turning point: whether the focus is on Coulomb and Poisson, or on Volta and his electric battery, or both, events after 1800 presented new opportunities and new challenges, marking a discontinuity with the earlier period.[1] A similar agreement exists concerning the beginning, around 1880, of what was called by contemporaries the "age of electricity."[2] The literature available on electrification in Western countries after 1880 has established the notion that the most significant technological developments associated with electricity began around that date, marking another discontinuity with previous events.[3]

A consequence of this agreed, if seldom problematized, periodization has been to convey the view that the tools appropriate for treating the history of electricity prior to 1800 are mainly those provided by the history of science, although the history of technology and economics should be brought in when addressing the period after 1880. Over the past few decades, a host of studies carried out according to methodologies inspired by cultural and social history have helped to reshuffle the traditional borders between the history of science and of technology; but the periodization of the history of electricity has not been affected accordingly. One consequence has been that the period from 1800 to 1880, while treated in a number of excellent studies,[4] has retained the status of a kind of magmatic interlude: a period when many crucial developments took place, whose connections with the rest of the story—the one before 1800 and the one after 1880—remain problematic.

As part of the reshuffling exercise alluded to, years ago, stimulated by John Heilbron's monumental history of the science of electricity up to 1800,[5] I chose Volta and the battery as a case study likely to shed some light on the interrelations of theory, scientific practice, technology, and culture in the history of electricity. After a full immersion in Volta's biography and his laboratory notes, I emerged with a partly unanticipated story.[6] I found, for example, that Volta's decision to build the battery in 1799 as well as some intriguing suggestions on how to build it, were inspired by the electric organs of the torpedo fish—the life sciences, and the broad concerns of natural philosophy, thus claimed a role—and by an apparatus that a chemist and instrument maker, William Nicholson, had imagined to imitate the fish with a mechanical-electrical machine. So, technology also claimed a role in the story, but it was a kind of technology that did not fit easily into the modernist perspective of many historians of technology. To my further surprise, I found that the intriguing debts to the fish and to Nicholson, which Volta had acknowledged in his publications, had disappeared from the accounts of the battery provided by generations of physics textbooks and histories of physics. According to my reconstruction, Volta's debt to the life sciences and to the "bricoleur" spirit of instrument makers was a reminder of the winding roads of scientific and technological developments circa 1800. As I am still reflecting on the possible implications of my story for the history of electricity, my contribution to the present volume is a sequel to that story, and it addresses questions such as the following: Assuming we can agree that the battery came on the scene in the strange and unexpected way I claim, why did it take about eighty years—*not* a quick development in the century of "progress"—to move from Volta's *fishy* battery[7] to electric lighting as an everyday marvel? Do the steps leading from the battery to the "age of electricity" reveal developments and unintended consequences comparable to those that led Volta to the battery? Is there anything interesting to be learnt from such considerations for the history of electricity, and for our understanding of the enterprise we now call science and technology?

My case study is provided this time by William Thomson and his "apparatus room" at the University of Glasgow.[8] The room was packed with batteries and other electric paraphernalia that Thomson had set up as a newly appointed professor of natural philosophy after 1846, when he was barely 22. From about 1857, the facility was known as "the laboratory," and Thomson and later historians regarded it as the first such teaching facility in the history of physics. During those same years, as is well-known, Thomson developed a theory of electric and magnetic phenomena to which a younger contemporary, James Clerk Maxwell, declared he owed most when introducing his own new approach to the science of electricity and magnetism.[9] As is also well-known, by the end of 1856 Thomson was one of the directors of the Atlantic Telegraph Company, which laid the first telegraph cables between Ireland and Newfoundland. By 1858, Thomson held a patent for telegraphy, which gave him an important position in the field for decades. Thus, in the dozen years following 1846, Thomson with his apparatus room showed that it was possible to move from the kind of "physical mathematics" in which Thomson himself had been trained as a student in Cambridge, to experimental physics and teaching, to industrial consultancy and patenting, and back again. The case will be used to highlight the web of knowing, doing, and patenting in which the science of electricity was woven half a century after the introduction of the battery, during the slow dawn of the age of electricity.

There is an abundant literature on Thomson and a slimmer one on his laboratory.[10] The novelty the present discussion can claim rests on the perusal of the wealth of manuscript resources available on the daily life of Thomson's apparatus room and on some broader interpretive issues associated with the notion of an "early age of electricity."[11]

William Thomson, electromagnetic theory, and telegraphy

I will quickly review a few aspects of Thomson's early theoretical work on electricity and magnetism.[12] This is to assess the roles theory and mathematics may have played in the process that led Thomson, in 1854, to engage in telegraphic research, becoming by the summer of 1858 one of the few winners in the story, otherwise doomed to failure, of the first Atlantic cables. As is well-known, after working more or less properly for a few weeks, the 1858 cable stopped letting messages through. For the record, it took several more attempts, and eight more years, to have permanent telegraphic communications established across the Atlantic.[13]

My claim is that although Thomson's mathematics and his electromagnetic theory helped to open the doors of the Atlantic Telegraph Company to him, it was his mirror galvanometer—an instrument he had developed in *loose* connection with his theoretical views, in an apparatus room where teaching, research, and industrial commitments merged every day—that was the winner of the 1858 transatlantic experiment. I will further claim that the story had an impact on Thomson's own perception of the relative merits of theory and instruments in his work: by 1858 he had learnt to trust his instruments as well as his mathematics.

Young Thomson had been trained at the University of Cambridge as a brilliant mathematician. In Cambridge in those days "mathematics dominated undergraduate studies to the almost complete exclusion of all other subjects."[14] The kind of physics allowed within that tradition was referred to by Thomson as "physical mathematics."[15] An example is the problem Thomson was expected to solve as a candidate during the Tripos examinations of 1845. The problem, as he phrased it, "suggested some considerations about the equilibrium of particles acted on by forces varying inversely as the square of the distance."[16] In addressing such problems young Thomson excelled; however, to solve these problems no acquaintance with experimental physics was expected. Thomson, however, left Cambridge for Glasgow in 1846, and he was quickly exposed to different models and needs.

Glasgow University in those days did not reserve for mathematics the place it enjoyed in Cambridge. The teaching of natural philosophy was intended for all the students enrolled in the university and no special mathematical training was required to join the class. These being the arrangements, as a candidate to the chair and then a professor of natural philosophy, Thomson had to convince people in Glasgow that, despite his training in Cambridge and his early mathematical publications, he was *not* "a mere mathematician."[17]

When a candidate for the chair in Glasgow, Thomson had visited Paris to become acquainted with leading philosophers there. Victor Regnault and Joseph Liouville were among them, and they were later to write letters in support of Thomson's application.[18]

On March 26, 1845, as a result of a four-hour conversation in Paris focusing on Faraday's objections to the French theory of electrical action at a distance,[19] Liouville, a mathematician, convinced Thomson to write a paper to help smooth out the conflict. As Liouville emphasized in the letter he wrote 16 months later to support Thomson's application for the Glasgow chair, in his work for that purpose—meanwhile published in French in Liouville's journal—Thomson had shown his ability to build on the plan set out in an earlier attempt by George Green at reconciliation between French and British traditions in the science of electricity.[20]

In Paris Thomson also learnt how important instruments and experimental demonstrations were in the business of natural philosophy. It was Faraday however—whom Thomson had first met in 1845, immediately starting a correspondence with him—who showed him a new balance between the theory of electricity and experimental physics, and it was Faraday who, most forcefully, in 1854 called attention to the connection between electrical theory and telegraphy when trying to explain the retardation of electric signals noticed in submarine telegraph cables.

Retardation phenomena in the signals sent through long underground and submarine cables were first reported in 1852.[21] Faraday's trials aimed at assessing the phenomena were carried out during 1853, and his reflections on the subject were published early the following year.[22] Faraday used the occasion to reassert ideas on induction, conduction, and insulation that he had published as early as 1838 without connection to telegraphy. His ideas suggested that, when explaining electric propagation and its speed, the surrounding medium was just as important as the conductor itself.

The correspondence between Thomson and George Gabriel Stokes, both trained in Cambridge, bears witness to the attention the two devoted to Faraday's 1854 publication on telegraphy and to its implications for physical theory. Practical problems were considered as well. At the meeting of the British Association during the summer the question of a telegraph cable across the Atlantic was discussed, and Thomson was quick to focus on that aspect when assessing Faraday's publication and his own possible role in connection with theory and telegraphy. Meanwhile, he thought it fit to translate into English and circulate, with additions, two papers on the "theory of electricity in equilibrium" that he had published in 1845 at Liouville's instigation.[23]

As Thomson saw things in October 1854, Faraday's approach to the problem of retardation, when treated in the mathematical way he and Stokes (unlike Faraday) were familiar with, allowed "answering practical questions regarding the use of the telegraph wire." And that was possible, according to Thomson, by using the "well known equation of the linear motion of heat in a solid conductor" developed by Fourier.[24]

From then onward Thomson's advancement in electromagnetic theory went hand in hand with his growing involvement in telegraphy. His well-advertised achievements in the theory of electricity gave him visibility among experts and the lay public interested in telegraphy. On May 24, 1855, Thomson submitted to the Royal Society a communication "On the Theory of the Electric Telegraph." The paper was in his own words "in an incomplete form," "as it may serve to indicate some important practical applications of the theory especially in estimating the dimensions of telegraph wires and cables required for long distances."[25]

Thomson's plan of publications focusing on theory must have been further stimulated by letters he received from Maxwell. Seven years junior to Thomson, Maxwell had also been trained as a mathematician in Cambridge and was somewhat unhappy with the mathematical theories of electricity then circulating. In February 1854 Maxwell sought the advice of Thomson on how best to approach the study of electricity. In the course of 1855 his pressure on Thomson on the subject of electricity became insistent. Maxwell realized that, since the 1840s, Thomson had set out several interesting pieces likely to form, when put together, a new theory of electricity and magnetism.[26] That must have temporarily reinforced Thomson's determination to brandish theory as his main credential in the public debate over the feasibility of the Atlantic cable. Accordingly, he intensified the publication of important theoretical papers under the auspices of the Royal Society, and a few years later, he would expound his theoretical views with powerful rhetoric.[27]

A determination to use theory and mathematics as major credentials in questions concerning submarine telegraphy is shown also in the public debate, which Thomson engaged in with Wildman Whitehouse, the chief electrician of the Atlantic Telegraph Company, in the autumn of 1856.[28] The clash, which included an exchange of letters in the widely read *Athenaeum*, was used by Thomson to foster his connections with the Company. In the same weeks Thomson submitted two more technically oriented papers on the same topic to the Royal Society.[29] The campaign—involving both the expert Royal Society and the "popular" *Athenaeum*—was successful: by December 1856 the Glasgow investors elected Thomson to the board of directors of the Atlantic Telegraph Company.[30] So it was that, in the following years, the professor of natural philosophy found himself on board the ships laying the first telegraph cables across the Atlantic. Thomson's perception, at that stage, of the combined pleasures of mathematical theory and telegraphy was conveyed in a letter to Helmholtz sent on December 30, 1856.[31] He conveyed a similar mood, just before the departure of the 1858 Atlantic expedition, in a letter to his brother James, an engineer and a patentee.[32]

However, the situation described with fervor to Helmholtz—the propagation of electricity through submarine wires "is the most beautiful subject possible for mathematical analysis"[33]— turned out to be an idealized one. The letter to James in fact bears evidence that having his own instruments made "to work the telegraph" was already a part of Thomson's strategy some time before he left Ireland.[34] Finally, during the summer of 1858 when facing the challenges of the new cable being laid across the Atlantic, Thomson realized that mathematics and his theoretical views were helping him, and the managers of the Atlantic Telegraph Company, less than he had expected. Instead, a set of instruments that he had developed in the apparatus room of the University of Glasgow under the combined pressures of teaching, research, and industrial ambitions, proved up to the challenge.

The new balance among mathematics, theory, instruments, and industrial requirements was conveyed in a speech Thomson delivered in Glasgow a few months after the failure of the 1858 attempts at establishing permanent telegraphic communications across the Atlantic. The key passage was reported by *The Engineer*, in the third person, as follows:

> An exact mathematical investigation of the circumstances showed that a sufficiently large size of conductor and insulating coat would entirely remedy the anticipated embarrassment [i.e., retardation]. But a cable of such dimensions as the calculations showed to be required for signaling through 2,000 miles, at the rapid rate of ordinary telegraph, would be too unwieldy and too costly to be thought of in a first attempt.
>
> Therefore the Atlantic Cable Company, in adopting the improvement that he [Thomson] suggested, did not carry it further than in making the quantities of copper and gutta percha in any part of their cable nearly double of those in an equal length of any previously constructed telegraphic line, so far at least as the first cable was concerned, and prudently, in his [Thomson's] opinion, they left the further mitigation of the anticipated slowness to be worked out by improvements in the methods and instruments to be used for the transmission and the receipt of messages.[35]

The "methods and instruments" alluded to were the product of Thomson's apparatus room.

Thomson's apparatus room

As the ships commissioned to lay the Atlantic cable were ready to set out from Ireland on a trial trip, late in May 1858,

> Thomson, waiting anxiously for the arrival of the latest instruments from Glasgow, was almost the last man to go aboard, a precious package being handed up to him at the last moment by his assistant Donald Macfarlane, who had brought it by express. It contained an object like a small brass pot standing on four legs—the "marine mirror galvanometer", constructed during the preceding fortnight.[36]

Three months later, when the cable began giving signs of its imminent failure, Thomson's galvanometer and the currents of his sawdust Daniell batteries were the only instruments able to send and receive clear enough messages across the Atlantic, including a message by the president of the United States in answer to Queen Victoria's greetings for the achievement.[37]

It had taken much longer than a fortnight, of course, to develop the new type of galvanometer and the other instruments that Thomson had prepared or conceived in connection with the Atlantic venture of 1858. Some of them had been refined over the past several months in competition with those prepared by Whitehouse for the same purpose. The Atlantic Cable Company, in fact, had refused to pay for Thomson's instruments, although they had paid, and dearly, for Whitehouse's.[38] Thomson in any case had protected his creatures with a patent; a patent which, in 1871, he was still able to get prolonged for another eight years.[39] His new galvanometer and the other twenty or so instruments mentioned in the patent were indeed the result of the expertise and know-how he had accumulated in the apparatus room of the University of Glasgow since 1846. That expertise had grown out of teaching, research, and—more decidedly from 1854—industrial commitments, merging every day in the process of competitive imitation in which Thomson was engaged at the local, national, and international levels.[40]

The sources available on the daily life of Thomson's apparatus room from 1846 to 1858, and beyond, are rich. They allow us to reconstruct the exchanges—which usually go unrecorded—between the professor, his "hands," his "computers," his instrument makers, and his personal secretaries, often coinciding in the very same person, such as the already mentioned Donald Macfarlane. Sources range from the correspondence Thomson exchanged with colleagues whose labs inspired him when planning his new facility, to the correspondence he entertained with assistants and instrument makers—especially during the long summer vacations, when the professor traveled a lot, while the assistants stayed on, carrying out the tests and experiments he had assigned to them. I will use some of these sources to capture the diverse—local and nonlocal— factors that shaped Thomson's apparatus room and the network of people that enabled him to build the expertise he used in the telegraph venture of 1858.

On taking possession of the chair 12 years earlier, Thomson had inherited the collection of old-fashioned instruments put together by his predecessors. Glasgow College had a long-established tradition in the field, as well as in experimental chemistry.[41] On appointing the new teacher, the professors had shown that they expected him to revive the tradition, attracting large numbers of students. Thomson accepted the challenge. By November 22, 1846, he reported having 93 students, including "private" students and students in the experimental class (more on this below).[42] It was typical of Glasgow University that at the beginning of the academic year students paid a fee directly to the professors whose class they wanted to attend. The fees being a substantial addition to the professors' salary, being a popular lecturer had an immediate impact on their earnings.

Demonstrative experiments, and careful planning of the experimental class, were central during Thomson's long career as a professor. Where informal advertising among the students was thought to be insufficient to attract a good number, recourse was made to paid advertisements in the local newspapers. A typical ad of this kind, in the *Glasgow Herald* of October 26, 1854, offered a "Descriptive Course of Mechanical and Physical Science, illustrated by experiment, every morning at 9 o'clock, commencing Tuesday, Nov. 7," for the duration of about a semester. The fee was £3.3s, and there was an assistant's fee of 3s. Tuesday's and Thursday's lectures, devoted to "Experimental Discoveries in Physical Science," could be attended separately; the fee for the experimental lectures being £2.2s, and the assistant's fee being 2s.6d. The announcement, in the same ad, of a course of lectures on the "Mathematical Principles of Mechanics, Every Forenoon, except Saturday, at 11 o'clock; and on Saturdays at 10 A.M.," mentioned no fee.

Considering that, till 1858, Thomson was paying his assistant entirely out of his own pocket, the system realized an interesting (if uneven) joint venture between the professor and the assistant, focusing on the production of brilliant experiments and demonstrations. Apparatus and manpower were essential to succeed. Money for the apparatus was provided by the University of Glasgow itself through a series of grants intended for the acquisition of new instruments, which Thomson often ordered from the best instrument makers in Paris and London. The manpower, as mentioned, Thomson himself provided.

Aware that in Glasgow, because of his training as a mathematician in Cambridge, his fame was that of a young man "too deep to have popular talent,"[43] already in view of his application for the Glasgow chair, Thomson had used Paris as a remedy. During the four and a half months he spent there in 1845, he exploited Paris as a shopping mall offering exposure to—and testimonies of accomplishment in—all sorts of experimental natural philosophy, serious and popular.

In Paris he attended, in order of decreasing popular appeal, the lectures of Pouillet, Dumas, and Pelouze, and finally he was admitted for a while as an assistant of sorts to the laboratory of the youngest and the least popular of all: Victor Regnault, recommended by some as the best physicist in town.

As seen through the eyes of Thomson, both serious and popular lecturers in Paris distinguished themselves by the abundant experimental apparatus they used, which was "exceedingly good and on a very extensive scale." Another connected feature, especially among popular lecturers, was that "all the things" were "prepared with great care beforehand," so that—Thomson noted—public performances were varied and they seldom failed.[44]

To prepare lists of the instruments Regnault and others used in their "cabinets de physique" was the next obvious step once Thomson was appointed to the chair. So prompted by his own personal motives, and by Glasgow's local opportunities and challenges, in Paris, young Thomson launched himself on a conscious process of competitive imitation that was to bind Glasgow and Paris together in the field of natural philosophy for years to come.

The most tangible dimension of the imitation process was the stream of instruments reaching Thomson's apparatus room from Paris over the subsequent years. The Paris instrument-makers involved in the game included Marloye (acoustics), Froment (electricity), Silbermann (thermometry), Frastré (thermometry again), Pixii (mainly electricity and magnetism), Golaz (mechanics), Duboscq (optics), and Ruhmkorff (electricity again).[45]

Thomson's ties with enterprising French instrument-makers were reinforced during subsequent visits he paid to Paris, for example in the summer of 1847, when he spent two days in Marloye's workshop, or again in the fall of 1850, when he saw some beautiful experiments conceived by Foucault performed in Dubosq's workshop.

One of the latter experiments is worth recalling in Thomson's own words: they convey an enthusiasm—to be found somewhere at the intersection between machines, manipulative performance, and physics teaching—that helps capture Thomson's own special place in the history of science and technology:

> A prism and lenses were arranged to throw upon a screen an approximately pure spectrum of a vertical electric arc between charcoal poles of a powerful battery, the lower one of which was hollowed like a cup. When pieces of copper and pieces of zinc were separately thrown into the cup, the spectrum exhibited, in perfectly defined positions, magnificent well-marked bands of different colours characteristic of the two metals. When a piece of brass, compounded of copper and zinc, was put into the cup, the spectrum showed all the bands, each precisely in the place in which it had been seen when one metal or the other had been used separately.[46]

Note that these demonstrations as witnessed by Thomson went on in Paris in the workshop of an instrument maker. Duboscq was probably the one performing. The demonstration had been devised by Foucault, himself an interesting blend of a philosopher and an experimenter, and the young professor of natural philosophy from Glasgow took notes. No surprise that Thomson's later public demonstrations with the mirror galvanometer and the mirror electrometer will show many similarities to this pattern of demonstration.[47]

In a few years, thanks to Thomson's frequent travels, the models he was inspired by included labs in the German states, Switzerland, and Scandinavia, as well as France and, closer to home, the Royal Institution, of course. Thomson's apparatus room in Glasgow was affected accordingly.

I have so far stressed the importance of teaching and demonstrations in the early life of Thomson's laboratory. The pattern adopted also implied employing one or two assistants on a regular basis, as Regnault did in Paris, and a few volunteer students helping with instruments and preparations.

The first of Thomson's assistants was apparently Robert Mansell, whom we find routinely signing the administrative records of Thomson's apparatus room in the 1840s. As an instrument maker, Mansell was good enough to build an "aether thermometer," which Thomson used for experiments leading to his demonstration of the effect of pressure in lowering the freezing point of water.[48] Despite Thomson's honorable mention of Mansell and his thermometer in a publication, by the summer of 1849 the assistant had left.[49]

The little we know of Mansell, however, gives an idea of his competences and social condition and shows that the assistant had ties with the telegraph business before the professor himself had. Writing to Thomson in the autumn of 1849, in the hope of being hired again, Mansell described the other options then open to him. The first was going abroad, considering the slight prospect of finding suitable employment at home. The second was to join the "wire-work" in which one of his former employers, a certain Wilson, was involved in Inverness. Those were the years of the expansion of telegraph lines in Scotland, and during the summer Mansell had found temporary employment with a brother of Wilson's. Many years later a G. D. Wilson advertised that he had made arrangements to get Greenwich time telegraphed to Inverness every day.[50]

Once Mansell had left, Donald Macfarlane's long season in the apparatus room began. Under Thomson's supervision, often exerted from afar, Macfarlane presided over the daily life of the laboratory for about two decades. He first appears in documents in 1850,[51] and was still working for Thomson in the 1870s, when he published under his own name a couple of papers in the *Proceedings of the Royal Society*. He had probably been a student of Thomson's father, a professor of mathematics in the same university. For sure, Thomson relied on Macfarlane in several capacities: as an excellent teaching assistant, supervising and marking students' laboratory work and exams; as a good "computer"; a keen performer of delicate experiments; a reliable proof reader; and a trusted administrator.

With Macfarlane in the apparatus room, from 1850 onward, Thomson could expand its scope: from a facility subservient to the pedagogy of experimental natural philosophy, favoring Thomson's own personal transition from a prevalent interest in physical mathematics to experimental work, to a research facility in tune with Thomson's vast national and international ambitions.

As to Thomson's national ambitions, they were channeled through the British Association, and then the Royal Society. The growing expertise being accumulated in the apparatus room proved an important asset in that direction too. Thomson's acquaintance with James Joule for example—they had first met at the Oxford meeting of the British Association in 1847[52]—led to joint research and experimental work for which the apparatus room, Macfarlane, and the help provided by several of Thomson's students proved crucial. By 1851 the Royal Society of London had granted Thomson and Joule jointly some money for research on "the relation of the molecular actions which formerly were attributed to imponderable fluids, but now are generally considered modes of operation or power." The rationale sustaining the grant was that Thomson's and Joule's joint research might "bring the whole doctrine of affinity within the range of calculation."[53] The research goals set in the statement supporting the grant mentioned Joule's demonstration that "heat has a definite equivalent of mechanical power, and also of electric current," while Thomson had applied the "resources of theory" to Joule's facts. From Thomson, an "experimental verification of his reasoning" was expected; from Joule, an "investigation of the change of volume which takes place in iron on which magnetism is induced, indicating, as it seems to do, a close connection between that energy, and the tensile and compressive forces of the metal."

If the stated goals were overly ambitious, the Royal Society's grant did sustain some important experimental work in Joule's house in Acton Square, Salford[54] and in Thomson's apparatus room. When compared to the money invested by the University of Glasgow and by Thomson personally in his teaching and experimental activities, the Royal Society's money was little more than symbolic; but the grant gave visibility to the work carried out with it, and it contributed tangibly to the esprit de corps of the team working in the apparatus room.

In those days, the main phenomena promising access to "molecular actions" were the variations observed in the currents sent with a voltaic battery through wires of different metals kept at different temperatures or under different conditions of mechanical stress. The key instruments used for such observations were thermometers, galvanometers, and hydromechanical apparatuses of various kinds needed to heat and cool differently various parts of the wires.

A list of thermometers, prepared by Macfarlane in May 1856, mentioned 13 such instruments available in Thomson's apparatus room: 5 produced in London and 8 in Paris.[55] Thermometer readings allowed the compilation of fine tables of quantitative data, the joy of Thomson qua mathematical physicist, and the pride of assistants or students such as a Charles A. Smith, who saw one of his tables of readings published under his name in the *Philosophical Transactions*.[56]

As to galvanometers, one made by Horne, Thornthwaite & Wood in London had entered the apparatus room in June 1848,[57] whereas a Coulomb torsion balance, built by Pixii in Paris, entered the room in spring 1849. Another galvanometer, made in Manchester by J. B. Dancer—the instrument maker working with Joule[58]—was bought in December 1851. From the autumn of that year, Ruhmkorff was providing "electro magnetic" instruments from Paris. James White, the Glasgow instrument-maker with whom Thomson would later establish a joint business, entered the list of suppliers of the apparatus room in April 1854, providing an "ivory reel for galvanometer."

Use of several types of galvanometers was standard practice in the kind of research carried out in the apparatus room in the early 1850s. Joule himself had been involved in improving the galvanometer for many years. By March 1852 he and Thomson were comparing several types of the instrument in view of their joint work for the Royal Society. By that date, as Joule reported, Thomson had improved the tangent galvanometer so that, "by making two circles of unequal force oppose one another," he obtained "the means of measuring powerful currents accurately, without needlessly increasing the size of the galvanometer."[59] Ambitious scientific goals and expediency went hand in hand in the collaboration established between Joule and Thomson, just like in the latter's apparatus room, which Joule visited from time to time.

From the summer of 1855 Thomson's personal acquaintance with Hermann von Helmholtz—they first met in Germany, where Thomson put pressure on Helmholtz to go to the British Association meeting in Glasgow the following September—offered new opportunities for improvement of the techniques associated with the galvanometer, which the German used for his electro-physiological researches. Helmholtz was known for his "two coils" tangent galvanometer. In a paper communicated to the Royal Society in November 1856, "on practical methods for rapid signaling by the electric telegraph," Thomson recommended the adoption of a form of Helmholtz's galvanometer.[60] In December of that same year, Thomson, as we know, was reporting with enthusiasm to Helmholtz about his involvement in the Atlantic cable venture, while taking the opportunity to enquire about a galvanometer produced by Siemens and Halske in Berlin.[61]

Side by side with the instruments, and with the vast network of national and international contacts Thomson maintained, another asset of the apparatus room was the kind of teamwork that had developed there since the professor and his assistant had started offering experimental classes for a fee.

We find teamwork carefully recorded in the manuscripts now preserved in the Glasgow University Library,[62] which Thomson himself selected and reorganized when writing the 101-page-long text of his Royal Society's Bakerian Lecture for 1856. The title of the lecture announced, and the content maintained, an interesting blend of high theory and industry—"On the electro-dynamic qualities of metals"[63]—while the acknowledgements on the first page offered a snapshot of the team that had been working in Thomson's apparatus room over the past several years. Thomson acknowledged "much valuable assistance in the various experimental investigations...from his assistant Mr. McFarlane, and from M. C. A. Smith, Mr. R. Davidson, Mr. F. Maclean, Mr. John Murray, and other pupils in his laboratory." Thomson relied on a similar team for his telegraph work in those same years.

We can get a glimpse of the social relations within the apparatus room through a symbolic episode. In 1863, on offering a present at the end of the academic year to Macfarlane—promoted on the occasion to the rank of "assistant professor of natural philosophy"—24 grateful students signed themselves "The Laboratory Corps of 1862–63."[64] Thomson knew that his students liked the apparatus room as a place for socializing as well as for experimental work.[65] Esprit de corps and competition ran high within the team, and could occasionally create tensions, like when an ambitious student such as the already mentioned Charles A. Smith tried to show that he was possibly a better mathematician than Thomson's assistant.[66]

In tune with the broad theoretical goals pursued under the Royal Society's grant, the Bakerian Lecture experiments on thermoelectric phenomena were used to open the way to a general theory of the relationships between heat and electricity. Most of the experiments reported in Thomson's lecture were aimed at assessing the "convective effect" generated by currents applied to conductors of different metals (often pairs of metals), kept at different temperatures. Thomson called the phenomena collectively "the electrical convection of heat."[67]

Thermometers, galvanometers, and well-coordinated teamwork played a key role in these experiments, as in the following one:

After about an hour and a half, the thermometer at the middle of the conductor indicated 170° Fahr. (76°.7 Cent.); and one of the brass bridges of the commutator was then lifted so as to break the circuit. Immediately the liquid mounted rapidly in each of the three glass tubes of the air-thermometers, and it was prevented from rising above a certain point in the middle one by completing the circuit again. The column of liquid was kept as steady as possible at this point in the middle air-thermometer by a person observing it, and making and breaking the circuit by means of the brass bridge, while two other persons noted the indications of the two lateral thermometers...Mr Joule assisted in this experiment, and was satisfied with the evidence it afforded in favour of the conclusion that *the Resinous Electricity carries heat with it in iron.*[68]

Thomson's interest in telegraphy prompted by Faraday's 1854 publication on retardation phenomena entered the apparatus room that we are describing easily. Similarly, in that same year, telegraph news—or "Telegraphic Intelligence," as it was called—found its way into the daily life of prosperous Glaswegians through the newly inaugurated telegraph terminal in the rooms of the Glasgow Athenaeum, which promised its members "Telegraphic News as it arrives."[69]

Wires of various metals had been tested by Thomson for their thermoelectric properties since 1851, and gutta-percha—the material then used as an insulator for telegraph cables worldwide—was available in the apparatus room in quantities: it was used for the pipes carrying the cooling and the heating liquids needed to keep the metals being tested at different temperatures.

Extensive experience in the apparatus room with testing wires of many kinds and the effects produced on the thermoelectric properties of conductors when using several, joined slips of metal instead of a single conductor,[70] was at the origin—together with Thomson's mathematical views on the best possible section for submarine cables—of his first attempt at a patent in the field of telegraphy in December 1854. Encouraged by Macquorn Rankine, one of his engineering colleagues in Glasgow,[71] during the autumn of that year, Thomson saw the implications that his theoretical interpretation of the problem of retardation, and his experimental work with conductors of various kinds, opened up for a patent in the field. The "invention" claimed in the provisional specification of that first, attempted patent consisted in "providing for each independent electric current in electrical conductors for telegraphic communication a strand, cord, or rope, consisting of several conducting wires in contact with each other," thus increasing the sectional area while improving flexibility.[72]

Thomson's 1854 provisional specification "did not proceed to the Great Seal." The obstacle was an earlier patent by Henry Vernon Physick, a London civil engineer with whom for a while Rankine and Thomson considered establishing contact with the aim of buying his patent.[73] Physick had proposed a multiwire cable, similar to the one suggested by Thomson, for the purpose of avoiding breaks, while making the different wires recognizable from their position and from the colors of the insulating materials. Physick did not mention retardation phenomena, nor their theory, as a motive behind his intended improvement.

Frustration with the 1854 attempted patent did not deflect Thomson from his plan to enter the business of telegraphy with the expertise accumulated through both his theoretical work and the practice, and teamwork provided by the apparatus room.

The same network of colleagues, correspondents, and friends on whom Thomson had relied to circulate and discuss ideas about thermoelectricity was easily adjusted to collecting information about retardation phenomena in telegraph lines. Pressed by Thomson, from February 1855 Joule was enquiring informally about the retardation phenomena observed along the lines between Manchester and London.[74] Joule also ordered from his suppliers samples of wires intended for Thomson's apparatus room.[75]

Galvanometers, we know, were already in the apparatus room in numbers when the season of systematic testing of telegraph wires began. We get a glimpse of the early typical testing of copper and telegraph wires in the apparatus room from a letter sent by Thomson—who was as usual on the move, on a journey to London to consolidate his network of scientific and recently added business contacts—to his assistant in Glasgow. The letter provides a glance into the combination of experimental setup, measuring techniques with the galvanometer, teamwork, and "rule of thumb" procedures in Thomson's early work on telegraphy:

Moffat, May 8, 1857

Dear Sir,

It is not unlikely you will get a set of samples of single copper wire from London / *added:* such as the strand for cable to [*be*] spun out of / to test, each sample 12 y.[ds] long.—If so, cut 144 inches (12 feet) from each to test for resistance and make the tested length...as nearly as you can to 142 6/7 inches (being 1/7 of 1000 inches). Then, taking one of M. Morrison's 1000 inch samples, if possible one of those referred to as ½ I in his book, find the resistance of each of the new samples in terms of it. Vary, as a check, by finding the resistance of each of the new samples in terms of one of themselves. Remember never to have more than two / *added:* of the 1 samples / in the circuit of the battery at once, & use one good cell (one of the large ones, with simple zinc will do perfectly) of Daniell.
You remember the plan of the experiment.

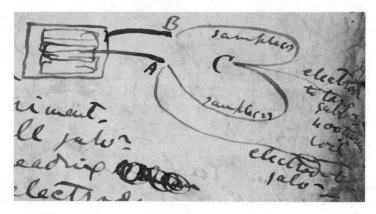

Figure 13.1 William Thompson's sketch of an experimental set up used to test samples of metal wires, 1857. By permission of University of Glasgow Library, Special Collections.

[*Here Thomson added the sketch reproduced above, showing the battery, with poles B and A, on the left, and an indication, on the right, that the electrodes leading to the tangent galvanometer—one with a 400 turn coil, not represented—should be connected in C and A*]

Take a full galv.[r] quadruple reading with the electrodes as shown. Then shift one electrode from A to B, leaving the other at C, & take full reading.

If mean defl.[n] [=*deflection*] in the case illustrated in diagram be d_2, & with the electrode transferred to A d_1, then

resist. of sample (1) / resist...(2) = tan d_1 / tan d_2

For this formula to be rigorous, the resistance in the cell & electrodes from it to A & B must be very small compared with that of the samples tested: but since in the present case the resistance of the galv.[r] coil is very great this condition might be considerably violated without introducing sensible error.

[...]

Send me results as soon as possible (giving me the nat. tang.[r] of the mean deflect.[ons]). If you post any day before 3.15 I should receive the same evening & despatch by night mail in time to reach London on the following forenoon.

Mr Morrison, or Mr Murray,[76] or any one competent who may [be] about, might give you assistance. Attend to the labels on the samples: & if you have M. Morrison book, I would like a copy of his last determinations sent along for comparison.

Yours truly
William Thomson[77]

Apart perhaps from the hurry, all the ingredients mentioned in this letter had been available in the apparatus room for several years.

As Crosbie Smith and Norton Wise have shown, Thomson's decision to take advantage of the telegraph frenzy of his days was rooted in circumstances, people, and interests linked to Glasgow.[78] Yet, if we watch closely at what was going on in the apparatus room and around it, it seems appropriate to emphasize also that the expertise that Thomson was able to throw into his new venture was molded by the competitive imitation game that he had been playing at the local, national, and international levels since 1846 as an ambitious professor of natural philosophy.

The web of knowing, doing, and patenting

In March 1852, we know, Joule reported that Thomson had introduced into his galvanometers a practical new feature, which enabled the instrument to be used for measuring both powerful and weak currents. The versatility of the new instrument was in tune with Thomson's mobility across teaching, research and, from 1854, industrial ventures.

We also know that Thomson's first attempt at entering the telegraph business with a patent, focusing on a cable combining several small-section copper wires, was linked to his mathematical model for submarine cables, connecting retardation phenomena with the cable's lateral dimension and length. There was no single or easy avenue, however, from a mathematical theory of cables to successful solutions of the problem of retardation. Failure to go on with the 1854 patent must have encouraged Thomson to try different paths, one of which focused on the galvanometer. By the end of 1856, as we have seen, Thomson was comparing different types of galvanometers developed by Helmholtz for electrophysiological research, and by Siemens and Halske for telegraphy.[79] In November he was still recommending the use of Helmholtz's galvanometer for telegraphy, "with or without modification."[80] Finally, sometime during 1857 or early 1858, in close interaction with James White—who would become Thomson's partner in several industrial ventures in subsequent years[81]—and with his assistant Macfarlane, he developed a galvanometer appropriate for rapid signaling with the smallest possible currents, which his conception of the "electric pulse" and of Fourier's harmonics recommended for long-distance, submarine telegraph operations.

In Thomson's mirror galvanometer two very light magnets replaced the needle used in other galvanometers. The magnets were attached to a silvered piece of microscope glass—this being White's suggestion—acting as a mirror, suspended to a fine platinum wire, later replaced by "a stout bundle of twenty or thirty silk fibres." By projecting the ray of a lamp on to the mirror, the deviations caused to the magnets by an electric current in a wire were observed, magnified, upon a scale, "the beam of light serving as a weightless index of exquisite sensitiveness."[82] Details of a couple of mirror galvanometers with such characteristics were included in Thomson's patent of 1858, together with a long list of other "inventions" that he had developed in the apparatus room.[83]

There is some irony in the fact that, although the Atlantic Cable Company had hired Thomson chiefly for his mathematical skills and academic prestige, during the 1858 expedition he was able to help mainly through the mirror galvanometer, which he had developed in interaction with the varied interests and the mixed population we have seen.

Amidst the long list of knowledge, management, money, personality, and public-relations issues that contributed to making the success and then the failure of the 1858 Atlantic cable an amazing story, and an endless source of controversy among experts and in the press, Thomson's mirror galvanometer was the clear winner. It is not a minor part of the story that the winner was partly unexpected by its own inventor who, until recently, had trusted his mathematics and his theory of electricity more than his instruments.

A hint of the shift in the hierarchy of ascribed values produced by these developments—as far as Thomson could perceive it in the middle of action—was conveyed in comments such as the one he made in writing from the telegraph station in Valencia to fellow experimenter Joule:

> "Telegraphic work,...when it has to be done through 2400 miles of submarine wire, and when its effects are instantaneous exchange of ideas between the old and new worlds, possesses a combination of physical and (in the original sense of the word) *metaphysical* interest, which I have never found in any other scientific pursuit."[84]

Concerning the *uncertainties* affecting "telegraphic work," they had already been clear enough to Thomson in November 1856. Then, announcing to the Royal Society "various practical applications of formulae" belonging to the "theory of the electric telegraph," he had ended his paper recalling how, in a short, air telegraph line even "a trickling of water along a spider's web" could, in the middle of a message, suddenly "throw all the indications into confusion."[85]

Thomson's perception of the uncertainties associated with telegraphic work increased, of course, after his direct experience on board the ships of the Atlantic expeditions of 1857 and 1858.[86] Thomson's 1858 patent, now remembered mainly for the marine mirror galvanometer, listed 21 different apparatuses or "parts" of the "invention." The 1858 patent was indeed like a net knit by Thomson with the experience he had accumulated in the apparatus room, within the board of directors of the Atlantic Cable Company, and on the ships: a net aimed at capturing whatever would turn out successful during the Atlantic experiment. Apparently, net throwing and taming uncertainty characterized Thomson's patenting strategies just like his knowing and doing as a natural philosopher.

Conclusion

Like the mirror galvanometer, I suggest, the age of electricity emerged out of a web of knowing, doing, and patenting such as the one sketched above: out of people, interests, goals, and instruments like those we have met within and around Thomson's apparatus room. Given the variety of humans and artifacts involved, and the distinctly porous walls of such places, it should come as no surprise that the road that led from the battery to telegraphy, and then to electric lighting and the "age of electricity," turned out so tortuous and slow. It is also not surprising that there was no single avenue leading from a mathematical theory of submarine cables to successful signaling across the Atlantic. Nor is it surprising, more generally, that the process was full of unexpected developments, and that it took some 80 years to unfold despite the auspicious, complacent context provided by the age of progress and by the market that telegraphy had created for field theory.[87]

With hindsight, it is revealing that Thomson's key asset in the 1858 Atlantic challenge was an instrument that, by its very name—the galvanometer—evoked the same fuzzy borders between different fields, between knowing and doing, and between the life sciences and the measuring of weak electricity, which had favored the introduction of the Voltaic battery some 60 years earlier.

Moreover, there is irony in the fact that, although the Atlantic Telegraph Company had hired Thomson mainly because of his prestige as an academic and his excellence as a mathematician, in the 1858 campaign he was able to help instead through an instrument the company had refused to pay, and that he had developed in interaction with the diverse goals and the mixed population we have encountered in the apparatus room.

To expose the deceptive notion of the "simplicity of nature," Georg Christoph Lichtenberg—a most perceptive eighteenth-century "electrician," and a key figure in several of John Heilbron's works on the history of electricity—wrote ironically that "the lofty simplicity of nature all too often rests on the plain simplicity of the one who thinks he sees it."[88] I suggest that we apply Lichtenberg's maxim also to the history of science and technology: the lofty simplicity of the history of science and technology all too often rests on the simplicity of those who think they see it. Adopting Lichtenberg's maxim as a caveat is especially convenient when trying to develop a new interpretive framework for the early age of electricity: that magmatic, unpredictable "interlude" that was nonetheless—or because of that?—a particularly creative period.

Notes

I wish to thank Mario Biagioli, Jessica Riskin, and an anonymous referee for valuable comments and criticism on earlier versions of this paper.

1. The classic study in this connection is John Heilbron, *Electricity in the 17th and 18th Centuries. A Study of Early Modern Physics* (Berkeley: University of California Press, 1979; new edition: Mineola, New York: Dover Publications, 1999).
2. The first book in English bearing that title was Park Benjamin, *The Age of Electricity. From Amber-Soul to Telephone* (New York: Charles Scribner's, 1886).
3. The classic study is Thomas P. Hughes, *Networks of Power. Electrification in Western Society, 1880–1930* (Baltimore: The Johns Hopkins University Press, 1983).
4. Literature includes books such as: Christine Blondel, *A. M. Ampère et la création de l'électrodynamique (1820–1827)* (Paris: Bibliothèque Nationale, 1982); *Faraday*

Rediscovered. Essays on the Life and Work of Michael Faraday, ed. David Gooding and Frank A. J. L. James (Houndmills: Macmillan, 1985); Crosbie Smith and M. Norton Wise, *Energy and Empire. A Biographical Study of Lord Kelvin* (Cambridge: Cambridge University Press, 1989); *The Uses of Experiment*, ed. David Gooding, Trevor Pinch, and Simon Schaffer (Cambridge: Cambridge university Press, 1989); Geoffrey Cantor, *Michael Faraday: Sandemanian and Scientist* (Houndmills and London: Macmillan, 1991); *Invisible Connections. Instruments, Institutions, and Science*, ed. Robert Bud, Susan E. Cozzens, and Roy F. Potter (Bellingham, Wash.: SPIE Optical Engineering Press, 1992): James R. Hofmann, *André-Marie Ampère* (Cambridge: Cambridge University Press, 1995); Iwan R. Morus, *Frankenstein's Children: Electricity, Exhibition, and Experiment in Early-Nineteenth-Century London* (Princeton: Princeton University Press, 1998); Olivier Darrigol, *Electrodynamics from Ampère to Einstein* (Oxford: Oxford University Press, 2000); *From Natural Philosophy to the Sciences. Writing the History of Nineteenth-Century Science*, ed. David Cahan (Chicago and London: The University of Chicago Press, 2003); Graeme Gooday, *Domesticating Electricity: Technology, Uncertainty and Gender, 1880–1914* (London: Pickering and Chatto, 2008); and papers like Christine Blondel, "Electrical Instruments in 19th Century France, between Makers and Users," *History and Technology* 13 (1997). 157–182 and Bruce J. Hunt, "Doing Science in a Global Empire: Cable Telegraphy and Electrical Physics in Victorian Britain," in *Victorian Science in Context*, ed. Bernard Lightman (Chicago and London: University of Chicago Press, 1997), 312–333.
5. Heilbron, *Electricity.*
6. Giuliano Pancaldi, *Volta. Science and Culture in the Age of Enlightenment* (Princeton: Princeton University Press, 2003; paperback 2005), esp. Chapter 6.
7. Simon Schaffer, "Exactly like a Stingray," *London Review of Books* 26, no. 11 (June 3, 2004): 28–29.
8. In the 1850s Thomson gave as the postal address of his assistant (more on him below): "Apparatus Room, College" (Glasgow): William Thomson to James David Forbes, May 27, 1856, St Andrews University Library, Papers of James David Forbes, Ms Deposit 7.
9. James Clerk Maxwell, *A Treatise on Electricity and Magnetism*, 2 vols. (Oxford: Clarendon Press, 1873), vol. 1, x.
10. Book-length studies on Thomson include: Silvanus P. Thompson, *The Life of William Thomson*, 2 vols. (London: Macmillan, 1910); Smith and Wise, *Energy and Empire*— the standard biography—and David Lindley, *Degrees Kelvin* (Washington, DC: Joseph Henry Press, 2004). On Thomson's laboratory the classic study is Crosbie Smith, "'Nowhere but in a Great Town': William Thomson's Spiral of Classroom Credibility," in Crosbie Smith and Jon Agar, eds., *Making Space for Science. Territorial Themes in the Shaping of Knowledge* (Houndmills: Macmillan Press and St. Martin Press, 1998), 118–146.
11. On the latter: Giuliano Pancaldi, "Interpreting the Early Age of Electricity," *2007 IEEE Conference on the History of Electric Power* (Newark, NJ, 2008), 212–221.
12. Literature on Thomson's theoretical work on electricity and magnetism includes: Jed Z. Buchwald, "William Thomson and the Mathematization of Faraday's Electrostatics," *Historical Studies in the Physical Sciences* 8 (1977): 101–136; M. Norton Wise, "The Flow Analogy to Electricity and Magnetism: William Thomson's Reformulation of Action at a Distance," *Archive for History of Exact Sciences* 25 (1981): 19–70; Ole Knudsen, "Mathematics and Physical Reality in William Thomson's Electromagnetic Theory," in P. M. Harman, ed., *Wranglers and Physicists. Studies in Cambridge Physics in the Nineteenth Century* (Manchester: Manchester University Press, 1985), 149–179; Smith and Wise, *Energy and Empire*, Chapters 7, 8, and 13; Darrigol, *Electrodynamics from Ampère to Einstein*, Chapter 3.

13. Bern Dibner, *The Atlantic Cable* (Norwalk, Conn.: Burndy Library, 1959).
14. Andrew Warwick, *Masters of Theory: Cambridge and the Rise of Mathematical Physics* (Chicago: University of Chicago Press, 2003), 37.
15. For Thomson's conception of "physical mathematics" in the period under review, see William Thomson, *Reprint of Papers on Electrostatics and Magnetism*, 2nd ed. (London: Macmillan, 1884, 2; 2n).
16. Thomson, in Smith and Wise, *Energy and Empire*, 215.
17. Thompson, *The Life of William Thomson*, 54.
18. On Regnault: Matthias Dörries, "Easy Transit: Crossing Boundaries between Physics and Chemistry in Mid-19th Century France," in Smith and Agar, *Making space for science*, 246–262. On Liouville: Jesper Lützen, *Joseph Liouville, 1809–1882, Master of Pure and Applied Mathematics* (New York: Springer-Verlag, 1990).
19. Thompson, *The Life of William Thomson*, 127.
20. George Green, *On the Application of Mathematical Analysis to the Theories of Electricity and Magnetism* (Nottingham: Wheelhouse, printed for the Author, 1828). On Green: I. Grattan-Guinness, "Green, George (1793–1841)," *Oxford Dictionary of National Biography* (Oxford: Oxford University Press, 2004), http://www.oxforddnb.com/view/article/11381. On Thomson as a mediator: Darrigol, *Electrodynamics from Ampère to Einstein*, 130–131, 136.
21. See: Bruce J. Hunt, "Michael Faraday, Cable Telegraphy and the Rise of Field Theory," *History of Technology* 13 (1991): 1–19, on p. 7.
22. Michael Faraday in *Proceedings of the Royal Institution*, January 20, 1854, and in the *Philosophical Magazine*, June 1854, as reprinted in Michael Faraday, *Experimental Researches in Electricity* (London: Taylor and Francis, 1855), vol. 3, 508–523. On the history of telegraphy: Jeffrey Kieve, *The Electric Telegraph. A Social and Economic History* (Newton Abbot: David & Charles, 1973) and Ken Beauchamp, *A History of Telegraphy* (London: The Institution of Electrical Engineers, 2001).
23. Hunt, "Michael Faraday, Cable Telegraphy and the Rise of Field Theory," 7.
24. Thomson to Stokes, October 28, 1854, in David B. Wilson, ed., *The Correspondence between Sir George Gabriel Stokes and Sir William Thomson*, 2 vols. (Cambridge: Cambridge University Press, 1990), vol. 1, 172.
25. William Thomson, "On the Theory of the Electric Telegraph," *Proceedings of the Royal Society of London* 7 (1854–1855): 382–399 (read May 24, 1855).
26. Maxwell to Thomson, February 20, 1854; November 13, 1854; May 15, 1855; and September 13, 1855: in *The Scientific Letters and Papers of James Clerk Maxwell*, ed. P. M. Harman, 2 vols. (Cambridge: Cambridge University Press, 1990), vol. 1, 237–238, 254–263, 305–307, 319–323.
27. William Thomson, "Royal Institution Friday Evening Lecture" (May 18, 1860), reprinted in William Thomson, *Reprint of Papers on Electrostatics and Magnetism* (London: Macmillan, 1884), 208–225, 224–225.
28. Bruce J. Hunt, "Scientists, Engineers and Wildman Whitehouse: Measurement and Credibility in Early Cable Telegraphy," *British Journal for the History of Science* 29 (1996): 155–169.
29. William Thomson, "On Practical Methods for Rapid Signalling by the Electric Telegraph"(received November 14, 1856) and "Second Communication" of the same (received December 11, 1856), *Proceedings of the Royal Society of London* 8 (1856–7): 299–303 and 303–307.
30. Thompson, *The Life of William Thomson*, 339.
31. Ibid., 336–337.
32. Thomson to James Thomson, April 19, 1858, ibid., 352.
33. Thomson to Helmholtz, December 30, 1856, ibid., 336.
34. Thomson to James Thomson, April 19, 1858, ibid., 352.

35. From Thomson's speech in Glasgow as reported in: "The Atlantic Telegraph," *The Engineer*, January 28, 1859.
36. Thompson, *The Life of William Thomson*, 355.
37. Thomson to F. -N. -M. Moigno, October 23, 1858, in ibid., 381.
38. See Thomson to Forbes, April 24, 1858, St Andrews University Library, Papers of James David Forbes, Ms Deposit 7, where Thomson writes that, in view of the Atlantic expedition, he is preparing "instruments, in the first place at my own expense." Three days earlier the Atlantic Cable Company had refused a grant of £2000 that Thomson had requested to cover the cost of constructing new instruments. A subsequent request for a contribution of £500 was passed by the board by a bare majority (Thompson, *The Life of William Thomson*, 353–354).
39. The Patent Office, Patent N° 329, 1858. The dates on the "Letters Patent" were February 20 and April 23, 1858. The "Specification in pursuance of the conditions of the Letters Patent" was added by Thomson on August 19, 1858, that is, after the completion of the second, temporarily successful expedition carried out during the summer, and barely twelve days before the cable stopped sending signals. On the 1871 renewal of the 1858 patent: Thompson, *The Life of William Thomson*, 619–621.
40. On the notion of competitive imitation: Pancaldi, *Volta*, 207–210.
41. James Coutts, *A History of the University of Glasgow* (Glasgow: Maclehose, 1909), 385–386; Jack B. Morrell, "Thomas Thomson: Professor of Chemistry and University Reformer," *British Journal for the History of Science* 4 (1969): 245–265.
42. Thomson to Forbes, November 22, 1846, St Andrews University Library, Papers of James David Forbes, Ms Deposit 7.
43. Thompson, *The Life of William Thomson*, 116.
44. Thompson, *The Life of William Thomson*, 117.
45. As listed in the typescript "Department of Natural Philosophy, Glasgow University (Sir William Thomson Collection)," by The Secretary, National Register of Archives (Scotland), Edinburgh, 1992, National Archives of Scotland.
46. William Thomson, *Popular Lectures and Addresses*, 3 vols. (London: Macmillan, 1894), vol. 2, 172–173.
47. See for example Thomson's Friday Evening Lecture at the Royal Institution, on May 18, 1860, "On Atmospheric Electricity," reproduced in W. Thomson, *Reprint of Papers on Electrostatics and Magnetism* (London: Macmillan, 1872), 208–226, on p. 221.
48. William Thomson, *Mathematical and Physical Papers*, 6 vols. (Cambridge, Cambridge University Press, 1882–1911), vol. 1, 165–169.
49. Mansell to Thomson, October 13, 1849, Cambridge University Library, Kelvin Papers, M74.
50. *Inverness Advertiser*, Friday, June 6, 1873. On telegraph lines in Scotland in the 1840s: William Fothergill Cooke, *Telegraphic Railways...with Particular Reference to Railway Communication with Scotland and to Irish Railways*, London, 1842.
51. William Thomson, Notebook 34 , Cambridge University Library, Kelvin Papers, on p. 159 (September 6, 1850). On Macfarlane: David Murray, *Memories of the Old College of Glasgow* (Glasgow: Jackson, 1927), 140–141.
52. Thompson, *The Life of William Thomson*, 205.
53. See Earl of Rosse, "Address" (November 30, 1852), in *Abstracts of the Papers Communicated to the Royal Society of London* (1850–1854), vol. 6, 233–266, on pp. 235–236.
54. On Joule's house laboratory, Donald S. L. Cardwell, *James Joule* (Manchester and New York: Manchester University Press, 1989), 90. On Joule's experimental practice: H. Otto Sibum, "An Old Hand in a New System," in *Invisible Industrialist:*

Manufactures and the Production of Scientific Knowledge, ed. Jean-Paul Gaudillière and Ilana Löwy (Houndsmill: Macmillan, 1998), 23–57 and H. Otto Sibum, "Narrating by Numbers: Keeping an Account of Early 19th-Century Laboratory Experiences," in *Reworking the Bench: Research Notebooks in the History of Science. Archimedes 7* (Dordrecht and Boston: Kluwer, 2003), 141—158.

55. Macfarlane to Thomson, May 16, 1856, Glasgow University Library, Kelvin Papers, Mc10.

56. William Thomson, "On the Electro-dynamic Qualities of Metals" (The Bakerian Lecture), *Philosophical Transactions of the Royal Society of London* 146 (1856), 706. For a list of the observers in the apparatus room and the dates of their observations: ibid., 705.

57. This and the following information are taken from the typescript "Department of Natural Philosophy, Glasgow University (Sir William Thomson Collection)", note 45, above.

58. On Joule and Dancer: H. Otto Sibum, "Shifting Scales. Microstudies in Early Victorian Britain," Paper held at the conference *Varieties of Scientific Experience*, Berlin, 1997, Max Planck Institute for the History of Science, Preprint No. 171.

59. See James P. Joule, "On the Heat Disengaged by Chemical Combinations" (letter to the *Philosophical Magazine*, dated March 30, 1852), reproduced in James P. Joule, *Scientific Papers* (London, Taylor and Francis, 1884), vol. 1, 205–207.

60. Thompson, *The Life of William Thomson*, 332–333.

61. Ibid., 335–337.

62. The bulk of the manuscripts describing work going on in the apparatus room in the period examined here—including many reports of experiments, mostly by Macfarlane—is kept in the Glasgow University Library, Kelvin Papers, T 202. For an accurate list of "bills , accounts and correspondence and papers concerning supply and repair of laboratory fittings and apparatus," see the typescript "Department of Natural Philosophy, Glasgow University (Sir William Thomson Collection)", note 45, above.

63. Thomson, "On the Electro-dynamic Qualities of Metals," Part III, 649–751, 649n.

64. Glasgow University Library, Kelvin Papers, Mc20.

65. William Thomson, "Bangor Address," in William Thomson, *Popular Lectures*, 3 vols. (London and New York: Macmillan, 1894), vol. 2, 488.

66. C. A. Smith to William Thomson, June 28, 1854, Glasgow University Library, Kelvin Papers, S 60.

67. Thomson, "On the Electro-dynamic Qualities of Metals," 674, 693.

68. Ibid., 692–693.

69. An ad in the *Glasgow Herald* for November 24, 1854 read: "'Telegraph News.' The proprietor of the Glasgow Athenaeum…has much pleasure in announcing to the Public that he has arranged with the Magnetic Telegraph Company, to supply the Room with the Telegraphic News as it arrives, on and after 1st December 1854…Athenaeum Reading Rooms, with Telegraphic Intelligence, Library & c. One Guinea a Year."

70. See: Thomson, "On the Electro-dynamic Qualities of Metals," 665; the experiments reported on "multiple sheet" conductors are dated November 1853.

71. On Rankine: Ben Marsden, "Rankine, (William John) Macquorn (1820–1872)," *Oxford Dictionary of National Biography* (Oxford: Oxford University Press, Sept 2004); online edn., May 2008, http://www.oxforddnb.com/view/article/23133.

72. The Patent Office, Patent 2547, 1854: William Thomson, William John Rankine, and John Thomson, "Provisional Specification." On the intellectual property regime that accompanied the early age of electricity: Mario Biagioli, "Patent Specification and Political Representation: How Patents became Rights," in M. Biagioli, P. Jaszi, M. Woodmansee, eds., *Making and Unmaking Intellectual Property* (Chicago and

London: University of Chicago Press, 2011), 25–39. On inventors and patents in Britain: Christine MacLeod, *Heroes of Invention: Technology, Liberalism and British Identity, 1750–1914* (Cambridge: Cambridge University Press, 2010). See also Brad Sherman and Lionel Bently, *The Making of Modern Intellectual Property Law: The British Experience, 1760–1911* (Cambridge: Cambridge University Press, 1999).

73. Rankine to Thomson, June 19, 1855, Glasgow University Library, Kelvin Papers, R1. On Physick's patent: *The Repertory of Patent Inventions* 25 (1855): 386–388.

74. Joule to Thomson, February 22, 1855, etc., Cambridge University Library, Kelvin Papers, Add. MS 7342, J192, J193, J194.

75. Joule to Thomson, January 1, 1855, Cambridge University Library, Kelvin Papers, Add. MS 7342, J189. On the history of wires and cables: Robert M. Black, *The History of Electric Wires and Cables* (London: Peter Peregrinus in association with the Science Museum, 1983).

76. A student named John Murray was mentioned in the Bakerian Lecture (note 63, above).

77. Thomson to Macfarlane, May 8, 1857, Glasgow University Library, Kelvin Papers, Mc12. By permission of University of Glasgow Library, Special Collections.

78. See Smith and Wise, *Energy and Empire*, Chapters 2 and 5; and Smith, "Nowhere but in a Great Town" (note 10, above).

79. Thompson, *The Life of William Thomson*, 335.

80. William Thomson, "On Practical Methods for Rapid Signaling by the Electric Telegraph," *Proceedings of the Royal Society of London* 8 (1856–1857): 299–303, on p. 301.

81. On White: T. N. Clarke, A. D. Morrison-Low, and A. D. C. Simpson, *Brass & Glass. Scientific Instrument Making Workshops in Scotland* (Scotland: National Museum of Scotland, 1989), 252–275.

82. Thompson, *The Life of William Thomson*, 348. On the development of Thomson's mirror galvanometer, ibid., 335–348.

83. William Thomson, The Patent Office, Patent N° 329, 1858 (2nd edition), 1-36, on p. 23, Part I, figs. 5–6.

84. Thomson to Joule, September 25, 1858, quoted in Thompson, *The Life of William Thomson*, 378.

85. Thomson, "On practical methods...(Second communication)," 303–307, on p. 307.

86. On Thomson onboard the ships: Dibner, *The Atlantic Cable*, 19–45.

87. On telegraphy providing "a *market* for field theory" see Hunt, "Michael Faraday, Cable Telegraphy and the Rise of Field Theory," 15.

88. Georg Christoph Lichtenberg, quoted in John Heilbron, "Wit and Wisdom," *Nature* 438 (November 3, 2005): 29.

Contributors

Ken Alder is the Milton H. Wilson Professor in the Humanities and professor of History at Northwestern University, where he directs the Science in Human Culture Program. He has written three books of history—*Engineering the Revolution* (1997, 2010), *The Measure of All Things* (2002), and *The Lie Detectors* (2007)—plus one novel, *The White Bus* (1987). Despite the argument of his article in this volume, he has nothing against international junkets.

Jim Bennett has just retired from the directorship of the Museum of the History of Science in the University of Oxford, where he was also professor of History of Science. He is the author of *The Divided Circle* (1987) and *Church, State and Astronomy in Ireland* (1990); the coauthor of *Sphaera Mundi* (1994) with D. Bertoloni Meli; *The Garden, the Ark, the Tower, the Temple* (1998) with S. Mandelbrote; and *London's Leonardo* (2003) with M. Cooper, M. Hunter, and L. Jardine; and the coeditor of *The Oxford Companion to the History of Modern Science* (2003) with J. L. Heilbron, editor in chief, F. L. Holmes, R. Laudan, and G. Pancaldi.

Mario Biagioli is Distinguished Professor in the STS Program and School of Law at University of California at Davis, where he directs the Center for Science and Innovation Studies. He is the author of *Galileo Courtier* (1993) and *Galileo's Instruments of Credit* (2006); the editor of the *Science Studies Reader* (1998); and the coeditor of *Scientific Authorship* (2003) with Peter Galison and *Making and Unmaking Intellectual Property* (2011) with Peter Jaszi and Martha Woodmansee. He is currently completing a book on plagiarism in science.

Hasok Chang is the Hans Rausing Professor of History and Philosophy of Science at the University of Cambridge. His research and teaching focus on the history and philosophy of chemistry and physics since the eighteenth century, and the general philosophy of science. He is the author of *Inventing Temperature* (2004) and *Is Water H_2O?* (2012).

Paula Findlen is the Ubaldo Pierotti Professor of Italian History at Stanford University where she has codirected the Center for Medieval and Early Modern Studies, the Program in the History and Philosophy of Science, and the Science Technology and Society Program. Her publications include *Possessing Nature: Museums, Collecting and Scientific Culture in Early Modern Italy* (1994); *Merchants and Marvels: Commerce, Science, and Art in Early Modern Europe*, coedited with Pamela Smith (2002); *Athanasius Kircher: The Last Man Who Knew Everything* (2004); and other books and essays on science, society, culture, and collecting in early modern Italy. She has recently been writing about the history of the Uffizi.

Tal Golan is associate professor in the Department of History, University of California, San Diego. He specializes in the history of science in the eighteenth and nineteenth centuries, and in the relations among science, technology, and law. He is now working on two projects: a history of statistical evidence and the relations between Zionism and science.

Michael D. Gordin is professor of History at Princeton University, where he teaches the History of Science and Russian History. He is the author of four books: *A Well-Ordered Thing: Dmitrii Mendeleev and the Shadow of the Periodic Table* (2004), *Five Days in August: How World War II Became a Nuclear War* (2007), *Red Cloud at Dawn: Truman, Stalin, and the End of the Atomic Monopoly* (2009), and *The Pseudoscience Wars: Immanuel Velikovsky and the Birth of the Modern Fringe* (2012).

Anthony Grafton teaches History and the History of Science at Princeton University. His work focuses on the history of scholarly and scientific practices and the history of books and readers in early modern Europe. Among his books are *Worlds Made by Words* (2009) with Joanna Weinberg; *"I Have Always Loved the Holy Tongue": Isaac Casaubon, the Jews, and a Forgotten Chapter in Renaissance Scholarship* (2011); and *The Culture of Correction in Renaissance Europe* (2011).

Matthew L. Jones teaches at Columbia University. He is the author of *The Good Life in the Scientific Revolution* (2006). He is currently completing a book on calculating machines from Pascal to Babbage, and, with the support of the Guggenheim and Mellon Foundations, he has just begun a study of data mining from 1963 to the present.

Daniel J. Kevles, the Stanley Woodward Professor of History at Yale University, has long written about issues in science, technology, and society, past and present. His works include *The Physicists: The History of a Scientific Community in Modern America; In the Name of Eugenics: Genetics and the Uses of Human Heredity;* and *The Baltimore Case: A Trial of Politics, Science, and Character.* He is currently writing a book on the history of innovation and intellectual property protection in plants, animals, and people.

Giuliano Pancaldi is professor of the History of Science at the University of Bologna. His books include *Darwin in Italy: Science across Cultural Frontiers* (1991) and *Volta: Science and Culture in the Age of Enlightenment* (2003).

Dominique Pestre is a social and political historian of nineteenth- and twentieth-century science, but he has written on more philosophical aspects and contemporary topics as well. He has coedited several volumes, notably *Science in the Twentieth Century* with John Krige, and *Dictionnaire culturel des sciences* with N. Wittkowski and J. M. Levy-Leblond. He has published *Physique et physiciens en France, 1918–1940; Louis Néel, le magnétisme et Grenoble;* and *History of CERN*, collaboratively, in three volumes. His most recent books are

Heinrich Hertz, L'administration de la preuve (2002) with Michel Atten; *Science, Argent et Politique* (2003); *Les Sciences pour la guerre, 1940–1960* (2004) with Amy Dahan; and *Introduction aux Science Studies* (2006).

Jessica Riskin teaches History at Stanford University. She is the author of *Science in the Age of Sensibility: The Sentimental Empiricists of the French Enlightenment* (2002) and the editor of *Genesis Redux: Essays in the History and Philosophy of Artificial Life* (2007). She is finishing a book about the history of mechanist-scientific accounts of life and mind entitled *The Restless Clock*.

Index

absolutism, 127, 132–4
Académie des sciences, 7–8, 22, 29, 125–6, 127, 136–9
accelerator physics, 151
ACLU, *see* American Civil Liberties Union (ACLU)
Adams, George, 44
admissability, 175–8
Ad Vitellionem paralipomena (Kepler), 109, 112–13
Aelian
  Varia historia, 198
aether thermometer, 271–2
Agar, John, 96, 97
Agent Orange, 168, 170–3
AIDS/HIV, 155–6
Alexander VII (Fabio Chigi), 209, 210, 214
Alexander the Great, 196
Allacci, Leone, 208, 209, 210
  Apes Urbanae, 206–7, 209
allegiances to nation-states, 22–3, 30
Allen v. United States, 168–70, 173
Almagest (Ptolemy), 195, 197
al-Suyuti, Jalaluddin, 193
Altemps, Marco Sittico, 238
American Civil Liberties Union (ACLU), 157–9
American College of Medical Genetics, 158, 160
animals
  animal-machine idea, 12, 229, 230, 233, 239–40
  automata, 229, 231, 235, 236, 240
anions, 48
Annales ecclesiastici (Baronius), 191
Annius of Viterbo, 191–2, 193
anti-aircraft warfare, 7, 84, 87–8
anti-submarine warfare, 7, 84, 88–9, 90
Apes Urbanae (Allacci), 206–7, 209
Apianus, Petrus, 10–11, 198, 199, 252
  Astronomicum Caesareum, 196–7
  Cosmographia, 261, 262

Instrument Buch, 255, *256,* 258
apparatus room, University of Glasgow, 13, 264, 268–78
Aquinas, Thomas, 233, 240n1
Archive Meter, 28, 31–2, 35
artisans
  and intellectual property, 126–7, 130–1
  and invention, 8, 130–1, 136
  payment of, 138
Artis historicae penus (Wolf), 187
asbestos, 168
Association for Molecular Pathology, 158
Astronomia nova (Kepler), 115, 116
Astronomicum Caesareum (Apianus), 196
astronomy
  and calculating machines, 126
  and chronology, 11, 187–8, 196–9
  heliocentrism, 205, 209, 210, 213, 214, 215
  and philosophy, 110–13, 116
  and sundials, 12, 249–52
  witnessing in, 8, 103–18
  see also cosmography
As You Like It (Shakespeare), 251
Athenaeum, 267
Atlantic Telegraph Company, 13, 264, 265, 267–9, 278, 279, 280
Atlas (Mercator), 190
Atomic Energy Commission (AEC), 150
atomism, 6, 65–6, 69–71, 72, 74, 75
Augustus, 198
automata, 229–40
  animals, 229, 231, 235, 236, 240
  animated paintings, 245n49
  and clocks, 229, 232, 245n49
  and humor, 232, 233, 239
  and organs, 229, 232
  and the Reformation, 233–4
  religious, 12, 230–5, 240n5, 244n46–7
  waterworks, 235–9, 247n73–4
Avogadro, Amedeo, 61

Printed in the United States of America